THE EQUATIONS OF

The Equations of Materials

Brian Cantor

OXFORD
UNIVERSITY PRESS

OXFORD
UNIVERSITY PRESS

Great Clarendon Street, Oxford, OX2 6DP,
United Kingdom

Oxford University Press is a department of the University of Oxford.
It furthers the University's objective of excellence in research, scholarship,
and education by publishing worldwide. Oxford is a registered trade mark of
Oxford University Press in the UK and in certain other countries

Published in the United States of America by Oxford University Press
198 Madison Avenue, New York, NY 10016, United States of America

British Library Cataloguing in Publication Data
Data available

Library of Congress Control Number: 2019945445

ISBN 978–0–19–885187–5 (hbk.)
ISBN 978–0–19–885188–2 (pbk.)

DOI: 10.1093/oso/9780198851875.001.0001

Printed and bound by
CPI Group (UK) Ltd, Croydon, CR0 4YY

Links to third party websites are provided by Oxford in good faith and
for information only. Oxford disclaims any responsibility for the materials
contained in any third party website referenced in this work.

To my partner, Gill, for all her love and support over many years

Contents

List of Portraits

Preface

This book describes some of the important equations of materials and the scientists who derived them. It is aimed at anyone interested in the manufacture, structure, properties and engineering application of materials, such as metals, polymers, ceramics, semiconductors and composites. It is meant to be readable and enjoyable, a primer rather than a textbook, covering a limited number of topics, and not trying to be comprehensive. It is pitched at the level of a final-year school student or a first-year undergraduate who has been studying the physical sciences and is thinking of specialising in materials science and/or materials engineering, but it should also appeal to many other scientists at other stages in their career. *The Equations of Materials* requires a working knowledge of school maths, mainly algebra and simple calculus, but nothing more complex. It is dedicated to a number of propositions, as follows:

1. The most important equations are often simple and easily explained.
2. The most important equations are often experimental, confirmed time and again.
3. The most important equations have been derived by remarkable scientists who lived interesting lives.

Each chapter covers a single equation and materials subject. Each chapter is structured in three sections: first, a description of the equation itself; second, a short biography of the scientist after whom it is named; and third, a discussion of some of the ramifications and applications of the equation. The biographical sections intertwine the personal and professional life of the scientist with contemporary political and scientific developments. A short bibliography is provided for further reading at the end of each chapter. Each chapter is self-contained, i.e. can be read without prior reading of any other chapters.

You might ask: *Why materials? Why equations? Why simplicity? Why experimental? Why biography?* Let me explain.

Why materials? Everything in our modern world (and, indeed, in the ancient world) is made up of complex materials. We could not have things such as fridges, cars, TVs, aeroplanes, MRI scanners, prosthetic limbs, books, phones or pretty much anything else without the materials from which they are made. We could not even have much simpler things such as needles and pins, bricks and mortar, or pots and pans. The development of humankind—including our politics, economy, art and culture—is, at its root, a continuing attempt to use materials in ever-more sophisticated ways to enhance the quality of human life, from early developments such as building mud huts and domesticating agricultural crops to modern inventions such as mass transport, keyhole surgery and internet communications. The fields of *materials science* and *materials engineering* are, therefore, essential to our way of living. To maintain and improve human health, wealth and well-being, we need to understand the physics, chemistry, biology and engineering applications of a wide range of materials, such as metals, polymers, ceramics, semiconductors and composites.

Why equations? Albert Einstein said, 'An equation is for eternity'[1] and Stephen Hawking said, 'Science is beautiful ... including the equations of physics'.[2] Philosophers have long agonised about how we *know* anything, how we can be sure that anything is true. This is the field of *epistemology*. And it has proved difficult to pin down truth. Historians, philosophers, artists, writers and poets all have their own corner of the truth, their own view of it, their own insights into human existence. But truth is elusive. Scientists catch their understanding of things via equations, which codify in a neat way literally millions and millions of experimental observations. Equations represent tentative scientific truths that have been repeatedly confirmed, millions and millions of times over, and are, therefore, highly likely to be true, but could still be found wanting if either the world takes a capricious turn or someone comes up with a weird new interpretation (and, from time to time, someone does). Equations are the essential stuff of science and engineering, our working tools. Richard Feynman said, 'Equations are the best method to obtain laws which are presently unknown'.[3]

Each chapter in this book deals with a different equation and a different related topic in materials science and materials engineering. However, exceptions prove the rule, and some chapters do not do this in the strict sense of the word. I have stretched somewhat the notion of an equation. Thus, the Bravais lattices are fundamental to crystallography, and there are 14 of them, so there is, effectively, a *Bravais equation* for the number of Bravais lattices N_L:

$$N_L = 14,$$

but this is not how we normally refer to this fact. Similarly, the Fermi level E_F is the maximum energy of the electrons in a material, and determines, fundamentally, its electrical behaviour, so there is effectively a *Fermi equation*:

$$E_F = E_{max};$$

but again, this is not how we usually refer to this fact. As a final example, the Burgers vector b is the defining parameter for dislocation defects, which control plasticity and flow in crystalline materials. Dislocations are topological defects that cannot end in a crystal, so the Burgers vector is conserved and there is effectively a *Burgers equation*:

$$b = \text{constant}$$

along a dislocation line, but again we do not normally describe this fact this way. I have included chapters on Bravais lattices, the Burgers vector and the Fermi level, finding this preferable either to using a more complex equation (but one more often referred to as an equation) or leaving out these aspects of materials. Bearing this in mind, the list of chapters, equations and materials topics is as follows:

1. Bravais Lattices: Crystals
2. Bragg's Law: Diffraction
3. The Gibbs Phase Rule: Phases
4. Boltzmann's Equation: Thermodynamics
5. The Arrhenius Equation: Reactions
6. The Gibbs-Thomson Equation: Surfaces
7. Fick's Laws: Diffusion
8. The Scheil Equation: Solidification
9. The Avrami Equation: Phase Transformations
10. Hooke's Law: Elasticity
11. The Burgers Vector: Plasticity
12. Griffith's Equation: Fracture
13. The Fermi Level: Electrical Properties

The list of chapters has a general logic of progression, beginning with structural topics such as crystals and phases, progressing through processing topics such as solidification and phase transformations, and finishing with material properties such as elasticity, fracture and electrical conduction. There are, of course, many other materials topics that I could have selected, each with their own characteristic equations. For instance, the list does not include anything about microscopy, extraction, fatigue or magnetic properties, which I suppose I am effectively saving for a possible second book. In the end the list is subjective and represents, to some extent, a personal preference. Each chapter has been written to be independent and self-contained, i.e. there is no cross referencing, and each chapter can be read without reading previous chapters. For simplicity and ease of reading, there are no footnotes.

Why simplicity? Confucius said, 'Life is simple',[4] and Isaac Newton said, 'Truth is to be found in simplicity'.[5] There are many complex equations in materials science and engineering, but the main phenomena and the most eternal truths (i.e. the least tentative) can be encapsulated in some beautifully simple equations. Hooke's law shows that extension is proportional to applied force, Fick's first law shows that atomic migration is proportional to concentration gradient, Griffith's equation shows that fracture is proportional to crack length, and Boltzmann's equation shows that atomic disorder is proportional to probability. It is much easier to learn a simple equation than a complex one, much easier to explain it and much easier to understand it (but don't be fooled: it is often much harder to discover it).

In the first part of each chapter I derive and explain the key features of a simple equation. After a short biography of the scientist after which it is named, I give a flavour of its scientific implications. The ramifications of simple equations often lead to great complexity, involving new and deep mathematical and scientific abstractions and explanations, which I do not cover in detail or even at all. Thus I do not deal with group theory for describing crystallographic structures, the wave equation or the reciprocal lattice for describing diffraction, tensor calculus for describing elastic or plastic stress–strain behaviour, Legendre transforms for describing thermodynamics, coincident site lattice theory for describing crystal boundaries, or quantum mechanics for describing electrons in atoms and materials. These are all important topics and beautiful examples of deep scientific understanding, but they are too detailed for a readable primer. Nevertheless, I do try to flesh out the significance of each equation, some of its applications, and some related phenomena. The interested reader whose appetite has been whetted is directed to more detailed texts in the bibliography at the end of each chapter.

Why experimental? I have been an experimental scientist throughout my career. I believe that, although theoretical science is important, it is sometimes given too much prominence. I believe fiercely in the fundamental significance of the *experimental method*. I was taught by my schoolteachers when I was growing up in Manchester, and my lecturers and professors when I was a student at Cambridge, to treat experimental facts as sacrosanct. Theories are mutable. They are just ideas, speculation that might or might not apply, and might at any time be superseded by better theories. When writing a scientific paper, I was taught anyone else should be able to follow the description of my experiments and I should be entirely confident they would get the same results, but I should see my theoretical explanation of the results as, at best, tentative, because someone else may come up later with improved explanations (and they often did). I have always tried to teach the same to my own students.

This is not just a question of temperamental preference. It is to do with a profound, underlying philosophy about how to make sense of the world. Science has been phenomenally successful in improving peoples' lives, and it has achieved this through a fierce commitment to the experimental method, not to having clever theoretical ideas. The world is a complicated place. There are many clever and widely believed ideas that have been proved wrong: the earth is not flat, the earth is not at the centre of the universe, a substance called *phlogiston* is not produced when something burns, there is not an ether to carry light waves, physical objects are not fully dense with matter, and so on. Self-evident truths have not stood up well to the scientific gaze. As Bertrand Russell remarked, 'Aristotle maintained that women have fewer teeth than men; although he was twice married, it never occurred to him to verify this statement by examining his wives' mouths'.[6] Erasmus Darwin spoke more tartly (but not intentionally, I think, about Aristotle), 'A fool is one who has never made an experiment'.[7]

I repeat, the world is a complicated place, and it has only given up some of its secrets by arduous and carefully controlled experiments that have yielded incontrovertible results and been reproduced time after time after time. When I was young, I used to think that science would solve all our problems, but I was wrong. In fact, there is a deep paradox: the more we research and understand things, the more there seems to be left to understand. It is a commonplace in academic circles to say that any good doctoral thesis throws up more new questions than it solves. The deeper we dig, the more difficult everything seems to become. I now think of the world as a vast sea of complexity and fog, with occasional tiny islands of clarity and understanding, hard to find and important to hold on to. These islands of knowledge include the beautiful equations of materials dealt with in this book. Some of them were discovered experimentally, such as the Arrhenius and Griffith equations, and some were discovered theoretically, but are proved almost every time an experiment is performed, such as Boltzmann's equation or the Gibbs phase rule. But they all contain deep experimental scientific truths.

Finally, *why biography?* As I said, I have been an experimental materials scientist and engineer all my career, using the Arrhenius equation, the Gibbs phase rule, Hooke's law, the Avrami equation, Griffith's equation and all the other equations and laws in this book on almost a daily basis. But I realised late on that I had, at best, only a fleeting knowledge of the scientists who had discovered these equations. I knew nothing much about their lives or their loves, their careers or their passions. And what a cornucopia I discovered when I began to investigate them for this book. I expected, at first, to write a short two- or three-paragraph thumbnail sketch of each scientist. But I discovered enormous themes of war, catastrophe, nation-building and scientific breakthrough. I had no idea that Arrhenius invented physical chemistry and derived the first theory of global warming, that Hooke's life was shaped by the English civil wars and the Great Fire of London, that Griffith turned down Whittle's initial plan to build a jet engine in the 1930s. I found that my scientists' lives were interwoven with great political events, such as the American and English civil wars, the French Revolution, the First and Second World Wars, the growth of the German state, and the Industrial Revolution. They linked with major scientific developments, such as the discovery of the structure of DNA, the laws of gravitation, splitting the atom, the discovery of electricity and nuclear energy, the development of aeroplanes and air flight, the founding of the Royal Society and the instigation of the Nobel Prizes. And they lived tumultuous lives—Boltzmann committing suicide, Bravais climbing mountains, Bragg fighting in the trenches of the Somme, Fermi building the atomic bomb. I have tried to capture some of this excitement in my short biographies. I had no alternative but to extend them to several pages each, typically a third or more of each chapter. I have included a portrait of each of the scientists, except for Hooke and Scheil, for whom no reliable portrait exists. Understanding a little of scientists' lives and humanising their endeavours enriches one's understanding of their scientific achievements. Or, at least, it has mine. I only wish I had understood this earlier and learnt more about the scientists when I was studying.

I made a decision early on when writing this book that I would not have the time, energy or indeed money to conduct primary research into the lives of the scientists about whom I was writing. I could not possibly visit all their home and/or working environments, seeking information on the ground. Instead, I would have to rely on existing biographies, memoirs and, occasionally, brief letters to archivists at organisations where they worked. Some of the scientists covered are very well known and some are not. For instance, there are many biographies of Robert Hooke and Enrico Fermi, but none about some others. In most cases, I have been able to find sufficient information to write, I believe, an interesting, reasonably complete, several-page biography, but in one or two cases this was particularly difficult. Melvin Avrami was a complete mystery until I discovered, quite by chance, with help from a Columbia University archivist in New York, that he suffered a personal crisis, changed his name and dropped out of science and life in general before returning to a renewed and somewhat different career. I failed to find much at all about Erich Scheil, despite help from an archivist at the Max Planck Institute in Berlin, and this is the one case where I was not able to write much of a biography at all. In some cases I have covered biographical information about other, related scientists; but, in the interests of focus and brevity, I tried to avoid doing this too much. I included what seemed natural and interesting, but it was difficult to be consistent and, in the end, the degree of focus is rather variable. For instance, I could have written a lot, but I didn't, about Newton in the chapter on Hooke, about Mach in the chapter on Boltzmann, or about von Laue in the chapter on Bragg. I did, however, write a bit about Whittle in the chapter on Griffith, about Edison in the chapter on Thomson, and about Mehl and Kolmogorov in the chapter on Avrami.

It is a great sadness to me, but my scientists are all male. I always feel uncomfortable in an all-male environment. This happens much less frequently than some years ago, but it is still not uncommon in fields such as engineering and business. The simple truth is that we cannot really be successful in endeavours that exclude or militate against half the population. We need all the outstanding engineers and business managers we can find, as well as doctors, politicians and schoolteachers, not to say butchers, bakers, candlestick-makers and everyone else. Unfortunately, until relatively recently, scientific investigation was largely a male preserve. In modern times, it is great to see an increasing number of outstanding women scientists and engineers, but I cannot change or avoid the fact that almost all of our historic materials science equations have been discovered by men. I hope and expect that a similar book written some years in the future will reflect a better gender balance.

I want to thank some people who have helped me enormously: Roger Doherty, Patrick Grant and Mark Miodownik, all good friends and excellent scientists, who read some of the chapters and made insightful and useful comments; Anita Lekhwani from Singer and Sonke Adlung from Oxford University Press, commissioning editors who supported me in my writing plans; Harriet Konishi, Saranya Jayakumar, and Cat Ohala for help in preparing the printed version of the manuscript; Simon Nobis from the Archiv der Max Planck Gesellschaft, Louisianne Ferlier from the Royal Society, and Jocelyn Wilk from the Columbia University Archives for providing helpful documentary material; Michelle Kenyon, my outstanding personal assistant, who sorted out all the technicalities; many, many colleagues at the University of Bradford and elsewhere who showed great fortitude and continued to encourage me when, in my enthusiasm, I occasionally burbled on at great length about what must have seemed to them some obscure scientist or scientific equation; and, most important, Gill Partridge, my loving partner, who works as a chef and businesswoman, who is well known as a beekeeper, but who is also an undiscovered but brilliant linguist and mathematician, who read all the biographies and was my most challenging but most helpful and supportive critic.

All in all, I have enjoyed writing this book enormously. I have learnt a lot in focusing on simple experimental equations, and in allying this with biographies of the key scientists who derived them. I hope you enjoy reading it.

Brian Cantor
University of Oxford
March 2020

References

1. Robert Jungk. *Brighter Than a Thousand Suns: A Personal History of the Atomic Scientists* (Houghton Mifflin Harcourt, Boston, 1958), 201.
2. BrainyMedia Inc. 'Stephen Hawking quotes'. 2019. https://www.brainyquote.com/quotes/ stephen_hawking_627081 (accessed 5 March 2020).
3. Richard Feynman. 'The development of the space-time view of quantum electrodynamics'. In: Stig Lundqvist (ed.). *Nobel Lectures: Physics 1963–1970* (World Scientific, Singapore, 1998), 177.
4. 'Forbes quotes: thoughts on the business of life'. 2015. https://www.forbes.com/quotes/1925/ (accessed 5 March 2020).
5. Martin Mulsow and Jan Rohls, eds. *Socinianism and Arminianism: Antitrinitarians, Calvinists and Cultural Exchange in Seventeenth-Century Europe* (Brill, Leiden, 2005), 273.
6. Bertrand Russell. *The Impact of Science on Society* (Simon and Schuster, New York, 1968), 7.
7. W. Stanley Jeavons. 'Experimental legislation'. *Popular Science* **16** (1880), 754.

1

Bravais Lattices

Crystals

---

1 Bravais lattices

Most materials are solid, and most solids are *crystalline*. This means that the atoms and molecules that make up the material are arranged in regular arrays. And this raises the simple question: how many different ways can space be filled with a regular array of atoms or molecules? The answer turns out to be 14. This was discovered by the French scientist Auguste Bravais, and the 14 different regular arrays are called the *Bravais lattices*. Effectively, there is a *Bravais equation* that gives the number of different Bravais lattices N_L as:

$$N_L = 14.$$

To explain why 14 is the right answer, we have to think about the *symmetry* of regular arrays of atoms or molecules. Only certain kinds of symmetry are consistent with a regular array that fills space, and the essential difference between different arrays is that they have different symmetry. To understand the symmetry of crystals, we need to introduce the notion of a *lattice*.

A typical lattice is shown in Figure 1.1. A lattice is defined as an array of points, all of which have exactly the same environment within a given crystal. The arrangement of atoms and molecules around each *lattice point* is exactly the same. The lattice, i.e. the set of lattice points, describes the regularity of the arrangement of atoms or molecules within the crystal. We can think of the crystal as being constructed by taking a local atomic or molecular structure surrounding a given lattice point and then reproducing the same atomic or molecular structure around all the lattice points, extending throughout the crystal.

Each lattice is defined by three basis vectors *a*, *b* and *c*, which are not co-linear (i.e. do not point in the same direction) and are not necessarily orthogonal (i.e. are not necessarily at right angles to each other), as shown in Figure 1.1. Each lattice point is given by a vector *R*, which is an integral combination of the three basis vectors:

The Equations of Materials. Brian Cantor. Oxford University Press (2020). © Brian Cantor.
DOI: 10.1093/oso/9780198851875.001.0001

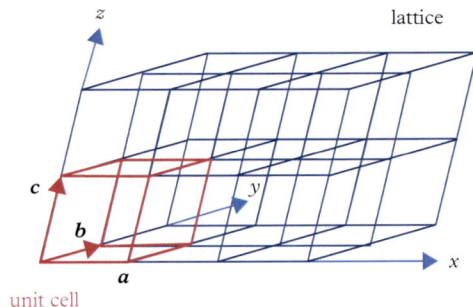

Figure 1.1 *A lattice composed of an array of lattice points, with basis vectors **a**, **b**, **c**, defining the unit cell*

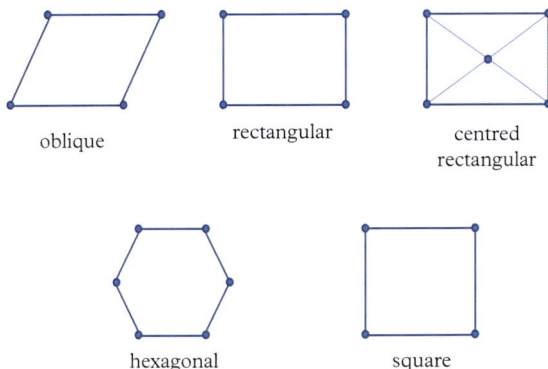

Figure 1.2 *The five Bravais lattices in two dimensions*

$$\boldsymbol{R} = n_1\boldsymbol{a} + n_2\boldsymbol{b} + n_3\boldsymbol{c},$$

where n_1, n_2 and n_3 are integers. The lattice is the set of all such vectors $\{R\}$, i.e. the set of all possible combinations of the integers n_1, n_2 and n_3. The *unit cell* of the lattice is defined by the basis vectors $\boldsymbol{a}$, $\boldsymbol{b}$ and $\boldsymbol{c}$, i.e. by the parallelepiped given by the eight lattice points with n_1, n_2 and n_3 equal to one or zero, as shown in Figure 1.1.

Lattices with different symmetry and structure are called *Bravais lattices*. Figure 1.2 shows the unit cells of the five different Bravais lattices in two dimensions, and Figure 1.3 shows the unit cells of the 14 different Bravais lattices in three dimensions.

Each Bravais lattice describes the *translational symmetry* of the regular array of atoms or molecules in a crystal. The lattice takes no notice of the detailed atomic or molecular structure, and each group of atoms or molecules is represented effectively by a single

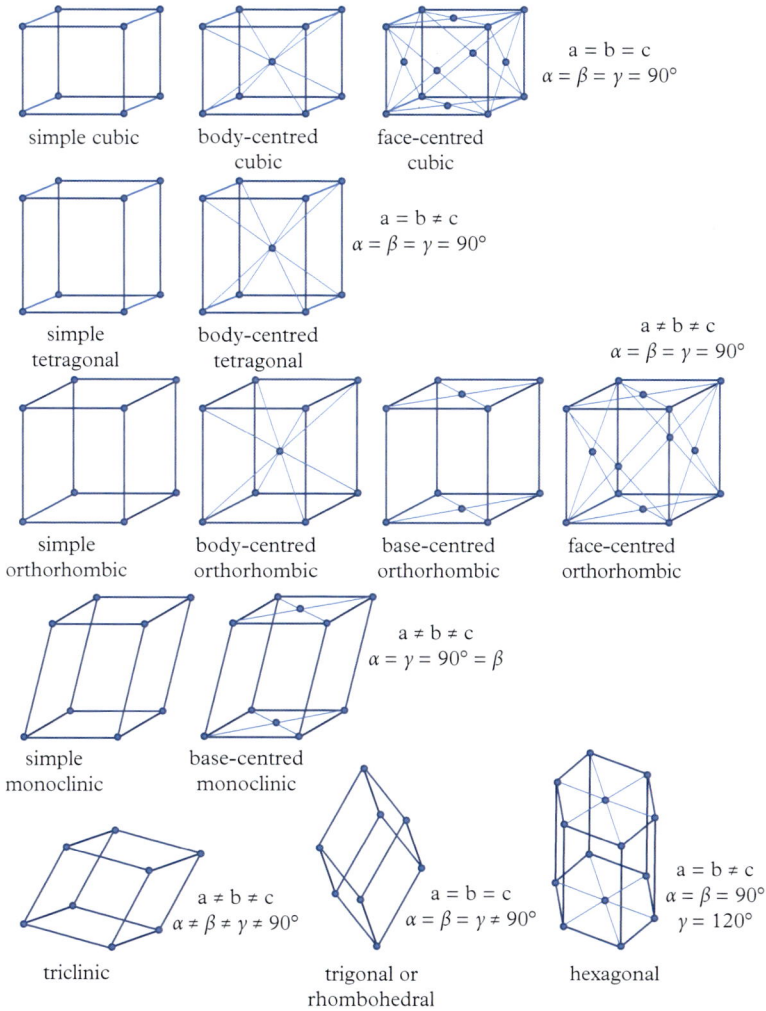

$a = b = c$
$\alpha = \beta = \gamma = 90°$

simple cubic   body-centred   face-centred
                 cubic   cubic

$a = b \neq c$
$\alpha = \beta = \gamma = 90°$

simple   body-centred
tetragonal   tetragonal

$a \neq b \neq c$
$\alpha = \beta = \gamma = 90°$

simple   body-centred   base-centred   face-centred
orthorhombic   orthorhombic   orthorhombic   orthorhombic

$a \neq b \neq c$
$\alpha = \gamma = 90° = \beta$

simple   base-centred
monoclinic   monoclinic

$a \neq b \neq c$ $a = b = c$ $a = b \neq c$
$\alpha \neq \beta \neq \gamma \neq 90°$ $\alpha = \beta = \gamma \neq 90°$ $\alpha = \beta = 90°$
                                                       $\gamma = 120°$

triclinic   trigonal or   hexagonal
                        rhombohedral

Figure 1.3 *The fourteen Bravais lattices in three dimensions*

lattice point. Translating the crystal by any of the vectors R connecting two lattice points reproduces the atomic or molecular arrangement exactly. The crystal is said to be *invariant* to such a translation. Each of the different Bravais lattices has a different translational symmetry, i.e. each Bravais lattice has a different structure of translational vectors $\{R\}$ that reproduce the atomic or molecular arrangement.

2   Auguste Bravais

Auguste Bravais

Mont Blanc is the highest mountain in Europe west of the Caucasus, 4,809 m above sea level. Anyone who has visited it knows that it dominates the Chamonix Valley, massive and glorious, higher than seems plausible, like some excessive and wildly imagined theatrical backdrop. It hurts the neck to lift one's eyes to the top, and the peak seems to defy one to consider scaling it, but at the same time calls one to try to do just that. The 18th-century Swiss geologist and meteorologist Horace Bénédict de Saussure was besotted by the idea of climbing Mont Blanc. He scaled many of the surrounding peaks but failed in several attempts at Mont Blanc. Finally, in 1760, he offered 'a very considerable reward' (there is some uncertainty about just how much) to whoever first got to the top, initiating a plethora of further failures. In the end, the first ascent was on 8 August 1786 by two very different local Chamonix men: Doctor Michel-Gabriel Paccard, an intellectual, trained as a medic in Turin, and passionate about natural history, botany and minerals; and Jacques Balmat, a macho mountain guide, hunter of chamois and collector of crystals.

The mountaineer Eric Shipton wrote, 'Theirs was an astounding achievement of courage and determination, one of the greatest in the annals of mountaineering. It was accomplished by men who were not only on unexplored ground but on a route that all the guides believed to be impossible'.[1] And according to the mountain writer and photographer Cyril Douglas Milner:

> The ascent itself was magnificent; an amazing feat of endurance and sustained courage, carried through by these two men, unroped and without ice axes, heavily burdened with scientific equipment and with long iron-pointed batons. The fortunate weather and a moon alone ensured their return alive.[2]

When they reached the top, Balmat was 'in a state of rapture . . . where no-one had as yet been, not even the eagle or the chamois'.[3] Despite the good weather, conditions were terrible and they were ill-equipped. They descended rapidly, fingers frostbitten, lips swollen, faces blistered and eyes inflamed. Paccard was blinded by the extreme brightness of the sun and snow, and had to be led by Balmat.

The two men later disagreed fiercely about the climb. Paccard was passionate about conducting scientific experiments on Mont Blanc, with a dream 'to carry a barometer to the summit and take a reading there'.[4] He said he took Balmat as his porter. Balmat was obsessed with the climb itself and had a nightmare about 'dropping down [from the rock] and grabbing a branch just before his death'.[5] He said he bumped into Paccard just as he was setting off, and persuaded him to come along and help. Balmat later bragged that he had set the route and that Paccard only went along for de Saussure's reward, and had, in any case, been too frightened to do the last pitch to the top. This story was publicised by the famous author Alexandre Dumas, who heard it during a drunken evening carousing with Balmat, and it became the accepted view, even though Paccard persuaded Balmat to sign a statement withdrawing it. In fact, they both set the route, Balmat took all the reward, and they both reached the top. In 1887, Chamonix put up statues of de Saussure and Balmat in the town centre. It took more than a hundred years for Paccard's role to be recognised fully, and a statue to him was not erected until 1986. Paccard married Balmat's sister and became a Justice of the Peace. Balmat continued climbing and died in 1834, falling off a cliff while prospecting for gold further north, near Sixt-Fer-à-Cheval, one of 'Les Plus Beaux Villages' in France. Paccard and Balmat's ascent of Mont Blanc is widely regarded as the event that sparked the modern sport of mountain climbing. Nowadays, more than 20,000 climbers reach the summit of Mont Blanc each year, helped, of course, by modern advantages such as impermeable plastic snow boots, high-quality climbing ropes and fixed pitons, weather stations, cable cars, overnight huts, geostatic navigation, smartphone communication, and well-equipped mountain rescue teams.

The year after reaching the summit, Balmat acted as a guide for a much larger scientific expedition designed to take de Saussure (finally) to the top of the mountain with a total of 20 scientists, guides and porters. This was the third ascent, since Balmat had already made a second earlier that year with two other Chamonix guides, Alexis Tournier and Jean-Michel Cachat, to check the route. de Saussure stayed for more than four hours at the top to conduct scientific studies, which he followed up with another expedition in 1788, when he stayed 21 days at the Col du Géant, just below the summit. His observations and experiments from a total of seven alpine journeys were published later, in four volumes entitled *Voyages dans les Alpes 1779–1796*.[6] More than half a century later, early in 1844, the French Ministre de l'Instruction Publique (Education Minister), Abel Villemain, and the French physicist, Claude Pouillet, asked the well-known explorer and Professor of Applied Mathematics and Astronomy at the Université de Lyon, Auguste Bravais, on behalf of the French Académie des Sciences to conduct another expedition to the top of Mont Blanc, following in de Saussure's footsteps and repeating his experiments with the benefit of modern equipment. Bravais' scientific curiosity and adventuring spirit were piqued. He read de Saussure's accounts avidly and agreed to accept the challenge together with his longstanding scientific partner and co-adventurer, Charles Frédéric Martins, and another scientific friend, the medical doctor Auguste Lepileur.

In the end, a convoy of 43 people set off for the summit of Mont Blanc on 31 July 1844, including the three scientists, three local guides, 32 porters to carry a bewildering array of (then) modern scientific equipment (barometers, magnetometers, hygrometers, thermometers, astrolabes, theodolites and the like), and two youngsters from the village who just wanted to tag along. They started up the Glacier des Bossons, following Paccard and Balmat's original route, reaching the Grands Mulets at the junction with the Glacier de Taconnaz after 15 hours of hard climbing, where they pitched tents. In the morning, they encountered, somewhat surrealistically, 80-year-old Jean-Michel Cachat, veteran guide from the second ascent, who had spied them setting off and felt he just had to hurry and join them. They continued to climb the icy rocks towards the Grand Plateau below the Dôme de Goûter, but the weather began to turn. After a brief pow-wow, most of the party decided discretion was the better part of valour and began to re-trace their steps, but the scientists and guides pressed on. Unfortunately, the weather deteriorated further and they were forced to pitch tents in drifting snow on a sloping ledge with little or no visibility in a fierce blizzard, hoping to shelter, protect the equipment and let the storm blow itself out. Charles Martins said,

> *Je n'avais pas compris comment des voyageurs pleins de vigeur et de santé péri à quelques pas de l'endroit où la tourmente était venue des surprendre; je le compris ce jour-là.* [I never understood how travellers full of vigour and health could perish just a few paces from being surprised by a storm; today I understand.][7]

Anyone who has overnighted halfway up a mountain in a severe storm knows just how he felt. They had no option but to retreat to Chamonix, leaving behind a cache of scientific instruments.

On 6 August, the weather lightened, and the next day they set off again, reaching their tents and the stashed equipment, this time with much more difficulty because of the problems posed by climbing on fresh snow and because of fatigue from their previous efforts. Unfortunately, the weather again deteriorated and they were again forced to retreat to Chamonix. Bravais said, '*Ce second échec ne nous découragea point; il fallait opposer la constance dans la résolution à l'inconstance du temps*' ['This second rebuff did not discourage us at all; we had the constancy of our resolve to overcome the inconstancy of the weather'].[8] On 28 August, they set off on a third attempt, and this time the weather remained calm. They reached the Grand Plateau for the third time, only then seeing the fabulous view in clear daylight for the first time, spent a day conducting experiments; slept in freezing, strength-sapping conditions, suffering from altitude sickness; and finally set off to reach the summit, where they again made a host of measurements: the geometric angles to peaks near and far, the boiling point of water, the air pressure, the strength of the sun's rays, the orientation of the earth's magnetic field, the altitude of the peak above sea level and so on. Finally, they left the summit, descending rapidly to the Grand Plateau, where they spent three days making more measurements before going down to Chamonix. Bravais wrote about his exploits and scientific results in a book entitled (unimaginatively) *Le Mont Blanc*,[9] the first full scientific exploration of the physical geography of the mountain. Most notably, he described and explained the phenomenon, which he had seen for the first time from the top of Mont Blanc, of the formation of halos around the sun caused by the refraction of its light through clouds in the atmosphere.

Auguste Bravais' father, Victor, was born in 1764 in Annonay, the largest town in the Ardèche region of France, in the southern part of the Massif Central. Victor's mother and father were Claire and François Bravais, the latter a consular clerk from Saint Péray, further south on the banks of the river Rhône, just across from the larger city of Valence in the Drôme. Victor went to school in Valence and then attended the Oratory College, usually coming top of his class. In 1782 he went to study medicine, chemistry and botany in Montpellier, near the Mediterranean coast. He was intelligent and adventurous and, not surprisingly, bored with the dullness of provincial life. He was fascinated by the expedition in 1785 of Jean-François de Galaupe, Comte de la Pérouse, sponsored enthusiastically by King Louis XVI, who wanted a Frenchman to follow in the footsteps of Captain Cook, circumnavigate the globe and outdo the exploits of the British explorer. de la Pérouse's two ships, *La Boussole* and *L'Astrolabe*, set sail from Brest in August 1785, steered initially south, rounded Cape Hope and, for a couple of years, zig-zagged bewilderingly to and fro across the Pacific, visiting an array of (then) exotic places— Easter Island, Maui, Mount St. Elias in Alaska, California, Macau, Manila, Jeju Island in Korea, Sakhalin Island and the Kamchatka Peninsula in Russia, Samoa and Tonga— before reaching Australia in January 1788. There he established its second European settlement on the north shore of Botany Bay (now the Sydney suburb of La Perouse, named after him), just five days after Captain Arthur Philip had landed at Botany Bay and established the first (penal) settlement at Sydney Cove. After taking on supplies, de la Pérouse set off again in March 1788 for New Caledonia, Santa Cruz, the Solomon Islands and the Louisiades, but disappeared with all hands and was never heard from again.

Victor Bravais was enthralled by this expedition, inspired by and wanting to follow its Chief Engineer, Paul Monneron, a fellow-townsman from Annonay. In 1791, Victor signed up, therefore, with Admiral Antoine Bruni d'Entrecasteaux, who was mounting an expedition to try and find out what had happened to de la Pérouse, but Victor's parents were worried about the uncertain and dangerous political situation in France and wouldn't let him go. In 1792, Victor again signed up with a similar expedition, this time mounted by Lieutenant Aristide Aubert du Petit-Thouars, but once again his parents intervened to prevent him. It was the height of the French Revolution, which had begun in 1789 with the overthrow of the monarchy, followed by extreme civic unrest, the establishment of a republic in 1792, King Louis' execution in 1793, the rise of Georges Danton and Maximilien Robespierre, and the establishment and dictatorship of the dreaded Committee of Public Safety and its Reign of Terror in 1793 and 1794. In 1788, a year before the beginning of the Revolution, Victor's father, François, was appointed as one of the representatives in the civic assembly, and then as Magistrate of the commune. He was exposed politically, fell rapidly foul of the extremists (as almost everyone did) and was imprisoned in the Ardèche town of Viviers in 1793. Victor fled to Paris after hiding under the protection of Monseigneur François Marbos, Constitutional Bishop in the nearby Drôme region, only returning later, after the so-called Reaction of Thermidor, the dissolution of the Committee of Public Safety and its replacement by le Directoire in 1795, when his father was re-instated. Victor was now resigned to a provincial life. He became a doctor at the local hospital and, in 1798, he married Adélaïde Amélie Thomé, the daughter of a local bookseller Joseph Thomé and the niece of Bishop Marbos, with a trousseau of 16,000 francs (equivalent to about 100,000 euros today). The marriage contract was signed at the offices of the local paper manufacturer in Annonay, Jacques-Etienne de Montgolfier (famous along with his brother Joseph-Michel for inventing the hot-air balloon), where the brother of the mayor, who was also the registrar, worked. In 1807, Victor bought a fine house with a beautiful garden and settled there with his family on the Rue Clochesang.

Victor and Amélie were blessed with 10 children during their first 14 years of marriage, although only five survived to adulthood. Auguste Bravais was the ninth, born on 24 August 1811 in Annonay. His mother died during childbirth the following year. Auguste and his older brothers and sisters, Louis, Jules, Camille and Marie, were all brought up by a young maid and nanny, Jeannette, hired by Amélie, in her own words '*digne pour elle d'être la gardienne*' ['worthy to be their guardian'].[10] Auguste was particularly close to his youngest sister, Marie, born just a year before him: '*sa camarade de jeux et surtout sa complice*' ['his playmate and, above all, his partner in crime'].[11] While the elder children went to school, Auguste and Marie were taught to read and write by Jeannette, and were given a love of life, nature, science and adventure by their father, Victor. Marie wrote in her diary: '*quand mon père avait le loisir . . . il donnait un goût très profond à Auguste pour l'histoire naturelle*' ['whenever my father could, he passed on to Auguste a deep taste for natural history'].[12] And according to Auguste's biographer, Marie-Hélène Reynaud, '*Victor emmène ses enfants dans ses longues promenades au cours desquelles il ramasse plantes et fleurs, minéraux et insectes*' ['Victor took his children on long walks, collecting plants, flowers, minerals and insects'].[13] Just north of Annonay is the great massif of Mont Pilat. One day, the Bravais children began to plan making the long and difficult climb to the summit, but excluded Auguste, then aged nine, as being too young. Early the next morning, Auguste, nothing daunted, took off by himself, reached the top after six hours of hard climbing, slept just below the summit so he could watch the sun rise over Mont Blanc and the Alps some 250 km (150 miles) further east, and returned slowly, collecting novel plants, stones and insects along the way. He had truly inherited his father's love of nature and adventure.

The executive council le Directoire was established by the Reaction of Thermidor and governed France from 1795 to 1799. It curbed the most extreme excesses of the

Reign of Terror, stopping mass executions and relaxing the worst forms of repression, but it was still a sorry affair, suspending elections, reneging on its debts, persecuting the church, suppressing public comment and pursuing revolutionary wars abroad. It was bedevilled by corruption and infighting, and it finally collapsed in a coup led by Napoléon Bonaparte, born in Corsica from a modest Italian background, who had risen rapidly through the ranks of the military in the early years of the Revolution, becoming a general and a war hero with his exploits, notably subjugating Italy and Austria. Napoléon became First Consul in 1799 and then Emperor in 1804, leading wars of conquest throughout Europe until repulsed in his invasion of Russia in 1812, forced to abdicate by the Allies (Austria, Britain, Portugal, Prussia, Russia and Spain) in 1814, and, after escaping from exile on the island of Elba, finally defeated by the Allies at Waterloo in 1815. At its height in 1812, the Napoléonic Empire extended throughout most of Europe, including all its central heartland, excluding only its outer fringes. Napoléon's defeat was followed by the restoration of the Bourbons under Louis XVIII and Charles X, their subsequent replacement with an Orléans monarchy under Louis-Philippe I, and, finally, after another revolution in 1848, the installation of Louis Napoléon (Bonaparte's nephew) who ruled first as President and then as Emperor Napoléon III until 1870.

During the revolutionary and Napoléonic wars, the Regency of Algiers benefitted enormously from supplying food to France, which was far too busy with war to grow its own. Much of these supplies were purchased on credit, i.e. the bills were not paid. The Dey (Lords) of Algiers began to run out of money and were forced to raise taxes, squeezing the local population, who responded with widespread piracy against European and US shipping in the Mediterranean, leading to the Barbary wars of 1801 to 1815 between (mainly) the United States and the Barbary states—the city states of Algiers, Tunis and Tripoli—which were nominally part of the Ottoman Empire, but in practice were ruled separately and acted largely independently. In 1827, Hussein Dey decided to cut through his financial problems and demanded from France the repayment of 28 years of food debt, but King Charles X refused. During tense negotiations, the Dey famously struck the negotiating French consul Paul Deval three times with his flyswatter (the so-called flyswatter incident). This *cause-célèbre* provided the pretext for Charles to break off diplomatic relations, blockade Algiers, destroy trading posts at the ports of Annaba and La Calle, and finally dispatch 34,000 troops to land at Sidi Fredj, 21 km west of Algiers on 14 June 1830. They took Algiers three weeks later, bringing to an end more than 300 years of Ottoman rule, launching French colonisation of North Africa and initiating France's worldwide second colonial empire.

Auguste moved to study at the Collège d'Annonay, where he finished in 1825 in brilliant fashion at the tender age of 14, with excellent grades throughout, winning the college prize in maths. He spent a year at the Collège Stanislas de Paris, one of the best high schools in France, studying rhetoric and philosophy, but he disliked classical subjects and secretly continued to study maths and science in correspondence with his old Annonay college professor Louis Raynaud. One of his classmates was the famous mathematician Évariste Galois, whom he beat in a college maths competition. In 1828, Auguste took the examination, at the unusually early age of 16, for entrance to the École Polytechnique, the grandest and most prestigious of the famous *grands écoles* in France.

He passed with flying colours, winning the *prix d'honneur de mathématiques*, and came in second overall. He was taught and examined by the famous French mathematician Siméon-Denis Poisson, who recommended him for doctoral studies. In 1831, Auguste graduated *couvert de lauriers* (covered with laurels) and decided, perhaps surprisingly, to join the navy. He wanted to travel and see the world, learn new things and study the natural world in regions beyond the Pilat Massif and the Ardèche. For the next six years, he held a number of commissions on ships in the Mediterranean, engaged actively in policing the French Barbary Coast of North Africa, at the same time pursuing ecological, botanical and biological scientific studies whenever possible.

Auguste set sail on 13 January 1832 from the French naval base of Toulon as a young naval recruit (*Élève de Première Classe de la Marine*), on the frigate *Finistère*. For five years, between 1832 and 1837, he served on a variety of naval brigs, steamships, frigates and corvettes, including the *Finistère*, *Loiret*, *Pélican* and *Salamandre*, cruising and patrolling up and down the Barbary Coast, landing frequently, either for minor military skirmishes or, somewhat incongruously, for environmental investigation of the local flora and fauna, and laying up occasionally for refit, rest and leave periods back at Toulon. After taking the relevant exams and passing with ease, he was promoted to Lieutenant de Frégate in 1834. The *Préfet Maritime* (Port Admiral) wrote to the *Ministre de la Marine* (Naval Minister) in Paris: '*Son examen est des plus brillants et depuis longtemps, aucun élève n'avait satisfait aussi complètement à pareille épreuve*' ['His exam results are among the most brilliant ever; not for many years has a student completed such a demanding test so well'].[14]

Bravais showed his bravery as a soldier on 12 August 1836, when the commander of the *Loiret*, the senior lieutenant and the ship's surgeon went ashore and were immediately surrounded, seized and bound as prisoners by soldiers of Emir Abd-el Kadeer, the current Dey, who happened to be in the area. Auguste saw these events from the ship and quickly rounded up 27 of the crew to land and confront the Emir's 200 chevaliers and 40 infantrymen, and rescue the three senior officers, two of whom had been badly injured. According to the commander's report:

> *Je ne terminerai pas sans dire la bravoure, le sang-froid et l'habileté de mon collègue, M. Bravais.... Ce courageux ami commendait les matelots qui volèrent à notre secours; il dispose si bien sa troupe, il fondit si vigoureusement sur les Arabes, qu'il les força, en un clin d'oeil, à prendre la fuite.* [I cannot finish without speaking of the bravery, assurance and skill of my colleague, M. Bravais .... He courageously commanded the men who flew to our rescue; he organised them so well, and he swooped down on the Arabs so vigorously, that in the blink of an eye he forced them to flee.][15]

But Bravais spent most of his time in scientific pursuits, supported fortuitously by the interest in science of the Admiral of the Fleet, Auguste Bérard. According to Reynaud, '*il* [Bravais] *recueille des insectes inconnus, des crustacés, des poissons, des mollusques, des plantes inconnus en Europe*' ['he collected insects, shellfish, fish and plants unknown in Europe'].[16] Auguste sent samples to his father and brothers, Camille and Louis, as well as to scientists throughout France, including the botanist Ernest Cosson, the entomologist Paul Gervais and the naturalist Jean-Victoire Audouin. The latter, for instance, named a new shellfish after him, *Estheria bravaisi*, which Auguste had found in a brackish lagoon in Arzew, near the Algerian port of Oran. He collaborated extensively with his brother Louis, who, like his father, was a doctor in Annonay. In 1835, they presented a joint paper to the Académie des Sciences.[17,18] The study was theoretical as well as experimental, describing the geometric characteristics of leaf positions along a stem as defined by infinite series with irrational coefficients. The brothers concluded, '*le géométrie nous donne l'explication de tous les systèmes connus de position des feuilles*' ['geometry can explain all known systems of leaf positions'].[19]

On 21 September 1836, Auguste arrived at Toulon. He had decided to concentrate on his scientific work and applied to be transferred, first to the *Depot des Cartes et Plans* (the Naval Hydrographic and Oceanographic Archives) in Paris and then to the Paris Observatory, but both requests were refused. Nevertheless, the two brothers, Auguste and Louis, were proposed by Messieurs Guillemin and Brongniard and accepted as members of the prestigious scientific *Société Philomatique de Paris*. In 1837, the *Ministre de la Marine* (Naval Minister) nominated Auguste to join *La Commission Scientifique du Nord*, the government's scientific expedition to the arctic on the ship *La Recherche*, led by M. Gaymard and captained by Victor Charles Lottin, with the objective of recovering the ship *La Lilloise*.

In 1833, the naval officer and round-the-world explorer Jules-Alphonse-René, Baron de Blosseville, had started out in the ship *La Lilloise* on an arctic expedition that mapped the coast of Greenland, was damaged by pack ice, put in to Iceland for repairs and re-embarked, but then disappeared completely, with its last sighting on 15 August 1833. Two years later, Admiral François Thomas Tréhouart set off in *La Recherche* to find *La Lilloise*, but was unsuccessful, discovering no trace of the previous expedition. Bravais was to join Gaymard, who was now leading a second attempt to find *La Lilloise*, at the same time as conducting geographic, hydrological, mineralogical, botanical, zoological and other scientific studies in Norwegian Spitzbergen and Lapland. This appealed very much to Auguste's adventurous and scientific spirit, carrying strong resonance with his father's previous, failed attempts to circumnavigate the globe, and he accepted with alacrity.

La Recherche set off from Le Havre on 13 June 1838, docked briefly at Trondheim and Hammerfest to take on provisions, and on 24 July reached the island of Spitzbergen in the Svalbard archipelago, at a latitude of 78°N, halfway between the northern tip of Norway and the North Pole, entering the eastern Bellsundet Fjord (Bell Sound), where it dropped anchor. The scenery was superb and Auguste was in his element. The ocean was covered with weird and wonderfully shaped icebergs; the mountains were snow covered, with glaciers coming right down to melt at the water's edge; there were birds everywhere, of all imaginable kinds (eider, gulls, cormorants, petrels and guillemots); wild, exotic vegetation clung to the narrow strip of land between mountains and sea; and novel fauna survived in the harsh, snowy wastes (polar bears, marmosets, lemmings and blue foxes). Bravais and his scientific colleagues, notably his close friend Charles Martins, spent the next two years exploring Spitzbergen and the most northern parts of Norway, Sweden and Finland. The mountain Bravaisberget in Svalbard is now named after him.

Before winter arrived, they had to leave Spitzbergen so as not to become icebound, and they sailed back to Hammerfest, close to the northernmost tip of Norway. They then sailed along the coast and penetrated the Altafjord down to Bossekop; then, south to Kautokeino and then Karesuando and Karesuvanto, either side of the river Muonio, the northernmost localities in, respectively, Sweden and Finland; then down the Muonio and Tornio Rivers past magnificent mountains, raging waterfalls and tiny isolated villages such as Pelle and Aavasaksa, to reach Haparanda on the northern coast of the Gulf of Bothnia at the top of the Baltic Sea. From there, they took the ferry to Stockholm and then back to Paris, arriving in January 1840. During the course of their explorations, they had investigated many diverse phenomena: the depth and temperature of the fjords; the orientation of magnetic north; the geometric and optical structure of the aurora borealis and its evolution with time and season; the terraces formed on the edge of the fjords by prehistoric shrinking of the shoreline; the different types of arctic insects, birds and animals, and their comparison with those of the similar snowy wastes of the Alps; the cranial sizes and shapes of the different Lapp races and nationalities; and the growth of trees such as the Scots pine as a function of latitude.

Back in Paris and at home in Annonay, Auguste set about turning literally thousands of pages of notes into a series of important research reports and publications. In January 1841, for his scientific and exploration work, he was made a *Chevalier de la Légion d'Honneur*. The following month he left his commission in the navy and turned to academic life as Professor of Applied Mathematics and Astronomy at the Université de Lyon. He continued with scientific experiments, explorations and expeditions. In 1841 and 1842 he organised summer visits with his friend Charles Martins and his brothers, Louis and Camille, to the Bernese Alps, staying in huts, hiking and climbing in the beautiful mountain scenery of the Eiger, Mönch and Jungfrau, and studying the physical geography of the Alps and comparing it with the different landscapes of the Massif Central and Lapland. They studied features such as the composition of the air and the speed of sound as functions of altitude and temperature. In 1843 Auguste suspended these pursuits. He was grief-stricken at the death, after a protracted illness, of Louis, his brother and compatriot-in-arms in many scientific investigations. In 1844 Auguste made his epic scientific ascent of Mont Blanc, the success of which was instrumental in his appointment the following year to succeed his old professor, Victor le Chevalier, as Professor of Physics at his alma mater, the École Polytechnique in Paris. In 1845 he was made a *Chevalier de l'Ordre de l'Epée de Suède* and also a *Membre de l'Académie de Lyon*. Even more important, in 1847, aged 36, he met and married Marie-Antoinette Eugénie Moutié, a rich, young Parisienne, 12 years younger than him. Shortly thereafter she gave birth to their only son.

Bravais settled into a happy family life, looking after his wife and growing son, and continuing to publish papers based on his many exploits in North Africa, Lapland and the Alps. All his life, his scientific interests were wide-ranging and eclectic. He made contributions in many fields: astronomy, geology, botany, biology, physics and chemistry. And he now began his most famous work: in crystallography. While walking and climbing on snow, and while studying the atmospheric effects of clouds, he had become fascinated by the beautiful hexagonal crystal structure of snow and ice particles. He began to ruminate on the variety of possible underlying geometries of regular arrays of atoms and molecules that must form the structure of all solid matter. In 1848, he presented two *mémoires* to the *Académie des Sciences*, entitled 'Sur les Propriétés Géométriques des Assemblages de Points Régulièrement Distribués dans l'Espace' ['On the Geometric Properties of Assemblies of Points Distributed Regularly in Space'][20] and 'Études Cristallographiques' ['Crystallographic Studies'].[21] This work was a tour de force. He defined a lattice as a regular array of points in space, noting:

> *les molécules sont disposées en files rectilignes . . . les centres sont équidistants entre eux sur chacune de ces files . . . laissant indéterminée la forme de la molécule* [the molecules are arranged in linear rows . . . their centres are equidistant from each other along each of the rows . . . independent of the structure of the molecule].[22]

He showed the 14 different ways of constructing an infinite lattice in three dimensions, each with its own characteristic symmetry, and the relationship of these 14 *Bravais lattices* to the seven crystal systems (defined by the principal symmetry of the crystal) and the 32 point groups (the different combinations of symmetry that can be arranged at each lattice point). The two *mémoires* were assessed by an Académie panel, chaired by the famous French mathematician Augustin-Louis Cauchy, and they were full of admiration.

At the height of his success, however, fate intervened and Bravais' fortunes took a dark turn. On 1 January 1853, his 88-year-old father, Victor, slipped on ice at the entrance to the church where he was going to celebrate the New Year and died in his bed a few days later. The following year, 1854, his brother Jules also died, suffocated and poisoned by a gas leak while he was working on installing the new gasworks in Dijon. And later that year, his young son also died, carried off by a bout of typhoid fever. Auguste was, naturally, devastated. And this was not the end of it. To cap it all, in 1855 he and his wife, Antoinette, began to notice signs of his own mental decline. According to his friend, the geologist Jean-Baptiste Élie de Beaumont, '*Sa mémoire lui faisait défaut; il ne retrouvait plus les idées ingénieuses qu'il lui avait confiées sans les écrire*' ['His memory was failing him; he couldn't remember any more the good ideas he hadn't written down'].[23] At just 44 years of age, Bravais had begun to develop what we now know as early-onset Alzheimer's disease, and he deteriorated very rapidly. He couldn't continue teaching and was given leave of absence later that year by the École Polytechnique. Doctors were unable to help and, early in 1856, his consultant, Doctor Fioville, pronounced, '*les traitements n'ont aucune chance de succès*' ['treatments have no chance of success'].[24] In April, Bravais left the École Polytechnique and, later that year, arranged for his wife to take full control of his affairs, because of his mental incapacity. According to his biographer, Reynaud,

> *Auguste souffre; quelques moments de lucidité succèdent de longues périodes de difficultés. Un jour, dans un accès de désespoir, il brûle tous ses papiers, ces études qui avaient exigé des années d'expériences et de travail.* [Auguste suffered; just a few moments of lucidity succeeding long periods of difficulty. One day, in a fit of despair, he burnt all his papers, the studies which had required years and years of hard work and experience.][25]

In a cruel irony, Bravais' scientific success was being recognised at the same time as his catalogue of personal tragedies. In 1854 he was made a *Membre de l'Institut* (*Académie des Sciences*), and in 1856 he was promoted to be an *Officier de la Légion d'Honneur.* His failing mental abilities forced him to retire to Versailles, where he stayed in a terrible state of agony and semi-consciousness for seven years, supported as best as could be managed by his wife, Antoinette, and his sister, Marie, until he finally passed away on 30 March 1863, at the age of 52. Reynaud captured the sadness of the scene: '*Auguste pousse un dernier soupire; la vie vient de quitter son corps; il y a bien longtemps que ses esprits s'en sont allés*' ['Auguste heaved a last sigh; life left his body; his spirit had long since departed'].[26] At his funeral, the physicist and engineer Louis Urbain Dortet de Tessan, who had served with Bravais off the coast of North Africa on the *Loiret,* made a speech on behalf of the *Académie des Sciences,* during which he said Bravais had achieved '*les triomphes immortels de la science sur la nature*' ['immortal triumphs of science over nature'].[27]

3   Crystal systems

The 14 Bravais lattices fall into seven *crystal systems,* which correspond to the seven different types of symmetry that are found in macroscopic crystals. The crystal systems as shown in Table 1.1 are defined by the relationship between the unit cell basis vectors a, b and c, as described by the scalar magnitudes of the vectors a, b and c, and the angles between the vectors α, β and γ (by convention, α is the angle between b and c, β is the angle between a and c, and γ is the angle between a and b). The 14 Bravais lattices map onto the seven crystal systems, as shown in Table 1.2, where P, I, F and C are called Hermann-Mauguin symbols, designating simple, body-centred, face-centred and base-centred unit cells respectively.

Table 1.1 *The Definition of the 7 Crystal Systems*

Crystal system	Unit cell size	Unit cell angles
Cubic	$a = b = c$	$\alpha = \beta = \gamma = 90°$
Tetragonal	$a = b \neq c$	$\alpha = \beta = \gamma = 90°$
Orthorhombic	$a \neq b \neq c$	$\alpha = \beta = \gamma = 90°$
Monoclinic	$a \neq b \neq c$	$\alpha = \beta = 90° \neq \gamma$
Triclinic	$a \neq b \neq c$	$\alpha \neq \beta \neq \gamma \neq 90°$
Hexagonal	$a = b \neq c$	$\alpha = \beta = 90°, \gamma = 120°$
Trigonal (or rhombohedral)	$a = b = c$	$\alpha = \beta = \gamma \neq 90°$

Table 1.2 *The Relationship between the 7 Crystal Systems and the 14 Bravais Lattices*

Crystal system	Bravais lattice	Hermann-Mauguin symbol
Cubic	Simple	P
	Body-centred	I
	Face-centred	F
Tetragonal	Simple	P
	Body-centred	I
Orthorhombic	Simple	P
	Body-centred	I
	Base-centred	C
	Face-centred	F
Monoclinic	Simple	P
	Base-centred	C
Triclinic	Simple	P
Hexagonal	Simple	P
Trigonal (or rhombohedral)	Simple	R

4 Point groups

The 14 Bravais lattices describe only the *translational symmetry* of an array of atoms or molecules in a crystal. They take no account of the structure and symmetry of the atoms or molecules themselves. They ignore the *point symmetry*, i.e. how the atoms or molecules are distributed around each lattice point. This raises a second simple question: how many different ways can a group of atoms or molecules be arranged around a given single lattice point? How many different kinds of point symmetry are there? The answer to this question is 32, and the 32 different arrangements are called the *crystal classes* or *point groups*.

To understand point symmetry, we need to think about the different kinds of symmetry that can operate at a point. To do this, we define a *symmetry element* as an operation that, when applied to the crystal, reproduces exactly the same arrangement of atoms or molecules. The crystal and the arrangement of atoms and molecules are said to be *invariant* under the operation of the symmetry element. There are four important kinds of symmetry element that can operate at a point: a *centre of inversion, mirror plane, rotation axes* and *rotation–inversion axes*. We consider these separately.

Centre of inversion

A centre of inversion reproduces the arrangement of atoms and molecules when everything is inverted or reflected in the central lattice point. For every atom or molecule on one side of the lattice point, there is an equivalent atom or molecule at the corresponding equidistant position on the other side of the lattice point. A centre of inversion is represented by the Hermann-Mauguin symbol $\bar{1}$.

Mirror planes

A mirror plane reproduces the arrangement of atoms and molecules when everything is reflected in the plane. In other words, for every atom or molecule on one side of the plane, there is an equivalent atom or molecule at the corresponding equidistant point of reflection on the other side of the plane. Mirror planes are represented by the Hermann-Mauguin symbol m.

Rotation axes

An n-fold rotation axis reproduces the arrangement of atoms or molecules by rotations about the axis of $(1/n) \times 360°$, $(2/n) \times 360°$, $(3/n) \times 360°$,...up to $(n/n) \times 360°$ (i.e. 360°). In a complete rotation about the axis, there are n positions that reproduce the arrangement exactly. For example, a 2-fold rotation axis reproduces the arrangement of atoms or molecules by rotations of 180° and 360°. Figure 1.4 shows examples of 1-fold, 2-fold, 3-fold, 4-fold and 6-fold rotation axes, which are called, respectively, a *monad, diad, triad, tetrad* and *hexad,* and are represented respectively (and fairly obviously) by the Hermann-Mauguin symbols 1, 2, 3, 4 and 6. It turns out that these are the only rotation axes we need to bother about, because rotation axes with $n = 5$ or $n \geq 7$ are incompatible with a regular array, i.e. incompatible with the translational symmetries of the 14 Bravais lattices.

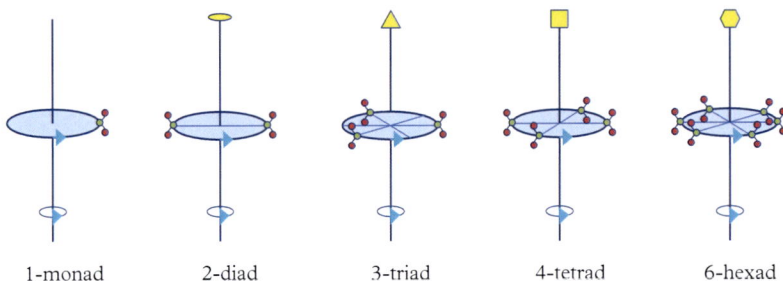

| 1-monad | 2-diad | 3-triad | 4-tetrad | 6-hexad |

Figure 1.4 *Rotation axes of symmetry: 1-fold, monad; 2-fold, diad; 3-fold, triad; 4-fold, tetrad; and 6-fold, hexad*

Table 1.3 *The Relationship between the 7 Crystal Systems and the 32 Point Groups*

Crystal System	Point Groups
Cubic	$23, m3, \overline{4}3m, 43, m3m$
Tetragonal	$4, \overline{4}, 4/m, \overline{4}2m, 4mm, 42, 4/mmm$
Orthorhombic	$mm, 222, mmm$
Monoclinic	$m, 2, 2/m$
Triclinic	$1, \overline{1}$
Hexagonal	$6, \overline{6}, 6/m, \overline{6}2m, 6mm, 62, 6/mmm$
Rhombohedral	$3, \overline{3}, 3m, 32, \overline{3}m$

Rotation–inversion axes

An *n*-fold rotation–inversion axis is one that reproduces the arrangement of atoms or molecules by rotations about the axis of $(1/n) \times 360°$, $(2/n) \times 360°$, $(3/n) \times 360°$, … up to $(n/n) \times 360°$ (i.e. 360°), followed by inversion in the lattice point. Once again, we only need to bother with 1-fold, 2-fold, 3-fold, 4-fold and 6-fold rotation–inversion axes, which are called, respectively, an *inversion monad, inversion diad, inversion triad, inversion tetrad* and *inversion hexad*, and are represented respectively by the Hermann-Mauguin symbols $\overline{1}, \overline{2}, \overline{3}, \overline{4}$ and $\overline{6}$. A rotation–inversion axis $\overline{1}$ is equivalent to a centre of inversion, and a rotation–inversion axis $\overline{2}$ is equivalent to a perpendicular mirror plane *m*.

The 32 point groups are obtained from all the different ways of combining these symmetry elements at a single lattice point, as shown in Table 1.3. By convention, a forward slash is used to denote that a mirror plane is perpendicular to a rotation axis. Otherwise, they are parallel.

5 Space groups

There are 230 different ways of combining point symmetry and translational symmetry, i.e. of arranging the 32 point groups across the 14 Bravais lattices. The 230 different arrangements are called the *space groups*. To describe the different arrangements fully, we need to introduce some new symmetry elements, *rotation screw axes* and *glide planes*, which are extensions of rotation axes and mirror planes when translation is also allowed.

Rotation screw axes

An *n*-fold rotation screw axis reproduces the arrangement of atoms or molecules by rotations about the axis of $(1/n) \times 360°$, $(2/n) \times 360°$, $(3/n) \times 360°$, ... up to $(n/n) \times 360°$ (i.e. 360°), followed respectively by translation along the axis by distances of x, $2x$, $3x$, ... up to nx. For example, a 2-fold rotation screw axis reproduces the arrangement of atoms or molecules (1) by a rotation of 180° and a translation by x, and (2) by a rotation of 360° and a translation by $2x$. Once again, we only need to bother with 1-fold, 2-fold, 3-fold, 4-fold and 6-fold rotation screw axes, which are called, respectively, a *screw monad, screw diad, screw triad, screw tetrad* and *screw hexad*, and are represented respectively by the Hermann-Mauguin symbols 1_x, 2_x, 3_x, 4_x and 6_x, where the subscript represents the fractional distance translated.

Glide planes

A glide plane reproduces the arrangement of atoms and molecules by reflection in the plane, combined with translation parallel to the plane by a distance x.

6   Crystal planes and directions

Crystals are *anisotropic*. This means their properties depend upon the *orientation* of the crystal. Their properties are different in different directions, and change as the crystal is rotated. This arises, fairly obviously, because the arrangement of atoms and molecules is different in different directions and along different planes within the crystal.

Directions and planes within a crystal are described using a notation system called *Miller indices*. Consider a general direction given by a vector R with its root at the origin of the Bravais lattice and its head at a point given by the coordinates x, y, z. The direction is then described by the Miller notation $[uvw]$, where the Miller direction indices u, v and w are related to the unit cell basis vectors a, b and c by

$$u = \frac{nx}{a}; v = \frac{ny}{b}; w = \frac{nz}{c},$$

with n selected as the smallest possible multiplier to make u, v and w all integers. In other words, $[uvw]$ are fractional coordinates x/a, y/b and z/c multiplied up to give the smallest possible set of integers. The notation $[uvw]$ is taken to refer to all parallel directions in the crystal. Negative intercepts lead to negative indices, which are written with an overline, e.g. $[\bar{u}vw]$.

Similarly, consider a general plane that intercepts the unit cell axes at x, y and z. The plane is then described by the Miller notation (hkl), where the Miller plane indices h, k and l are related to the unit cell dimensions by

$$h = \frac{na}{x}; k = \frac{nb}{y}; l = \frac{nc}{z},$$

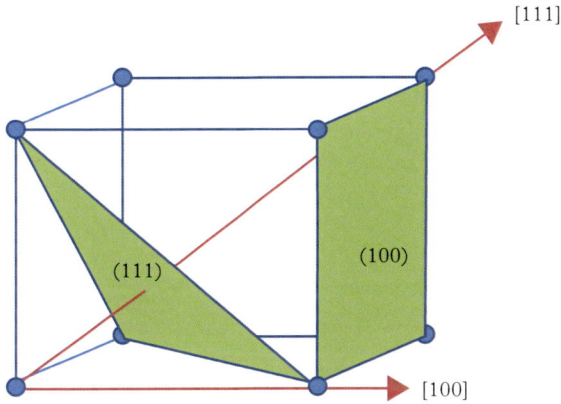

Figure 1.5 *Directions [100] and [111], and planes (100) and (111) in the unit cell of a simple cubic Bravais lattice*

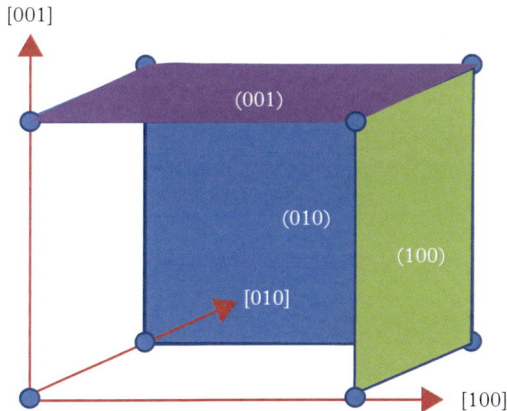

Figure 1.6 *Equivalence of [100], [010] and [001] directions, and (100), (010) and (001) planes in the unit cell of a simple cubic Bravais lattice*

with n selected as the smallest possible multiplier to make h, k and l all integers. In other words, (hkl) are inverse fractional intercepts a/x, b/y and c/z multiplied up to give the smallest possible set of integers. The notation (hkl) is again taken to refer to all parallel planes in the crystal. Negative intercepts lead to negative indices, which are again written with an overline, e.g. $(\bar{h}kl)$.

Figure 1.5 shows the directions [100] and [111], and the planes (100) and (111) in the unit cell of a simple cubic Bravais lattice. In cubic systems, a plane and its normal have the same Miller indices, i.e. $[hkl]$ is perpendicular to (hkl), but this is not true in general for other crystal symmetries.

Different directions and planes within a crystal are often equivalent, i.e. they have the same atomic or molecular structure. This is because they are related by operation of one or more of the crystal's symmetry elements. A set of equivalent directions is written <*uvw*>, and a set of equivalent planes is written {*hkl*}. Thus, in a cubic system, <100> refers to the set of all cube directions, [100], [$\bar{1}$00], [010], [0$\bar{1}$0], [001] and [00$\bar{1}$], which are all equivalent, with the same atomic or molecular structure; and {100} refers to the set of all cube faces, (100), ($\bar{1}$00), (010), (0$\bar{1}$0), (001) and (00$\bar{1}$), which again are all equivalent, with the same atomic and molecular structure. This is shown in Figure 1.6 for directions [100], [010] and [001], and planes (100), (010) and (001).

7   Stereograms

The technique for representing the *orientation* of a crystal is called *stereographic projection*, and the resulting representation is called a *stereogram*. The crystal is placed at the centre of a large sphere, a crystal plane normal is projected up to intersect with the Northern Hemisphere, this (first) intersection is re-projected down to the South Pole, and the subsequent (second) intersection with the equatorial plane is taken to represent the plane and its orientation, and is called the *pole* of the plane. Figure 1.7 shows stereographic projection from a (111) plane to create a (111) pole, and Figure 1.8 shows a typical

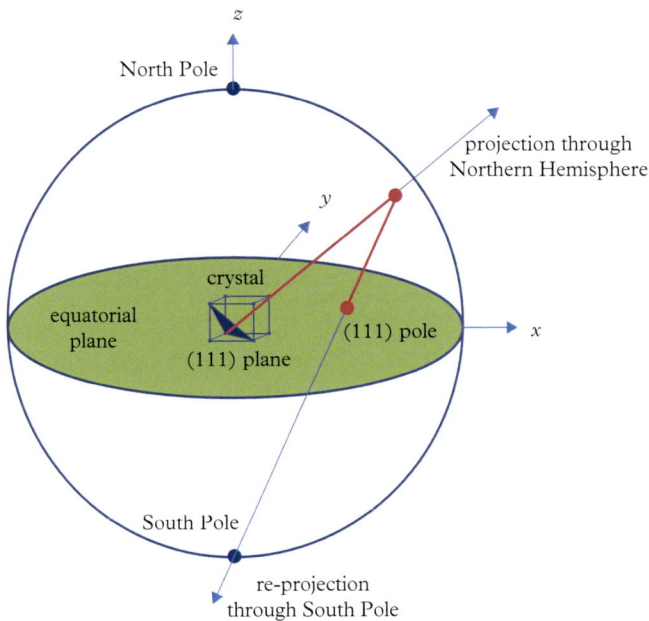

Figure 1.7 *Stereographic projection of a (111) plane in a cubic crystal*

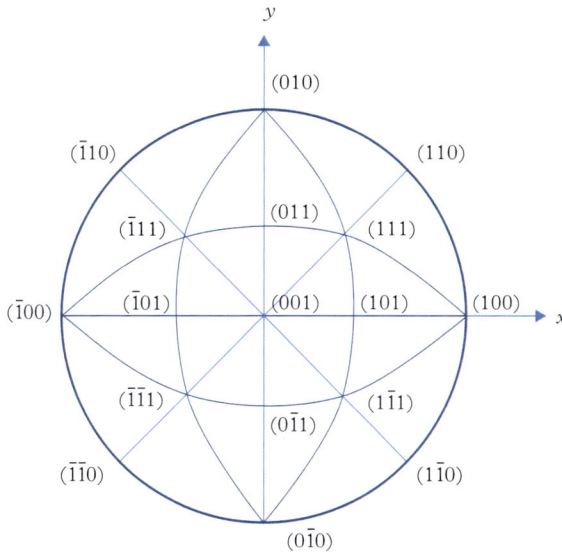

Figure 1.8 *Stereogram for a cubic crystal with [001] along the z-axis and at the North Pole*

stereogram for a cubic crystal with [001] along the z-axis, i.e. directed towards the North Pole.

8   Non-crystallographic materials

The only rotation axes compatible with the translation symmetry of a Bravais lattice are 1-, 2-, 3-, 4- and 6-fold, as shown in Figure 1.4. These are called *crystallographic symmetries.* Operation of these rotation axes at a given lattice point is compatible with the same operation at all other lattice points. Other rotation axes such as 5-fold, 7-fold, 8-fold, etc. are called *non-crystallographic symmetries.* Operation of these rotation axes at a given lattice point is incompatible with the same operation at other lattice points. For many years it was believed, therefore, that crystals could not be found with rotation axes such as 5-fold, 7-fold, 8-fold, etc. However, *quasicrystals* have been discovered, which appear to exhibit non-crystallographic symmetries such as 5-fold, 7-fold and 8-fold rotation axes.

In some solid materials, the atoms or molecules are not arranged in regular arrays. These materials are called *non-crystalline* or *amorphous.* Most polymeric materials are amorphous because their long-chain molecules easily become entangled and disorganised rather than aligned in a crystallographic array. Similarly, some ceramics are amorphous because their molecules easily form extended disorganised networks rather than become aligned in a crystallographic array. Similar effects are found, more rarely, in some metallic alloys.

....

9 REFERENCES

1. Eric Shipton. *Mountain Conquest* (American Heritage Publishing, New York, 1966), 28.
2. Cyril Douglas Milner. *Mont Blanc and the Aiguilles* (R Hale, London, 1955), 176.
3. Phil Edwards. 'Mont Blanc's ascent and the crazed crystal hunter who made it'. 8 August 2015. https://www.vox.com/2015/8/8/9119081/mont-blanc-first-ascent (accessed 3 December 2019).
4. Gaston Rebuffat. *The Mont Blanc Massif: The 100 Finest Routes,* Jane Taylor and Colin Taylor, trans. (Oxford University Press, New York, 1975), 12.
5. T. Louis Oxley. *Jacques Balmat or The First Ascent of Mont Blanc* (Kirby and Endean, 1881; re-printed by others including Trieste Publishing, Trieste, 2017, and Leopold Classic Library, Victoria, 2015).
6. Horace Bénédict de Saussure. *Voyages dans les Alpes* (Samuel Fauche, Neuchatel, 1779; and Barde, Manget, Geneva, 1786).
7. Marie-Hélène Reynaud. *Auguste Bravais: De La Laponie au Mont Blanc* (Editions du Viverais, Annonay, France, 1991), 138.
8. Ibid. 141.
9. Auguste Bravais. *Le Mont Blanc* (Wentworth Press, Sydney, 2016).
10. Reynaud, n 7, 22.
11. See ibid.
12. Ibid. 26.
13. Ibid.
14. Ibid. 36.
15. Ibid. 41.
16. Ibid. 34.
17. Auguste et Louis Bravais. 'Sur la disposition générale des feuilles curvisériées autour des tiges des végétaux'. *Annales des Sciences Naturelles* **VII** (1836), 42
18. Auguste et Louis Bravais. 'Essai sur la disposition générale des feuilles rectisériées'. In: *Congrès Scientifique de France 6th session* (Clermont-Ferrand, 1838), 53.
19. Ibid.
20. Auguste Bravais. 'Sur les propriétés géométriques des assemblages de points régulièrement distribués dans l'espace'. *C. R. des Séances de l'Académie des Sciences* **XXVII** (1848), 601.
21. Auguste Bravais. 'Études crystallographiques'. *Journal de l'École Polytechnique Paris* **XXXIV** (1851), 176.
22. Reynaud, n 7, 166.
23. See ibid, 182.
24. Ibid.
25. Ibid. 184.
26. Ibid. 186.
27. Ibid. 187.

....

10 BIBLIOGRAPHY

An Introduction to Crystallography. F. C. Phillips (Wiley, New York, 1972).

Auguste Bravais: De La Laponie au Mont Blanc. Marie-Hélène Reynaud (Editions du Viverais, Annonay, France, 1991).

Crystallography and Crystal Defects (2nd edition). Anthony Kelly and Kevin M. Knowles (Wiley, New York, 2012).

Structure of Metals (3rd edition). C. S. Barrett and T. B. Massalski (Pergamon, Oxford, 1980).

2

Bragg's Law

Diffraction

1   X-ray diffraction

When a beam of X-rays strikes the surface of a solid material, it behaves in a surprising way. An intense beam of reflected X-rays is seen at just a few special angles of incidence, with little or no reflected X-rays at all other angles of incidence. It turns out that this unusual behaviour is our essential tool for working out the complex atomic and molecular arrangements in different materials.

Figure 2.1 shows an arrangement for measuring the reflection of X-rays from a solid surface; Figure 2.2 shows a schematic of the resulting variation of X-ray intensity I with angle of incidence θ. High-intensity beams are reflected only at characteristic angles. The same basic behaviour is seen in whatever material is used for the solid surface and whatever source is used for the X-rays, but the exact angles of the intense, reflected beams depend on the material and the source.

The reason these results are surprising is because we are familiar with light beams, which produce reflection at all angles of incidence. Light and X-rays are both electro-magnetic radiation, but they have different wavelengths, and this is why their behaviour is so different.

Light beams have wavelengths in the range 0.2 to 0.8 μm, whereas X-ray beams have much smaller wavelengths, in the range 0.1 to 1.0 nm, close to the atomic size, which is typically 0.3 to 0.5 nm. X-rays are not really reflected by the surface in the way that light is. Instead, they are *diffracted* by the atoms and molecules at and just below the solid surface. Diffraction arises when waves interact with objects that are similar in size to the wavelength. The reflected waves from the different objects then add up in a complex way to produce a *diffraction pattern* of strong beams at characteristic angles, as shown in Figure 2.2.

The Equations of Materials. Brian Cantor. Oxford University Press (2020). © Brian Cantor.
DOI: 10.1093/oso/9780198851875.001.0001

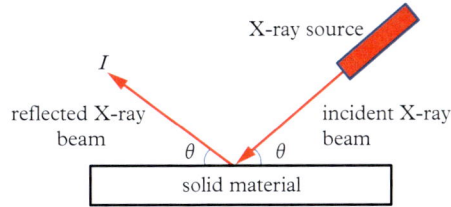

Figure 2.1 *Reflection of X-rays from a solid surface*

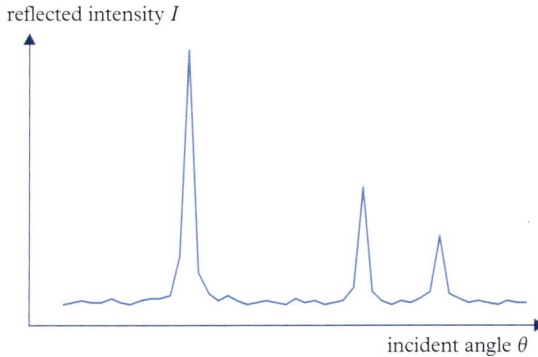

Figure 2.2 *Reflected X-ray intensity versus incident angle for copper X-rays and solid silicon*

2 Interference

To explain the surprising behaviour of X-ray reflection from a solid surface, we need to understand the *interference* of waves—in other words, how beams of waves add up when they interact. The disturbance caused by a wave is measured by its *amplitude A*. The strength or *intensity I* of the wave is measured by the square of its amplitude: $I = A^2$. The overall effect of superposing two beams of waves is to produce a disturbance that cannot be obtained simply by summing their amplitudes or intensities. This is because the overall disturbance depends not only on the strength of the two waves, but also on their relative *phase* (i.e. the relative position of their maxima and minima). This effect is known as *interference* between the two waves.

Figure 2.3 shows waves from two X-ray sources separated by a distance a, both emitting X-rays with the same amplitude A, intensity $I = A^2$, and wavelength λ, in the same direction along the x-axis. The *path difference* between the two waves is a and the *phase difference* between the two waves is defined as $\varphi = 2\pi a/\lambda$.

The disturbances caused by the two waves individually, y_1 and y_2, are given by

$$y_1 = A \sin \frac{2\pi x}{\lambda}$$

$$y_2 = A \sin \frac{2\pi (x - a)}{\lambda}.$$

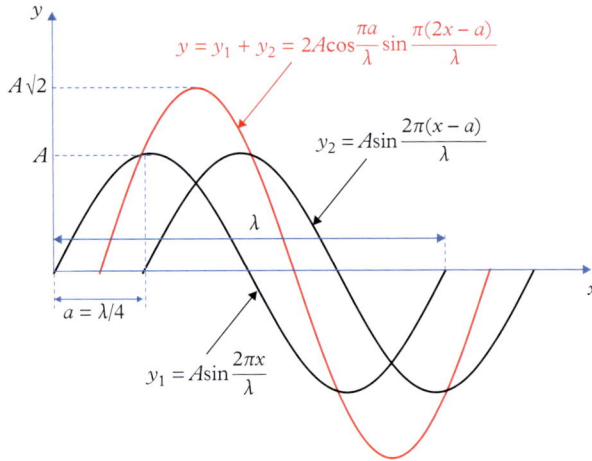

Figure 2.3 *Superposition of two waves, travelling in the same direction and separated by a distance a*

The overall disturbance y caused by superposition of the two waves is given by the standard trigonometric rule for summing sine waves,

$$y = y_1 + y_2 = A \sin \frac{2\pi x}{\lambda} + A \sin \frac{2\pi (x - a)}{\lambda}$$

$$= 2A \cos \frac{\pi a}{\lambda} \sin \frac{\pi (2x - a)}{\lambda},$$

which is a wave of amplitude $2A \cos (\pi a / \lambda)$ and phase $\pi a / \lambda = \varphi / 2$ (i.e. a phase and, therefore, maxima and minima halfway between the phases, maxima and minima respectively, of the two superposed waves). The combined intensity $I_{combined}$ of the two superposed waves is

$$I_{combined} = 4A^2 \cos^2 \frac{\pi a}{\lambda} = 4I \cos^2 \varphi / 2.$$

Figure 2.3 shows how the disturbance caused by the superposition of two waves depends on path distance a and phase difference φ. If the two waves start at the same position, the path difference a and the phase difference φ are both zero. The waves are called *in phase*, and they superpose with *constructive* interference. The peaks and troughs of the two waves are coincident, so they add up simply with an overall amplitude of $2A$ and intensity of $4I$. If the second wave starts at a distance $\lambda/2$ after the first wave, the path difference is $a = \lambda/2$ and the phase difference is $\varphi = 2\pi a / \lambda = \pi = 180°$. These waves are called *out of phase* and they superpose with *destructive* interference. The peaks of each wave are coincident with the troughs of the other and vice versa, so they negate each other with no overall disturbance and zero overall amplitude and intensity. In Figure 2.3, the second wave starts at a distance $\lambda/4$ after the first wave, the path difference is $a = \lambda/4$, and the phase difference is $\varphi = 2\pi a/\lambda = \pi/2 = 90°$. The waves are partially out of phase and they superpose with an amplitude of $2A\cos(\pi/4) = A/\sqrt{2}$ and intensity $I/2$.

3   Bragg's law

The angles of incidence θ for intense X-ray reflection are found to be given by *Bragg's law*:

$$2d\sin\theta = n\lambda,$$

where d is the spacing of planes of atoms in the solid, λ is the X-ray wavelength, and n is any integer.

Bragg's law can be derived as shown in Figure 2.4. X-rays reflected from the different planes of atoms in a solid material travel a different distance. When the path difference is an integral number of wavelengths, waves reflected from adjacent planes of atoms are in phase and interfere constructively, producing an intense, reflected X-ray beam. However, when the path difference is not an integral number of wavelengths, waves reflected from the different planes of atoms are not in phase and interfere destructively, with no reflected X-ray beam. The condition for constructive interference and a strong, reflected X-ray beam is, therefore,

$$\text{path difference} = 2\text{BC} = 2d\sin\theta = n\lambda.$$

The strong reflected beams found in an experiment, such as that shown in Figure 2.1, can be determined from Bragg's law. For a crystal structure such as simple cubic, adjacent rows of atoms are identical for all crystal planes, and Bragg's law predicts an intense beam for each value of n with each set of planes $\{h\,k\,l\}$. It is convenient to treat the nth-order diffracted beam from the $\{h\,k\,l\}$ planes as the first-order diffracted beam from the $\{nh\,nk\,nl\}$ planes (equivalent to planes parallel to $\{h\,k\,l\}$ with a spacing d_{hkl}/n). Diffracted beams are predicted when

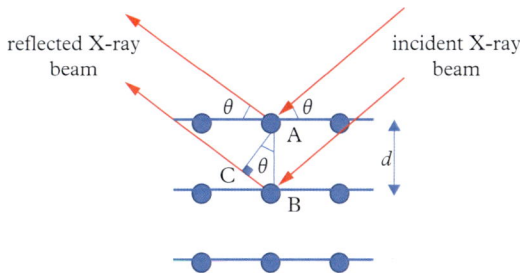

Figure 2.4 *Bragg's law obtained from the path difference between waves reflected from adjacent atomic planes*

$$\sin\theta = \frac{n\lambda}{2d_{hkl}} = \frac{\lambda}{2d_{hkl}}; \frac{\lambda}{d_{hkl}}; \frac{3\lambda}{2d_{hkl}}; \frac{2\lambda}{d_{hkl}}; \ldots; \text{ etc.}$$

For non-simple crystal structures such as face-centred cubic (fcc) or body-centred cubic (bcc), adjacent rows of atoms are not identical, leading to destructive interference and systematic absences of some of the diffracted beams. Diffraction is seen only when the crystal planes fulfil an additional criterion:

$$\text{fcc}: h, k, l \text{ all even}$$
$$\text{bcc}: h + k + l = \text{even.}$$

4   Lawrence Bragg

Lawrence Bragg

Science writer Lewis Thomas wrote, 'The greatest single achievement of nature to date is surely the invention of the molecule DNA'.[1] DNA—or deoxyribonucleic acid, to use its full name—is a polymeric (i.e. long-chain) molecule with two parallel helical chains (the famous *double helix*) of alternating sugar (deoxyribose, $C_5H_{10}O_4$) and phosphate (PO_4^-) groups, with pairs of *nucleobases* (adenine, cytosine, guanine and thymine, or just A, C, G and T) connecting the two chains regularly at right angles. Every human cell contains

46 *chromosomes*, and each chromosome is a DNA molecule of between 50 million and 200 million base pairs in length, so the total number of base pairs is more than three billion. The order of the bases contains all the information for cell growth, replication and functionality. Essentially, all instructions for human life are written from four letters—A, C, G and T—along the extended helical ladder of DNA. Reproduction takes place by unzipping the chains so that two identical molecules with the same order of base pairs can be created.

The structure of DNA was elucidated by Francis Crick and James Watson in 1953, working in the crystallography group in the Cavendish Laboratory in Cambridge under the leadership of Lawrence Bragg. Crick and Watson were an unlikely pair. The opening words of Watson's famous book *The Double Helix* are: 'I have never seen Francis Crick in a modest mood'[2]; the opening words of Crick's planned (but uncompleted) book *The Loose Screw* are: 'Jim [Watson] was always clumsy with his hands. One only had to see him peel an orange'.[3] Yet, when repairing to the Eagle pub in Cambridge after making their breakthrough, Watson remembered Crick announcing, 'We have found the secret of life'.[4] And Crick recalled telling his wife that evening, 'We seem to have made a big discovery'.[5] She later confided to him, 'You were always coming home and saying things like that, so naturally I thought nothing of it'.[6] Biochemist Erwin Chargaff commented acidly, 'Never before have two pigmies cast such a long shadow'.[7]

The story of Crick and Watson's discovery of the structure of DNA is remarkable. Late in 1951, a scientific conflab was underway at the Cavendish Laboratory. Maurice Wilkins and his colleagues Willy Seed, Rosalind Franklin and Ray Gosling had travelled up by train from King's College, London. Crick and Watson wanted to show off their new three-stranded helical model for DNA. Franklin was scathing, dismissing the new structure as completely implausible: there was, as yet, no direct X-ray evidence for a helical structure; the X-ray patterns were not compatible with phosphate groups on the inside, between the three strands of molecular backbone; magnesium ions used to bind the phosphate groups could not exist without hydration; and the amount of included water was far too small. She thought model-building was a waste of time. Progress would come only from further painstaking X-ray crystallography, which she was already doing. According to Gosling: 'Rosalind's view [was] that one could build models until the cows came home but it would be impossible to say which were nearer to the truth'.[8]

When Lawrence Bragg heard of the meeting, he was furious that his lab had looked so foolish. According to Watson, he found Francis Crick difficult: loud and bumptious, with a braying laugh, always acting as though he knew more than everyone else, even though he had not yet completed his own doctorate. This was the last straw. Bragg banned further DNA model-building. Crick and Watson were to return to the research work they were supposed to be doing: studying the structure of the proteins haemoglobin and myoglobin. At that time, most scientists still believed that proteins, rather than DNA, contained the key to life. In any case, there was a tacit agreement with King's that DNA was their baby. Crick and Watson complied overtly, putting their models out of view, but continued to discuss DNA and monitor developments by keeping in touch with Wilkins.

Two years later, Wilkins showed Franklin's latest results to Watson, the famous photo 51 of the B form of DNA. By now it was much clearer that DNA, rather than proteins, contained the key to life. In 1952, Alfred Hershey and Martha Chase at the Carnegie Institution in Washington, DC, had used isotope markers to show that, when a T_2 bacteriophage reproduces, phosphorus atoms from its DNA are passed on, but sulphur atoms from its proteins are not. The doyen of structural chemistry, Linus Pauling at Caltech in Pasadena, had published a successful single-stranded α helical structure for the polypeptide chains in proteins and was working on something similar for DNA. Franklin's new photo contained the crucial clues that a double-stranded, base-pair helix was the correct structure. Bragg was peeved they had been pipped to the post on polypeptides, and agreed to lift the ban on DNA modelling. They were all keen to beat Pauling, who was bound to find the answer sooner or later. Crick and Watson were elated and rushed ahead with their modelling. Their seminal paper on DNA was published in *Nature* on 25 April 1953.[9] Lawrence Bragg's invention—with his father, William Bragg— of the field of X-ray crystallography, which had already been so influential in discovering the structure of almost all physical matter, had now found its apogee in uncovering the most important structure of all biological matter. Crick and Watson had achieved one of the greatest breakthroughs in 20th-century science. For this, promoted strongly by Bragg, they shared with Wilkins the 1962 Nobel Prize for Medicine (sadly, Franklin died from cancer in 1958 so was not eligible).

William Lawrence Bragg ('Lawrence' or 'Willy') was born on 31 March 1890 in Adelaide, South Australia. His father, William Henry Bragg ('William'), was the Elder Professor of Mathematics and Experimental Physics at the University of Adelaide, and his mother was Gwendoline Todd, water colourist and daughter of Sir Charles Todd, Superintendent of Telegraphs, Government Astronomer, and Postmaster General for South Australia. (The town of Alice Springs was named after Todd's wife, Alice, while he was constructing the 3,200-km transcontinental Overland Telegraph, one of the greatest feats of Australian engineering in the 19th century.) Twenty-five years later, Lawrence and his father, William, would receive the Nobel Prize for Physics for their work on X-ray crystallography, the only case of a joint award to a father and son. Lawrence remained the youngest-ever recipient until the award of the Nobel Peace Prize in 2014 to Malala Yousafzai for her struggle for children's education against the Taliban in North Pakistan.

Bragg's father, William, was born in 1862 and was brought up in England, in Wigton, Cumbria, and then in Market Harborough, Leicestershire. William's father was a merchant sailor and farmer, and his mother was a clergyman's daughter. William Bragg studied at King William's School on the Isle of Man, excelling in maths, tennis and debating; becoming Head Boy; and, at age 17, winning an exhibition (junior scholarship) to Trinity College, Cambridge. Trinity was the epicentre of physics in Britain, the home of Sir Isaac Newton, James Clerk Maxwell, Lord Rayleigh and the then-Head of the Cavendish Laboratory and discoverer of the electron, J. J. Thomson. William Bragg graduated from Cambridge in 1884 as *third wrangler* (third-best student) in maths, and the following year was offered and accepted the Chair at Adelaide. According to William,

I was going along King's Parade to attend a lecture by J. J. Thomson at the Cavendish, and was joined on the way by the lecturer himself. He asked me [about] the Adelaide post. . . . I was astonished at the question: it had not occurred to me that anyone so young might be eligible . . . . I asked J. J. whether I might have any chance and he said he thought I might. So when the lecture was over I went and telephoned an application.[10]

He was just in time; it was the last day for applications. In the end, Bragg was appointed after an interview in London with the departing Professor Horace Lamb, J. J. Thomson and the Agent-General for South Australia, Sir Arthur Blythe.

William Bragg departed for Adelaide on P&O's finest liner, the 5,013-ton *Roma*, on 14 January 1886. He said, somewhat self-deprecatingly, 'I had never studied physics or chemistry [so] I tried to learn some on the way out'.[11] He was 24 years old, far from being a man of the world, with a character that was 'humble, private and self contained'.[12] After all, he had trained at St William's and Trinity, both closed and inward-looking clerical, all-male establishments. He arrived to take up his professorship in a young university with no equipment, no facilities and no money, in a rumbustious and muscular colonial city, in the middle of an economic crisis, with a drought and failing farms, mines and banks. The challenges were great, but Bragg's diffidence, good nature and intrinsic ability saw him through. He found Australia 'like sunshine and fresh invigorating air ... [with people] so open and kind and good natured'.[13] During the next two decades, he built laboratories for his students, quadrupled their number, initiated teacher training at the University, set up electrical engineering courses to support the introduction of electricity to Adelaide, oversaw the development of a School of Mines, helped establish the Australasian Association for the Advancement of Science, gave many successful public lectures and became a well-respected research scientist. He married Gwendoline in 1889, and they had three children: Lawrence (or Willy), Bob and Gwendy.

On 8 November 1895, Wilhelm Conrad Röntgen was experimenting with an electrical discharge tube at the University of Würzburg in Bavaria when he noticed a faint shimmering from a bench a few feet away from the tube. He had discovered X-rays. He ate and slept in his laboratory for the next few weeks while he studied their properties, quickly finding they passed through solid materials, notably the body, yielding an image of the underlying skeleton. For his discovery of X-rays, Röntgen received the first-ever Nobel Prize for Physics in 1901. William Bragg was excited by Röntgen's discovery. In May 1896, he demonstrated the new technology in Adelaide: 'Excellent photographs were achieved ... of William's hand, Alfred Lendon's foot and Dr Swift's knee and wrist, clearly showing the separate bones'.[14] Later that year, six-year-old Lawrence fell off his tricycle and smashed his elbow. His father X-rayed the elbow to monitor its recovery. As Lawrence later said, 'I must have been one of the first patients to be X-rayed in South Australia ... my father set up a tube worked by an induction coil ... I was scared stiff by the fizzing sparks and the smell of ozone'.[15]

In 1897 William asked for and was granted a year's leave of absence to visit his family in England and to 'hear about the latest discoveries in electromagnetic waves and radio, X-rays, natural radioactivity ... cathode rays and the electron'.[16] He was enthused and, on his return, set about initiating research into the nature of these different kinds of rays. This was a hotly contested topic: were they waves or particles? In fact, it took many more years of experimental research and theoretical understanding of quantum effects and wave–particle duality to identify α-rays and β-rays as beams of helium ions and electrons respectively, and γ-rays and X-rays as electromagnetic radiation with different wavelengths. Bragg was strongly of the view that radiation consisted of streams of particles. He discovered the *Bragg ionisation curve* describing the way α-rays are absorbed as they travel through matter. Light and X-rays act as waves, and lose less and less energy as they are gradually absorbed by a gas or solid. On the other hand, α-rays are helium ions and act as particles, so they slow down, increase their interactions and lose increasing amounts of energy. The resulting *Bragg peak* occurs just before they are fully absorbed. In 1907 Bragg was elected Fellow of the Royal Society (FRS). His fellow Adelaide professor George Henderson left a message on his desk: 'Hurrah, hurrah for Bragg and his University. Splendid'.[17] That same year Bragg was offered and accepted the Cavendish Chair of Physics at Leeds University, which he took up two years later.

Lawrence Bragg had a largely happy childhood in Adelaide. In his autobiography, he describes vacations in coastal resorts such as Port Elliot, Noarlunga and Yankalilla, and inland hills such as Mount Lofty. He developed a love of gardening, which stayed with him throughout his life. 'I remember vividly seeing the green tip appear from a daffodil bulb ... . I had a prolific peach tree in my garden, but the pride and glory was an immense vine'.[18] He went to Queen's Prep School until he was 11, and then St Peter's College. Aged just 15, he started at the University of Adelaide, graduating with a first-class degree in maths three years later, i.e. at an age when most students are just starting their undergraduate studies. He was academically gifted but emotionally underdeveloped. 'I was a misfit at school, being so very immature in some ways and so precocious in others'.[19] He claimed his emotional age when he started university was only 12, and described how he was at times bullied by his schoolmasters and ostracised by his classmates. Nevertheless, he excelled academically and enjoyed cycling, rowing, golf and debating. As well as gardening, his main hobby was collecting shells, building up a collection of about 500 species: 'I found a new species of cuttlefish which was named *Sepia braggi* by Dr Verco in Adelaide'.[20] He graduated in 1908 with a first-class degree in maths, and travelled to England with the family the following year to register, like his father before him, as a student at Trinity College, Cambridge. The Head of the Cavendish was still J. J. Thomson. Lawrence graduated in 1912 with another first-class degree, this time in physics.

Earlier that year, a young *privatdozent* (non-salaried lecturer) at the Institute of Theoretical Physics in Munich, Max Laue, was discussing the structure of solid crystals with Paul Ewald, a student of the Head of the Institute, Arnold Sommerfeld. For many years it had been thought that crystals might consist of repeating groups of atoms, but there was no direct proof. Ewald mentioned that the repeat distances would be on the order of Ångstroms (1 Ångstrom, or Å, is 0.1 nm), and Laue had a brainwave. If X-rays were indeed waves, their wavelengths would also be of the order of Ångstroms, so they should be diffracted by crystals in the same way as light waves are diffracted by regular arrays of slits. Sommerfeld would not let Laue and Ewald try the experiment, but they persuaded Paul Knipping, an ex-student of Röntgen, and Walter Friedrich, Sommerfeld's assistant, to help them do it surreptitiously. After several abortive attempts, on the evening of 21 April 1912, they successfully obtained a photograph of spot patterns from X-rays impinging on a zinc blende (ZnS) crystal, which seemed to confirm Laue's diffraction hypothesis. Oxford crystallographer Mike Glazer reported, 'Ewald once told me that in fact Friedrich and Kipping had secretly stolen Röntgen's apparatus in order to carry out the experiment'.[21] Laue received the 1914 Nobel Prize for this discovery.

That same year, on 26 June in Leeds, William Bragg received a remarkable letter from his friend the Norwegian physicist Lars Vegard: 'Recently, certain new curious properties of X-rays have been discovered by Dr Laue … . As I thought the matter would interest you, I asked Dr Laue … to give me one of his photographs to send you'.[22] He explained that Laue considered the spot patterns to be caused by diffraction of waves, with the crystal acting as 'a kind of grating',[23] but the results could not be explained fully. There were three difficulties: applying the diffraction condition for gratings in three dimensions was complex, the spots seemed too sharp, and they became elliptical depending on the position of the photographic plate. Lawrence was home from Cambridge for the summer and he discussed the problem with his father while they were on holiday at Cloughton on the Yorkshire coast. William was still wedded to a particle theory of X-rays, and Lawrence suggested the particles might have been channelled between the atomic rows in the crystal, but this explanation also didn't seem to work.

Lawrence went back to Cambridge, where he had begun his research work, and continued to nag away at the problem, later explaining,

> I can remember the exact spot on the backs [the fields adjacent to the River Cam, just behind some of the Cambridge colleges] where the idea suddenly leapt into my mind that Laue's spots were due to the reflection of X-ray pulses by sheets of atoms in the crystal.[24]

Reflection from planes of atoms led directly to Bragg's law, which is much simpler to use than the complexity of three one-dimensional grating equations (the *Laue equations*), though the two approaches can be shown to be mathematically equivalent. Laue had made critical errors in trying to match up his photographic spots with diffraction theory. He had assumed the X-ray beams were excited in the crystal with a single characteristic wavelength, and he was using an incorrect simple cubic crystal structure for zinc blende. When Lawrence Bragg used his new law with *bremsstrahlung* (i.e. white X-rays with a range of many wavelengths, like white light), and the correct fcc crystal structure, he got perfect agreement with the photograph, and was also able to explain the variation of sharpness and ellipticity of the spots.

He presented his results to the Cambridge Philosophical Society on 11 November 1912, and set about building an X-ray system to repeat Laue's experiments. The Cavendish was not well set up for experimental work:

> I had to manage with bits of cardboard and drawing pins … . I got so excited I worked the coil too hard and burnt out the platinum contact … the head mechanic who doled out the stores was very angry … it cost 10 shillings and to 'larn' me he made me wait one month for a replacement.[25]

Despite such problems, Lawrence proceeded to use his X-ray diffraction camera to determine a wide range of crystal structures: Laue's zinc blende (ZnS) and copper sulphate ($CuSO_4$); mica (a mixed complex silicate); alkali halides such as potassium and sodium chloride (KCl and NaCl respectively); other ionic compounds such as fluorspar (CaF_2), calcite ($CaCO_3$) and iron pyrites (FeS_2), and, triumphantly, diamond carbon. As he said later, 'It was a wonderful time, like discovering a new goldfield where nuggets could be picked up on the ground'.[26]

In the meantime, William at Leeds was developing the first X-ray spectrometer using the angles of diffraction from a crystal together with Bragg's law to investigate the wavelengths of emitted X-rays. In an ionisation tube, electrons from a cathode are transmitted to an anode, where they excite X-rays, with a mixture of *bremsstrahlung* and characteristic X-rays with specific wavelengths associated (as demonstrated later) with particular electronic transitions within the atoms of the anode material. William was able to determine the characteristic X-ray spectra of a wide range of anode materials, such as osmium, iridium, platinum, palladium, rhodium, copper, nickel, silver and tungsten.

Within a couple of years, this tremendous outpouring of scientific discovery from father and son was brought to a shuddering halt by dark external events: the onset of the First World War. On 28 June 1914, the Archduke Franz Ferdinand, heir to the throne of the Austro-Hungarian Empire, was assassinated in Sarajevo, shot in the street by the Yugoslav nationalist Gavrilo Princip. According to the historian Zbynek Zeman, 'the event [at first] almost failed to make any impression whatsoever … the crowds in Vienna listened to music and drank wine as if nothing had happened'.[27] Not surprisingly, however, the Emperor Franz Joseph was profoundly shocked and upset. In fact, the assassination brought to a head a series of simmering international tensions, and set light to a tinderbox. After a month of intense diplomatic manoeuvring between Austria–Hungary, Russia, Germany, France and Britain, Austria–Hungary issued an ultimatum to Serbia, 10 demands engineered to be undeliverable, and on 28 July declared war. Russia mobilised in support of Serbia, so in a pre-emptive strike, Germany declared war on Russia on 1 August, invaded Luxembourg on 2 August, and declared war on France on 3 August and Belgium on 4 August. That same day, Britain declared war on Germany, having issued an ultimatum to keep Belgium neutral. One of the deadliest periods of human conflict ever had been set in train. The war spread rapidly to 40 countries worldwide. Four-and-a-bit years later, more than nine million combatants and seven million civilians had died, and the political map of Europe and, to a fair extent, the rest of the world had been re-drawn. The Austro-Hungarian and Ottoman Empires collapsed; the Russian Empire was emasculated, leading to the Leninist–Marxist revolution of 1917; Japan expanded its influence dramatically throughout East Asia; and the United States emerged as one of the world's superpowers.

The onset of war disrupted scientific collaboration and created bad blood. The *Manifesto of 4 October: To the Cultural World*[28,29] was signed by 93 German intellectuals, including Friedrich Wilhelm Ostwald, Röntgen and Max Planck, insisting that Germany did not start the war and did not violate Belgian neutrality. The British response, *A Reply to German Professors*,[30,31] was signed by 117 British scholars, including William Bragg, Lord Rayleigh and J. J. Thomson, rejecting German accusations that England started the war. The German physiologist and pacifist Georg Friedrich Nicolai prepared an *Appeal to Europeans*,[32,33] disavowing the belligerence, but only three others (including Albert Einstein) signed it, and Nicolai was demoted and vilified for his pains. In June 1914, just before the outbreak of the war, William Bragg had been invited as an honoured guest to attend the Deutsche Physikalische Gesellschaft annual meeting and 'kindly speak about your latest research work concerning the dispersion of X-rays … . Why not bring your son over? We would be delighted to make his acquaintance'.[34] Needless to say, the visit never took place.

Instead of going to Germany, Lawrence was commissioned on 26 August 1914 as a 2nd Lieutenant in the Leicestershire Royal Horse Artillery and was posted to Diss in Norfolk for training: 'I was very much a fish out of water … . The officers were hunting men, who talked and thought horse to the exclusion of most other interests'.[35] He spent the best part of a year in Diss, bored to death and plagued by colds and flu. In July 1915 he was sent to the front, seconded to the Royal Engineers to develop sound-ranging, i.e. using a microphone array to pinpoint the position of enemy artillery. Days later, he received a letter: his brother Bob, with the allied forces in Gallipoli, had been hit by a shell and died the following day. As his biographer John Jenkin said, 'Lawrence's emotions were on an unprecedented roller-coaster ride'.[36] The sound-ranging section moved to Ypres. Their problem was how to differentiate the low-frequency noise of a field gun from the much louder shock waves from exploding shells under difficult field conditions: 'The weather has become atrocious, wet every day. The mud is unbelievable'.[37] Surreally, letters arrived to say that he and his father had been awarded the Nobel Prize: 'the friendly priest, in whose house we were quartered, brought up a bottle of Lachryma Christi [Italian red wine] from his "cave" [wine cellar] to celebrate'.[38] Lawrence's sound-ranging was vital for the British victories at Cambrai in 1917 and Amiens in 1918. By the end of the war there were 40 sound-ranging sections on the battlefront, each with about 50 staff and equipment. Lawrence was promoted to Captain and then Major, mentioned in dispatches three times, and awarded the Military Cross and made an Officer of the Order of the British Empire (OBE). He emerged from the war mature and self-confident, but he had seen great horror. His daughter said later that she 'sometimes woke … to hear him banging his head … trying to exorcise the scenes of war that came to him in the night'.[39]

Instead of going to Germany, William spent the last few months of 1914 on a lecture tour to the United States and Canada. On his return, he accepted an offer from University College, London, of the Quain Chair of Physics, which he took up in September 1915. He also joined the Board of Invention and Research (BIR), set up by the British government to provide scientific advice to the Royal Navy. The BIR's largest and most important section was on submarines and wireless telegraphy. Britain was heavily dependent on imports, and therefore shipping, for foodstuffs and raw materials, and German submarines were blockading Britain and attacking the merchant fleet. The work was initially spearheaded by the New Zealand physicist Ernest Rutherford, an old friend of William's. Rutherford came up with the idea of using acoustic detection, leading to the development of hydrophones and, later, asdic and sonar, but he became distracted by his ongoing fundamental research in Manchester, famously splitting the atom in 1917. While Lawrence was using sound to detect artillery in France, William was using sound to detect submarines under the waters around Britain. The work was bedevilled by petty disputes between the BIR and the Admiralty, between civilian and navy scientists, and between British and French systems. Substantial numbers of hydrophones were used, but '[of] all the U-boats destroyed by patrols during the war, only four were definitely sunk because of hydrophone contact'.[40] The answer to the submarine threat proved to be convoys with armed escorts, introduced in 1917, at which point marine losses reduced dramatically. Nevertheless, William was made a Commander of the Order of the British Empire (CBE) in 1917, and received a knighthood in 1920.

The Royal Swedish Academy of Sciences had suspended their Nobel award ceremonies during the war. They invited all prize winners for the years 1914 to 1919 to give their lectures in Stockholm at a grand ceremony in 1920. William and Lawrence didn't go, unhappy because of the presence of so many German scientists (including Laue and Planck). William never did give a Nobel lecture: 'the pain of the war had destroyed the joy of the award, and his previous internationalism was also a casualty',[41] but Lawrence did finally go to Stockholm to give his lecture in 1922. In 1923, William became Fullerian Professor of Chemistry and Director of the Davy Faraday Laboratory at the Royal Institution, where he continued to lecture and research until his death. In 1931 he was awarded the Order of Merit, the highest honour available in Britain, a personal gift of the King, restricted to 24 members. On 24 March 1942, aged 80, William went to bed feeling unwell and died two days later. According to Edward Andrade, 'Bragg's nature was simple, straightforward and tenacious … . He was a very great experimenter, who never wasted time on the trivial … . His work like his personality was simple yet profound, sincere and compelling'.[42]

After the war Lawrence Bragg returned to Cambridge briefly, just long enough to meet and propose to his future wife, Alice Hopkinson, a history student at Newnham College, although initially she turned him down. In 1919 J. J. Thomson resigned from the Cavendish to be replaced by Rutherford, and Lawrence was offered and accepted Rutherford's previous position as Langworthy Professor of Physics at Manchester. He found life difficult:

> I was only 29 when I took up my duties as Head of the Physics Lab at Manchester, and was handicapped by not having served my apprenticeship in a junior position in a department. Further, we had forgotten most of our physics during the war … . It was not made easier by a vile series of anonymous letters … [that] abused us bitterly as incompetent and useless … it drove me into what was really a nervous breakdown … . I recovered when the letters began to attack my father and Rutherford as well.[43]

Scientifically, for the next almost two decades, Lawrence determined the crystal structures of a series of complex spinels, ordered alloys and other inorganic materials. In 1921 he was elected FRS and finally married Alice and eventually had four children. In 1937 he became, briefly, Director of the National Physical Laboratory near London, before taking over the following year from Rutherford as Professor of Physics and Head of the Cavendish at Cambridge. This was again difficult: 'the Cavendish Laboratory had become a leading centre for nuclear physics … . The selection of a non-nuclear physicist could be seen as jettisoning Rutherford's legacy'.[44] However, Lawrence successfully re-structured the Cavendish to focus on a wider variety of physical phenomena. He initiated his research on haemoglobin with Max Perutz, who had fled from the Nazis in Vienna.

Lawrence's work was again interrupted by a major world war, and he spent time as the British Scientific Liaison Officer in Canada, as an advisor to the British Navy on asdic, and to the Army on sound-ranging. He was knighted in 1941. After the war, he set up in the Cavendish the Medical Research Council Unit to study biological molecular structures, which led to the discovery of the structures of DNA, haemoglobin and myoglobin. In 1953 he left the Cavendish and, following his father, became Fullerian Professor of Chemistry and Director of the Davy Faraday Laboratory in the Royal Institution from 1954 to 1966. Yet again there were difficulties, with major fights with the previous, autocratic Director, Edward Andrade. Like his father, Lawrence was a superb lecturer to children: they both presented the Royal Institution's Christmas Lectures for children on many occasions and to great acclaim. He retired in 1966 and, in 1971, survived a cancer operation, but suffered a relapse and died on 1 July 1971, aged 82.

Lawrence Bragg loved gardening all his life, and Crick recounts a story that, when he moved to the Royal Institution, he missed his garden and

> arranged that for one afternoon a week he would hire himself out as a gardener to an unknown lady living in the Boltons, a select inner-London suburb. He respectfully tipped his hat to her and told her his name was Willie.... all went well till one day a visitor, glancing out of the window, said to her hostess 'my dear what is Sir Lawrence Bragg doing in your garden'?[45]

Throughout his career, Lawrence was plagued by doubt and irritation because his father seemed to get most of the credit, even though Bragg's law had been solely his idea. He was attacked for being a classical physicist, a dinosaur in an age when the frontiers of science were being re-drawn by the exciting and bizarre revolutions of quantum mechanics and relativity. He was an outsider, a colonial, uncomfortable at High Table in Cambridge colleges. He was not a theoretical physicist, nor a mathematical physicist, but he was a brilliant experimentalist. As the physicist Brian Pippard put it, 'For Lawrence, beauty and economy were the touchstones of a physical argument or an experiment, and unless one sympathised with his quest for these ideals one missed his intellectual power and subtlety'.[46] Watson said, when first meeting Bragg, that he 'reminds you of the Colonel Blimp type ... sitting in London clubs like the Athenaeum',[47] but he discovered differently. Bragg liked Watson, who was also an outsider, and agreed to write the foreword to *The Double Helix*,[48] praising Watson generously and graciously, even though Bragg was heavily criticised therein. In 1966 Lawrence was awarded the prestigious Copley Medal of the Royal Society, and the citation said, 'Three new subjects, mineralogy, metallurgy and, now, molecular biology, all first sprang from your head, firmly based on applied optics'.[49]

5   X-ray scattering

Scattering by an electron

Bragg's law is very successful at explaining the characteristic scattering angles seen for X-ray diffraction from solids. However, X-rays are not, of course, scattered directly by

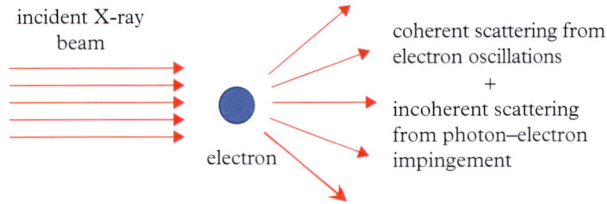

Figure 2.5 *Coherent and incoherent X-ray scattering by an electron*

planes of atoms, as implied by the derivation of Bragg's law in Figure 2.4. In fact, X-rays are scattered by their interactions with electrons.

There are two types of X-ray scattering by electrons:

(1) coherent or elastic scattering, when the scattered and incident X-rays are in phase and have the same energy; and

(2) incoherent or inelastic scattering, when the scattered and incident X-rays are not in phase and have different energies.

This is shown schematically in Figure 2.5. A beam of X-rays can be regarded as either an electromagnetic wave or a stream of photons. Its wave nature is responsible for coherent scattering, and its particle nature is responsible for incoherent scattering.

Coherent scattering

Consider coherent scattering first. X-rays are electromagnetic waves, i.e. they consist of oscillating electric and magnetic fields. When an X-ray beam strikes an electron, the electron's charge causes it to oscillate at the same frequency ν as the incident beam, with $\nu = c/\lambda$, where c is the speed of light. As it oscillates, the electron accelerates and decelerates, causing it to re-emit X-rays in all directions, again with the same frequency ν as the incident beam, in phase and coherent with it.

The intensity I of coherent X-rays scattered by a single electron was first determined by J. J. Thomson, who showed that it depends on the distance r from the electron and the scattering angle θ:

$$I = I_o \frac{e^4}{r^2 m^2 c^4} f(\theta),$$

where I_o is the incident intensity, e and m are the electron charge and mass respectively, c is the velocity of light, and $f(\theta) = \sin^2\theta$ or $(1 + \cos^2\theta)/2$ for polarised or unpolarised X-rays respectively.

The intensity of the scattered X-rays decreases with distance r from the electron, following an inverse square law, and is strongest in the forward or backward directions compared with at right angles to the incident beam. The amplitude of coherent X-rays scattered by a single electron is, of course, given by $A^2 = I$ or $A = \sqrt{I}$. It is the coherent scattering that causes interference and diffraction, leading to strong X-ray peaks at characteristic angles, as shown in Figure 2.2 and as described by Bragg's law.

Incoherent scattering

Now consider incoherent scattering. As explained earlier, an X-ray beam can also be regarded as a stream of photons moving at the speed of light c. When a photon strikes an electron directly, it loses energy to the electron and is deflected from its path. Losing energy leads to a corresponding decrease in frequency and increase in wavelength, since the X-ray energy E is given by the Planck-Einstein equation

$$E = h\nu = \frac{hc}{\lambda},$$

where ν and λ are the frequency and wavelength respectively, h is Planck's constant and c is the speed of light.

The more directly an X-ray photon strikes an electron, the greater its energy loss, the bigger the scattering angle θ, the lower the scattered X-ray frequency ν, and the higher its wavelength λ. In other words, the scattered photons are incoherent and have a range of energies $E < E_{incident}$, frequencies $\nu < \nu_{incident}$ and wavelengths $\lambda > \lambda_{incident}$. This effect is known as *Compton scattering*. The intensity of Compton scattering depends on the probability of direct photon–electron strikes and is, therefore, quite low. Incoherent Compton scattering is responsible for the non-zero background away from the characteristic X-ray diffraction angles, as shown in Figure 2.2.

6   The structure factor

When an X-ray beam is scattered by an atom, the resulting beam is the sum of the scattering from all the electrons in the atom. The coherent scattering from all the electrons has the same amplitude and intensity, given by Thomson's equation. In the forward direction, scattering from all the electrons is in phase, but this is not so for all other directions.

The atomic scattering factor f is defined as the ratio of the amplitudes scattered by an atom and an electron:

$$f = \frac{A_{atom}}{A_{electron}} = \frac{A_{atom}}{\sqrt{I}}.$$

In the forward direction $f = Z$, where Z is the atomic number (the number of electrons in the atom), since all the electrons scatter in phase. Because of the differences in phase in other directions, f decreases with scattering angle θ and increases with X-ray wavelength λ, approximately proportional to $\sin\theta/\lambda$. X-ray handbooks tabulate values of atomic scattering factor f for all atom species as a function of $\sin\theta/\lambda$, so that X-ray amplitudes and, therefore, intensities can be calculated in practical situations.

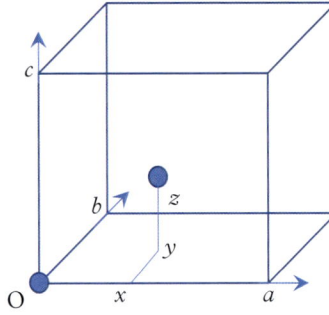

Figure 2.6 *Two atoms with coordinates (0 0 0) and (x y z)*

When an X-ray is scattered by a crystal, the resulting beam is the sum of the scattering from all the atoms in the crystal. Consider the nth atom in the unit cell, as shown in Figure 2.6, with coordinates $(x_n \ y_n \ z_n)$ and fractional coordinates $(u_n \ v_n \ w_n)$, where $u_n = x_n/a$, $v_n = y_n/b$ and $w_n = z_n/c$, and a, b and c are the unit cell dimensions. Strong scattering takes place in directions given by Bragg's law, i.e. at a series of angles θ corresponding to reflecting atomic planes $\{h \ k \ l\}$. Scattering by the nth atom in the direction θ has an amplitude and phase given by

$$\text{amplitude} = f_n$$
$$\text{phase} = \varphi_n = 2\pi \ (hu_n + kv_n + lw_n) .$$

To sum the different waves from all the atoms in the unit cell, it is convenient to use a complex representation for the wave scattered by the nth atom, as follows:

$$\psi_n = f_n \exp (i\varphi_n) = f_n \cos \varphi_n + if_n \sin \varphi_n$$
$$= f_n \exp 2\pi i (hu_n + kv_n + lw_n)$$
$$= f_n \cos 2\pi \ (hu_n + kv_n + lw_n) + if_n \sin 2\pi \ (hu_n + kv_n + lw_n)$$

The resulting wave scattered by the crystal is obtained by summing over all atoms in the unit cell, and is called the *structure factor F*:

$$F_{hkl} = \sum_1^N \psi_n = \sum_1^N f_n \exp 2\pi i (hu_n + kv_n + lw_n),$$

with an amplitude given by its modulus $|F|$:

$$|F| = \sqrt{FF^*} = \frac{A_{\text{unit cell}}}{A_{\text{electron}}} .$$

7 Fourier transforms

Consider an incident X-ray beam scattered by two electrons at points O and P, separated by a vector r within a solid material, with the incident and scattered directions represented by the unit vectors k_o and k, as shown in Figure 2.7. The path difference and, therefore, the phase difference between the two scattered waves are given by

$$\text{path difference} = \text{PA} - \text{OB} = r.k - r \cdot k_o = r \cdot (k - k_o)$$

$$\text{phase difference} = \left(\frac{2\pi}{\lambda}\right) r \cdot (k - k_o) = r \cdot q,$$

where the vector q is defined as $q = (2\pi/\lambda)(k - k_o)$.

The scattered wave in Figure 2.7 is given by

$$F(q) = \exp(ir \cdot q),$$

where the amplitude has again been normalised to the amplitude scattered from a single electron. X-ray scattering from the whole material is obtained by integrating for all values of r within it,

$$F(q) = \int \rho(r) \exp(ir \cdot q) \, dr,$$

where $\rho(r)$ is the density distribution of electrons throughout the material and $F(q)$ is called the *Fourier transform* of $\rho(r)$.

Fourier transformation is clearly a generalisation of the structure factor calculation for diffraction from a crystal to a more general body containing any distribution of electrons and atoms. When we obtain an X-ray diffraction pattern, we are effectively measuring

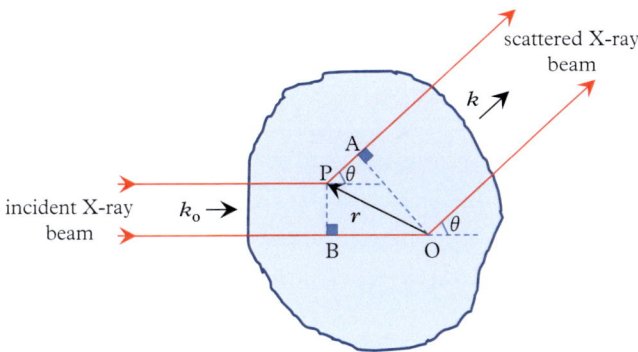

Figure 2.7 *X-ray scattering by two electrons at O and P*

$F(q)$. The corresponding distribution of electrons and, therefore, atoms and molecules can, in principle, always be obtained be inverting the Fourier transformation:

$$\rho\,(r) = \int F(q)\exp\,(-ir\cdot q)\,dq.$$

This is effectively what we did for crystal structures in the last section. Unfortunately, it is not always straightforward to perform an inverse Fourier transformation. Instead, we often have to guess the electronic and atomic structure and determine whether its Fourier transform agrees with the observed diffraction pattern.

8   Electron diffraction and neutron diffraction

Wave–particle duality was first discovered for electromagnetic waves such as light and X-rays. A light beam or X-ray beam exhibits the characteristic wave behaviour of interference and diffraction when interacting with a suitably sized grating or regular array of objects, but it also exhibits characteristic particle behaviour, e.g. in the photoelectric effect, when photons in the beam eject electrons from a solid surface. Subsequently, wave–particle duality was also discovered for beams of fundamental particles such as electrons and neutrons, both of which are also used in diffraction experiments, like X-rays, to study the structure of materials.

Louis de Broglie showed that the wavelength λ of a beam of particles of mass m and velocity v is given by

$$\lambda = \frac{h}{mv},$$

where h is Planck's constant. Electrons have a mass m and a charge e. If they are accelerated through a voltage V to reach a velocity v, their energy is

$$E = eV = \frac{1}{2}mv^2$$
$$\therefore v = \sqrt{\frac{2eV}{m}}$$

and their wavelength is, therefore,

$$\lambda = \frac{h}{\sqrt{2meV}}.$$

Electron diffraction is usually achieved using an electron microscope. The electrons are accelerated by voltages of 100 to 200 kV, corresponding to wavelengths of 3 to 4 pm, i.e. much smaller than typical X-rays. This allows fine-scale crystallographic information to be obtained by electron diffraction, which can also be correlated with microstructural information from electron imaging to study, for example, the different phases and crystals in a material, and the nature of defects such as *dislocations* and *grain boundaries*.

Neutron diffraction is usually achieved with a neutron beam from a nuclear reactor, which makes experiments difficult and expensive. Neutrons are scattered by the atomic nuclei rather than the electrons in a material, with neutron wavelengths ranging widely depending on the nature of the nuclear reaction used to generate the neutrons and the extent of moderation (slowing down) of the neutron beam. Neutron diffraction can be used to determine very fine-scale crystallographic details.

··

9 REFERENCES

1. Lewis Thomas. *The Medusa and the Snail: More Notes of a Biology Watcher* (Viking Press, 1979; re-printed Penguin, 1995), 27.
2. Alexander Gann and Jan Witowski (eds.). *The Annotated and Illustrated Double Helix: James D. Watson* (Simon and Schuster, New York, 2012), 3.
3. Francis Crick. 'The DNA double helix: the untold story'. *Liquid Crystals Today* 12 (2003), 1, note 4.
4. Gann and Witowski, n 2, 209.
5. Francis Crick. *What Mad Pursuit: A Personal View of Scientific Discovery* (Basic Books, New York, 1988), 77.
6. Ibid.
7. Crick, n 3, note 2.
8. Gann and Witowski, n 2, 91.
9. J. D. Watson and F. H. C. Crick. 'Molecular structure of nucleic acids: a structure for deoxyribose nucleic acid'. *Nature* **171** (1953), 737.
10. John Jenkin. *William and Lawrence Bragg, Father and Son* (Oxford University Press, Oxford, 2008), 61.
11. Ibid. 70.
12. Ibid. 85.
13. Ibid.
14. Ibid. 146.
15. A. M. Glazer and Patience Thomson, eds. *Crystal Clear: The Autobiographies of Sir Lawrence and Lady Bragg* (Oxford University Press, Oxford, 2015), 48.
16. Jenkin, n 10, 159.
17. See ibid. 223.
18. Glazer and Thomson, n 15, 58.
19. See ibid. 57.
20. Ibid. 63.
21. Ibid. ix.
22. Jenkin, n 10, 326.
23. See ibid. 327.
24. Ibid. 330.
25. Ibid. 334.
26. Ibid. 340.

27. H. P. Wilmott. *World War I* (Dorling-Kindersley, London, 2003), 26.

28. Jürgen von Ungern-Sternberg and Wolfgang von Ungen-Sternberg. *Der Aufruf 'An die Kulturwelt!': das Manifest der 93 und die Anfänge der Kriegspropaganda im Ersten Weltkrieg* (Verlag, Stuttgart, 1996).

29. G. F. Nicolai. *The Biology of War*. C. A. Grande and J. Grande, trans. (Century, New York, 1918), xi–xiii.

30. 'A reply to the German professors', *The Times* 21 October 1914, 10.

31. 'A reply to the German professors'. 1914. http://www.ersterweltkriegheute.de/1914/10/21/reply-to-the-german-professors (accessed 6 March 2020).

32. Siegfried Grundman. *The Einstein Dossiers* (Springer, Berlin, 2005).

33. S. L. Wolff. 'Physicists in the Krieg der Geister: Wilhelm Wien's proclamation'. *Historical Studies in the Physical and Biological Sciences* **33** (2003), 345.

34. Jenkin, n 10, 349.

35. Glazer and Thomson, n 15, 87.

36. Jenkin, n 10, 372.

37. See ibid. 374.

38. Glazer and Thomson, n 15, 90.

39. Jenkin, n 10, 385.

40. See ibid. 394.

41. Ibid. 401.

42. Ibid. 428.

43. Glazer and Thomson, n 15, 98.

44. Jenkin, n 10, 432.

45. Crick, n 5, 53.

46. J. M. Thomas and David Philips, eds. *Selections and Reflections: The Legacy of Sir Lawrence Bragg* (Northwood Science Reviews, Northwood, 1990), 97.

47. Gann and Witowski, n 2, 37.

48. James D. Watson. *The Double Helix* (Weidenfeld and Nicholson, London, 1968).

49. David Philips. 'William Lawrence Bragg 1890–1971'. *Biographical Memoirs of Fellows of the Royal Society* 25 (1979), 131.

··

10 BIBLIOGRAPHY

Crystal Clear: The Autobiographies of Sir Lawrence and Lady Bragg. A. M. Glazer and Patience Thomson, eds. (Oxford University Press, Oxford, 2015).

Elements of X-Ray Diffraction (international edition). B. D. Cullity and S. R. Stock (Pearson, Harlow, 2014).

The Double Helix. James D. Watson (Weidenfeld and Nicholson, London, 1968). [re-printed as *The Annotated and Illustrated Double Helix: James D. Watson*. Alexander Gann and Jan Witkowski, eds. (Simon and Schuster, New York, 2012).]

What Mad Pursuit: A Personal View of Scientific Discovery. Francis Crick (Basic Books, New York, 1988).

William and Lawrence Bragg, Father and Son. J. Jenkin (Oxford University Press, Oxford, 2008).

X-Ray Diffraction by Disordered and Ordered Systems. D. W. L. Hukins (Pergamon, Oxford, 1981).

3

The Gibbs Phase Rule

Phases

---

1 The Gibbs phase rule

Most materials are made up by mixing a series of *components*. For instance, most simple steels are iron–carbon alloys Fe-C, i.e. they are made up of a mixture of the two components: iron Fe and carbon C. Another example is window glass, which is also a mixture of two components: silica SiO_2 and sodium oxide Na_2O. Yet another example is simple solder, which is a lead–tin alloy Pb-Sn, i.e. a mixture of the two components lead Pb and tin Sn. Notice that a component can be an element such as iron or a compound such as silica.

Most materials consist of regions of space that are homogeneous in structure and properties. Each region with its own structure and properties is called a *phase*. The question arises: how many different phases can there be in a given material? Careful microscopic studies of Fe-C steels show that they usually contain two different regions, i.e. two phases. The same is found for solders, but window glass is found to consist of a single phase. Stainless steels are a mixture of at least three components—iron Fe, chromium Cr and nickel Ni—but consist of a single phase. Clearly, the number of phases is not necessarily the same as the number of components.

The answer to the question is that the number of phases p in a material containing c components is given at *equilibrium* by the *Gibbs phase rule*:

$$p + f = c + 2,$$

where f is called the *number of thermodynamic degrees of freedom*.

We can understand this as follows. In any phase at equilibrium, the temperature, pressure and composition must all be constant and uniform. Otherwise, there would be a flow of heat, volume (expansion or contraction) or matter within the phase. If the temperature varied from place to place, the phase would not be at equilibrium, and heat would flow from the hotter to the colder place until equilibrium was established at an intermediate constant temperature. Similarly, if the pressure varied from place to place, the phase would again not be at equilibrium, and the higher pressure would expand against the lower pressure until equilibrium was established at an intermediate constant pressure. And, finally, if the composition varied from place to place, the phase would

The Equations of Materials. Brian Cantor. Oxford University Press (2020). © Brian Cantor.
DOI: 10.1093/oso/9780198851875.001.0001

again not be at equilibrium, and matter would flow from the higher concentration to the lower concentration until equilibrium was established at an intermediate constant composition.

The number of variables needed to specify the condition or state of each phase is, therefore, $c - 1$ independent composition variables (the composition of all the components must add up to 100%, so it is only necessary to specify $c - 1$ composition variables) plus the temperature and pressure, i.e. $(c - 1) + 2 = c + 1$. The total number of variables $n_{\text{variables}}$ is, therefore, the number of phases p multiplied by the number of variables needed to specify each phase $c + 1$:

$$n_{\text{variables}} = p(c + 1).$$

But at equilibrium, the temperature, pressure and chemical potential of each component must not only be constant in each phase, but must also be equal in all phases. Otherwise, there would be a flow of heat, volume (expansion or contraction) or matter from one phase to another, in the same way as described previously within each phase. The total number of equations, $n_{\text{equations}}$, connecting the variables is, therefore, the number of quantities that must be equal in all phases $c + 2$ multiplied by the number of equations needed for equality across all phases $p - 1$:

$$n_{\text{equations}} = (p - 1)(c + 2).$$

The number of degrees of freedom (i.e. the number of independent variables) f is then the number of variables minus the number of connecting equations:

$$f = n_{\text{variables}} - n_{\text{equations}}$$
$$= p(c + 1) - (p - 1)(c + 2) = (pc + p) - (pc + 2p - c - 2) = c - p + 2$$
$$\therefore p + f = c + 2,$$

which is the Gibbs phase rule. At constant (e.g. atmospheric) pressure, the phase rule becomes $p + f = c + 1$. And at constant (e.g. atmospheric) pressure and constant (e.g. room) temperature, it becomes $p + f = c$.

What does the Gibbs phase rule tell us? It is usually stated as $p + f = c + 2$, but it is sometimes more convenient to think of it as determining the number of degrees of freedom in a material, i.e. to write it as

$$f = c - p + 2.$$

We can consider several simple cases for a single-component material, i.e. $c = 1$:

1. If we have a single-component material and one phase, there are two degrees of freedom, $f = 2$. This means that a single-component, single-phase material such as, for instance, solid pure copper can be varied in two independent ways, i.e. it can exist over a range of temperatures and pressures.

2. If we have a single-component material with two phases, then there is only one degree of freedom, $f = 1$. This means that a single-component, two-phase material such as, for instance, a mixture of solid and liquid copper can only be varied in one independent way, i.e. the temperature and pressure cannot be changed independently. In other words, the melting point T_m must vary in a defined way with pressure P. This relationship is called the *Clausius-Clapeyron equation*,

$$\frac{dP}{dT_m} = \frac{L}{T_m \Delta v_m},$$

where L is the latent heat of melting and Δv_m is the change in volume between liquid and solid.

3. If we have a single-component material with three phases, then there are no degrees of freedom, $f = 0$. This means that a single-component, three-phase material, such as a mixture of solid, liquid and vapour cannot be varied at all, i.e. it can only exist at a fixed pressure and temperature called the *triple point*.

2　J. Willard Gibbs

J. Willard Gibbs

The biography of Gibbs by Lynde Phelps Wheeler begins:

> The outward life of Josiah Willard Gibbs was singularly uneventful...He was a participant in no events of historical importance. He took no leading part in any of the movements of the age. He traveled less and lived at home more than the great majority of people of similar means. He neither sought nor occupied positions of influence even in his own scientific world. He was never in the public eye. And yet his life was one of high adventure such as few have achieved. He explored the far horizons of the hitherto unknown.... He mapped a major scientific continent. He rebuilt and completed one of the great edifices of 19th century science.[1]

What a remarkable contrast! A man whose personal life was the ultimate in quiet and unruffled dignity, yet whose scientific achievements were monumental.

Josiah Willard Gibbs the younger was born on 11 February 1839 in New Haven, Connecticut, fourth child of five and the only son of Josiah Willard Gibbs the elder and Mary Anna van Cleve. He came from a long and illustrious line of scholars, clerics, and military, mercantile and civic leaders on both his father's and mother's sides. His antecedents include Major Simon Willard, pioneer settler and commander-in-chief of the British forces in 1634; his son, the Reverend Simon Willard, pastor of the Old South Church in Boston and Vice-President of Harvard; his son, Josiah Willard, tutor at Harvard, commander of a ship sailing between Boston, Europe and the West Indies, and Secretary of Massachusetts; Colonel William 'Tangier' Smith, Governor of Morocco in 1675, President of Brookhaven Council and Chief Justice of New York; Dr John van Cleve, physician and Professor of Chemistry at Princeton; and William Churchill Houston, lawyer and jurist, Professor of Maths and Natural Philosophy at the College of New Jersey, and Delegate to the Constitutional Convention of 1787.

Gibbs' father was Professor of Philology and Sacred Languages, as well as College Librarian, at Yale University in New Haven, making scholarly contributions to the fields of philology, theology, archaeology and grammar. He received an honorary degree from Princeton in 1853. The young Willard Gibbs grew up in the 'pleasant, unhurried life in faculty circles in the New Haven of the 1840s and 1850s'.[2] Social life consisted of faculty parties and picnics, winter skating, lectures, concerts, and convivial suppers and dances, in stark contrast to the social turmoil of the times, leading inexorably to the American Civil War, which began in 1861. The northern states were industrialising rapidly, building railways to push economic development westwards, whereas the southern states were clinging to a more traditional and, for some, gracious lifestyle based on agriculture and slavery. North and South were set on an ultimately bloody collision course. Yale and New Haven professors and scientists contributed key engineering inventions, such as Samuel Morse's telegraph and Charles Goodyear's vulcanization of rubber for tyres. At the same time, New Haven was a prominent centre of anti-slavery agitation—protesting, for instance, vehemently against the accession of pro-slavery Kansas.

The young 'Willy' Gibbs was influenced more by the scientific than the political issues of the day, choosing engineering as his field of study. He went to Hopkins School in 1848, and then on to Yale College in 1854, where he studied until he completed his doctorate in 1863. He was a brilliant scholar, shy and industrious, keen on walks in the forest rather than more muscular leisure pursuits. He was prone to ill health, and was treated for early-stage tuberculosis and poor eyesight, which probably contributed to him not volunteering to fight in the Civil War. In his last year at Hopkins, the 'Baldwin' papers in the Yale Library record that, aged just 15, he developed a crush on a young girl, Fanny Storer, and gave her 'a most beautiful bunch of pond lilies . . . but Fanny likes his flowers better than she likes him'.[3] This is his only recorded dalliance with the opposite sex, and, whether it was Fanny's fault or a consequence of his studiousness, he remained a bachelor for the whole of his life. There is a story that he was so shy that he worried about how he would cope with the demanding oral entrance exam for Yale. He approached a friendly professor to discuss and try to familiarise himself with the kind of questions he would be asked. At the end of their discussion, the professor announced that Gibbs had answered so well as to be accepted directly, without needing to take the formal exam itself.

Gibbs' younger sister, Eliza, his mother and his father all died during his time as a student, and one of his elder sisters, Emily, died shortly afterwards. These unhappy family events must have been unsettling, as must have been the ongoing Civil War of 1861 to 1865, but they did not ultimately affect his successful academic progress. He won scholarly prizes every year of his undergraduate studies for both Latin and maths. Science and engineering were quite new in the United States, and were more applied than was common in Europe at that time. Gibbs' doctoral thesis was on the mechanics of gear design, entitled: 'On the Form of the Teeth of Wheels in Spur Gearing'.[4] When it was completed, he was the first person in the United States to receive a doctorate in engineering. Following his graduation, Gibbs took up a three-year tenure as a tutor in Latin at Yale. At the same time, he continued his engineering studies, inventing a *center-vent* hydraulic turbine, patenting an inertial railway car braking system, and giving a paper at the Connecticut Academy of Sciences and Arts on a new and rationalised system of mechanical units.

In 1866, Gibbs and his two elder sisters, Anna and Julia, used some of the money from their father's estate to embark on an extended European tour. They spent the next three years visiting, mainly, Paris, Berlin and Heidelberg. They enjoyed, no doubt, occasional sightseeing and visits to concerts, but the trip was not a holiday, and Gibbs continued to work assiduously to enhance his mathematical and scientific abilities. He attended classes by many of the foremost European scholars of the time: Joseph Liouville and Jean-Marie Duhamel in Paris; Karl Weierstrass and Leopold Kronecker in Berlin; and Gustav Kirchoff, Georg Cantor, Robert Bunsen and Hermann von Helmholtz in Heidelberg. His notebooks indicate that he was reading widely, consuming the works of the great French mathematicians Pierre-Simon Laplace, Joseph-Louis Lagrange, Siméon-Denis Poisson and Augustin-Louis Cauchy, the Germans Carl Friedrich Gauss and Carl Gustav Jacobi, and others such as William Hamilton, Hendrik Lorentz and William Thomson (Lord Kelvin). In 1867 in Berlin, Julia married Gibbs' ex-classmate, the librarian at Yale, Addison van Name, and they returned to the New Haven family house, to be joined two years later by Gibbs and his other sister Anna.

The family settled into an untroubled lifestyle in their New England backwater. They were reasonably well off, and refurbished the house fully, including such 'modern' accoutrements as gas lighting, running water and external paving. They purchased a horse and buggy, and a cutter (a horse-drawn sleigh) for the winter snows. Gibbs spent leisure time on walking and horse-riding trips to the nearby Adirondack hills. But, major developments were taking place in the world of higher education. Science was on the march. Alexander von Humboldt's vision of great research (as well as teaching) universities was taking hold. In 1874 the Cavendish laboratory was established at Cambridge University, with James Clerk Maxwell as its first Director; in 1871, the Victoria University in Manchester was set up to deliver technical education to the industrial powerhouse of northern England; and Charles Eliot was appointed President of Harvard in 1869, introducing the 'elective' system to broaden the curriculum to include science and engineering, and kick-starting the transformation of a sleepy provincial college into a dynamo of science and research.

At Yale, similar developments were triggered in 1871 by the imminent retirement of the long-standing traditionalist President Theodore Dwight Woolsey. A review of the university's needs identified fund-raising from alumni, a new library and new professorships as key requirements for the future. Almost immediately, indeed before this programme could be fully developed, a Yale Corporation meeting in 1871 recorded that 'Mr Josiah Willard Gibbs of New Haven was appointed Professor of Mathematical Physics, without salary, in the [new] Department of Philosophy and the Arts'.[5] This was an impressive advancement. Gibbs had limited claims to a professorship, even one without salary, having only published one patent and delivered one presentation to the local Academy. Lynde Phelps Wheeler speculates that Gibbs must have had an influential backer, and suggests that this was probably the Yale astronomer and mathematician Hubert Anson Newton, but in truth the reason for Gibbs' sudden preferment remains unclear.

In any case, whether it was a consequence of three years of scientific schooling in Europe, the confidence boost of being appointed early to a prestigious chair at Yale, or liberation from the drudgery of teaching Latin to undergraduates, the following decade proved incredibly productive for Gibbs' scientific creativity. His focus had moved from developing technical engineering innovations to fundamental explanation of engineering phenomena. By 1878 he had published three important papers that, in the end, constituted more than a third of his total career output. The first two developed geometric and graphical methods to represent and explain thermodynamic behaviour. Amongst other things, they introduced his famous function, *Gibbs free energy*, the maximum available work for a system at constant pressure and temperature, which is minimised when the system reaches equilibrium. Maxwell commented on these papers: 'many questions as to the behaviour of bodies which had seemed almost impossible of solution, received simple and direct answers'.[6] He was so impressed that 'he constructed a model of the thermodynamic surface of water, a plaster cast which he sent to Gibbs'.[7]

Gibbs' third paper, 'On the Equilibrium of Heterogeneous Substances',[8,9] was published in two parts, in 1876 and 1878, in *Transactions of the Connecticut Academy*. Unlike the other two papers, it was entirely analytical, and it was truly monumental: a gigantic monograph of more than 300 pages, a magnum opus, one of the great works of science, on a par with Newton's laws of gravitation or Maxwell's electromagnetic theory. It was described later as 'the Principia of thermodynamics'.[10] Thermodynamics had arisen during the course of the 19th century from efforts to try to understand the behaviour of *heat engines*, which were at the heart of the Industrial Revolution. The underlying problem was the relationships between heat, motion and work for an engine's *working fluid*—e.g. water, in the case of a steam engine. At the core of the problem is the transformation between different kinds of energy, in particular the microscopic kinetic energy of the atoms and molecules (heat), and the macroscopic kinetic energy of the motion of a piston (work) via the expansion energy of the fluid. This proved (perhaps surprisingly) to be a very complex problem, occupying many of the finest scientists of the time. The *first and second laws of thermodynamics* state respectively that energy is conserved and entropy (heat divided by temperature) is either constant or increases. Gibbs' genius was to integrate all previous work into a single, elegant, mathematically beautiful framework. He combined the first and second laws into a *fundamental equation of thermodynamics*, enabling electrical, magnetic, gravitational and other forms of energy to be included in a natural way. Most significantly, the resulting clarity of description allowed Gibbs to include chemical energy and thus cover more than one substance. At a stroke, this brought all of chemistry into the domain of thermodynamics, providing an underlying theory for when and how materials and chemicals can react and transform one into another.

With only 100 or so members and a subscription of $5, the Connecticut Academy of Arts and Sciences could not publish Gibbs' papers without making a special request for extra funds from the Yale faculty and local business leaders. This shows the strength of local support for Gibbs. The President of the Academy, the zoologist Addison Verrill said: 'no man who ever lived could understand Gibbs' papers except Maxwell'.[11] Gibbs received rapid recognition from the wider American scientific community. In 1879, a year after publishing his monograph, he was elected to the US National Academy of Sciences; in 1880 he was elected to the American Academy of Arts and Sciences of Boston; and in 1881 he was awarded their prestigious Rumford Medal. In 1879 he was offered a professorship at Johns Hopkins University in Baltimore, at a salary of $3,000 per annum. He declined, preferring not to disturb a settled family life and working pattern. He was, no doubt, influenced by a counter-offer from Yale, also of a salary (at last), albeit a lower one of $2,000 per annum. Recognition in Europe was slower. Gibbs sent copies of his galley proofs to Maxwell, Kelvin and Clausius, initiating a healthy correspondence with the leading European physicists. But Gibbs' style was austere and his papers were difficult to read: most chemists did not have the maths to understand them, and most mathematicians did not have the practical know-how to appreciate their significance. He was championed in Britain by Maxwell, but appreciation elsewhere awaited translation of his work into German by Wilhelm Ostwald in 1892, and into French by Henry Louis Le Châtelier in 1899.

For the next two decades, Gibbs continued to pursue elegant mathematical solutions to important scientific phenomena. Maxwell's *A Treatise on Electricity and Magnetism*[12] was published in 1873, with his *electromagnetic theory* based on Hamilton's *quaternion* maths. Gibbs invented the modern form of *vector algebra* to provide a more compact description of the associated electromagnetic *vector fields*. His pamphlet on vectors was published in 1881 and 1884,[13–15] introducing the now widely used *scalar dot product, vector cross product*, and ∇ (*del*) and ∇^2 (*del-squared*) *operator* notations. In five papers from 1882 to 1889,[16] Gibbs showed how optical phenomena such as *dispersion, birefringence* and *refraction* arise naturally from Maxwell's equations, independent of microscopic assumptions about the properties of the so-called *ether* (the medium that was thought, erroneously, to carry light waves). In two short letters to *Nature* in 1898,[17] he explained the *Gibbs phenomenon*, the overshoot associated with fitting a *Fourier series* to a discontinuity or a sharp step, which leads to ringing tone artefacts in *signal processing*. He also worked on other topics, such as the use of vectors to describe planetary orbits, and the soaring flight of unpowered planes.

In 1902 Gibbs published *Elementary Principles in Statistical Mechanics*[18] as his contribution to Yale's bicentenary. This was another tour de force. Gibbs invented *statistical mechanics* to explain the behaviour of large assemblies of particles, putting thermodynamics on a firm microscopic basis. Atomic and molecular theory had gradually gained ground in the 19th century. The *Maxwell-Boltzmann distribution* of particle velocities described well the behaviour of an *ideal gas*, i.e. one in which the atoms or molecules are infinitesimally small and do not interact except during elastic collisions, which allow them to exchange energy. But further progress was stalled because of a poor understanding of atomic and molecular structures in real gases. Gibbs concentrated on properties that could be deduced independent of structure. He defined an assembly of particles as the set of their instantaneous positions and velocities, and constructed an *ensemble* of all possible such sets, occupying a region in *phase space*. He then analysed their statistical probability and evolution with time. This showed that equilibrium corresponds to a particular fixed region in phase space. The distribution of atoms or molecules over the different energy levels determines the *partition function*, from which all the thermodynamic behaviour can then be derived.

From 1867, when he returned with his sister Anna from Europe, until his death in 1903, aged 64, from an acute intestinal obstruction, Gibbs continued to live in New Haven, initially with his two sisters and brother-in-law (Anna sadly fell ill in 1895 and died in 1898), rarely venturing outside New England. He remained a man of dignified gentility and modesty. One of his ex-students, Edward Bidwell Wilson, described him as 'not an advertiser for personal renown, nor a propagandist for science...he had no striking ways, he was a kindly gentleman'.[19] Another ex-student, Lynde Phelps Wheeler, said, 'he had a simple, contemplative, serene nature undistracted by personal ambition and unconcerned with rewards'.[20] And yet another ex-student, Henry A. Bumstead, referred to him as 'unassuming in manner, genial and kindly ... never showing impatience or irritation, devoid of personal ambition ... the ideal of the unselfish, Christian gentleman'.[21,22] Gibbs' niece, Theodora van Name, was quoted by Josephine Newton (Bumstead's daughter) as saying her uncle 'was the happiest man she ever knew'.[23]

Gibbs was indeed unassuming. As he said himself in his typical self-deprecating way: 'If I have had any success in mathematical physics it is, I think, because I have been able to dodge the mathematical difficulties'.[24,25] Although he corresponded profusely with many of the greatest scientists of the age, he largely avoided scientific and other organisations, in part because he disliked travelling away from New Haven and in part because he was uncomfortable pronouncing on scientific (or other) issues until he had a chance to think things through fully. He declined attendance at prestigious conferences such as the British Association for the Advancement of Science and the German Mathematical Society, even when invited by as great a figure as Ludwig Boltzmann. He declined submitting contributions to prestigious publications such as a special volume in honour of Lorentz, or another to celebrate George Stokes' jubilee as President of the Royal Society, despite invitations from Kamerlingh Onnes and J. J. Thomson respectively. Nevertheless, he was awarded a raft of academic prizes and fellowships, culminating with the Copley Medal of the Royal Society in 1902 (the greatest scientific honour before the launch of the Nobel Prizes), and was praised by Albert Einstein as 'the greatest mind in American history'.[26,27] Robert Millikan said that Gibbs 'did for statistical mechanics and for thermodynamics what Laplace did for celestial mechanics and Maxwell did for electrodynamics, namely, made his field a well-nigh finished theoretical structure'.[28]

All Gibbs' scientific contributions share certain characteristics: practical engineering relevance; a precise mathematical formalism; clarity and consistency of notation; integration of a large body of work into a single logical framework; and a determination to make as few assumptions as possible (hence his avoidance of the then-unknown details of atomic and molecular structure). These characteristics have ensured that his work has stood the test of time through to the present day, unaffected by seismic revolutions in 20th-century science such as quantum mechanics and relativity. His theories have a simplicity, a purity and a feeling of inevitability. The Gibbs phase rule is a beautiful example. It is one of the simplest equations possible, derived in just a few lines in the middle of Gibbs' enormous monograph. Yet its implications are profound: fundamental and (almost) philosophical, in determining something so basic as the number of phases (or different kinds of 'stuff') that can exist in any given material; yet highly practical, guiding metallurgists, ceramicists, geochemists, mineralogists and crystallographers to discover new materials with exciting and valuable properties.

3 Phase diagrams

Phase diagrams show the occurrence of different phases in a material as a function of varying temperature, pressure and composition.

Figure 3.1 shows a one-component phase diagram for water H_2O. The phases are plotted on a graph of pressure versus temperature. The phase diagram shows

(1) single-phase regions for solid ice, liquid water and gaseous steam, where both pressure and temperature can be varied independently;

(2) lines of two-phase equilibrium between ice and water, water and steam, and ice and steam, where pressure and temperature cannot be varied independently, and the Clausius-Clapeyron equation describes the equilibrium temperature as a function of the pressure; and

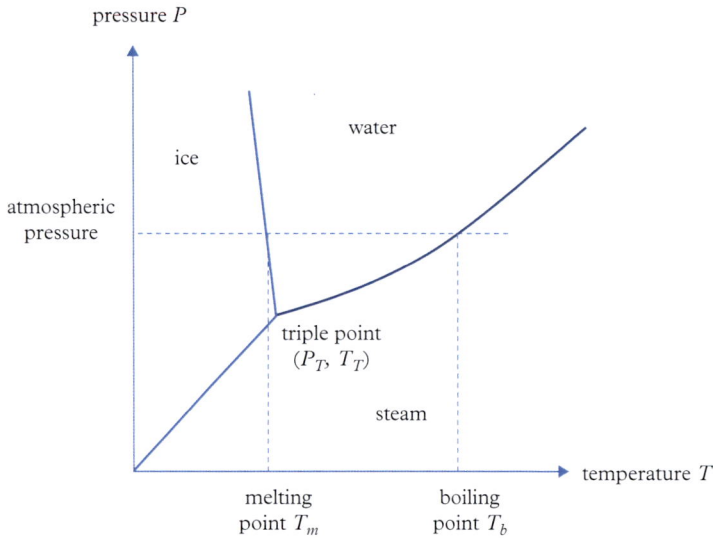

Figure 3.1 *One-component phase diagram for water (H₂O)*

(3) a triple-point mixture of ice, water and steam, which is only possible at a fixed pressure and temperature. The triple point pressure is P_T and the triple point temperature is T_T.

Figure 3.2 shows a two-component phase diagram for mixtures of face-centred cubic (fcc) copper Cu and fcc silver Ag. The phases are plotted on a graph of temperature versus composition, at constant (atmospheric) pressure. The phase diagram shows

1. a single-phase region of Cu–Ag *liquid solution* at high temperature, where both temperature and composition can be varied independently;
2. *terminal solid solutions* of silver in fcc copper α-Cu(Ag) and copper in fcc silver β-Ag(Cu), at low temperature and, respectively, high and low copper content, again where both temperature and composition can be varied independently;
3. lines of equilibrium called *solvus lines*, which are the solid solubility limits of the α and β phases, where temperature and composition cannot be varied independently, and there is an equation describing the maximum solubility as a function of temperature;
4. two phase regions of mixtures of α + liquid, β + liquid and α + β phases, where temperature and composition cannot be varied independently, because these regions consist of horizontal *tie lines*, describing the compositions of the two phases (tie lines are explained later in this chapter); and

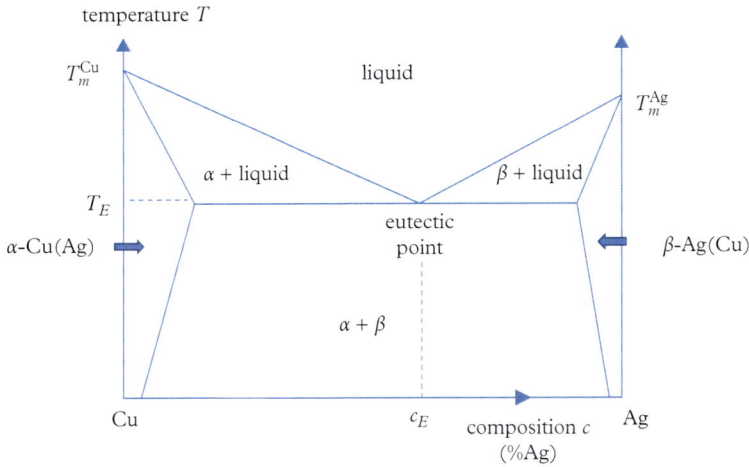

Figure 3.2 *Two-component phase diagram for copper Cu and silver Ag*

5. a triple-*eutectic point* mixture of α, β and liquid, which is only possible at a fixed temperature and composition. The *eutectic temperature* is T_E and the *eutectic composition* is c_E.

4   Invariant reactions

The triple point in a one-component material as shown in Figure 3.1, and the eutectic point in a two-component material as shown in Figure 3.2, are examples of *invariant points* with no degrees of freedom. This can be seen for the triple point by taking $c = 1$ and $p = 3$ in the Gibbs phase rule:

$$f = c - p + 2 = 0$$

and for the eutectic by taking $c = 2$ and $p = 3$ in the Gibbs phase rule at constant (atmospheric) pressure:

$$f = c - p + 1 = 0.$$

At the eutectic point in Figure 3.3, we say that there is a *eutectic reaction*. *Eutectic solidification* is when liquid of eutectic composition solidifies on cooling at or below the eutectic temperature to form a two-phase mixture of α and β solid solutions:

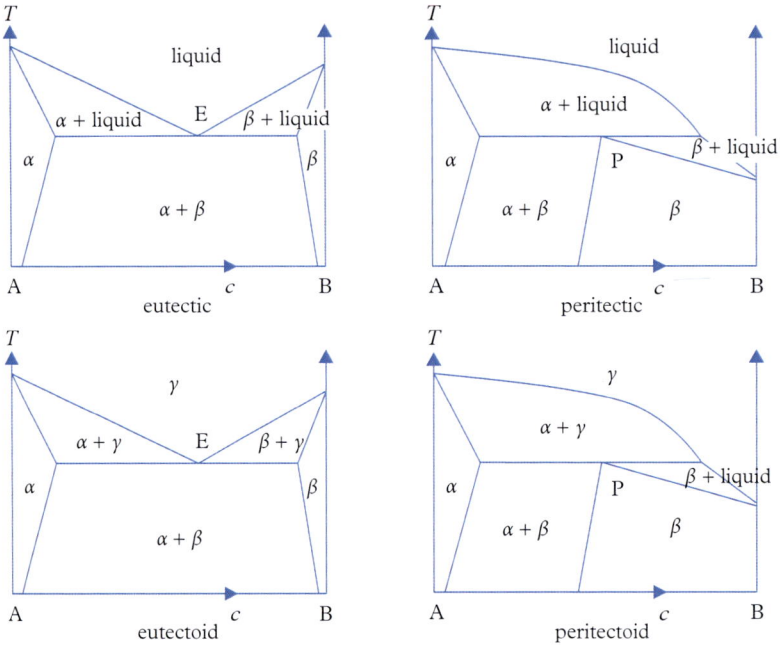

Figure 3.3 *Invariant reactions in a two-component A–B material*

$$L \to \alpha + \beta.$$

Eutectic melting is the reverse reaction, when a two-phase mixture of α and β solid solutions melts on heating at or above the eutectic temperature to form a liquid of eutectic composition:

$$\alpha + \beta \to L.$$

A eutectic reaction is an example of an *invariant reaction*. There are many other examples, such as *peritectic, eutectoid* and *peritectoid reactions*:

$$\text{peritectic: } L + \alpha \to \beta$$
$$\text{eutectoid: } \alpha \to \beta + \gamma$$
$$\text{peritectoid: } \alpha + \beta \to \gamma$$

Phase diagrams with eutectic, peritectic, eutectoid and peritectoid points are shown in Figure 3.3. Eutectics and peritectics are examples of invariant solidification and melting reactions. Eutectoids and peritectoids are examples of invariant solid-state reactions.

5   The lever rule

The Gibbs phase rule at constant (atmospheric) pressure is

$$f = c - p + 1.$$

In a two-component material at constant (atmospheric) pressure, a single phase has, therefore, two degrees of freedom. The temperature and composition can be varied independently, as shown by the two-dimensional areas occupied by the single-phase liquid, α, β and γ regions in Figures 3.2 and 3.3. On the other hand, three phases have no degrees of freedom, with invariant points, as shown for the eutectic, peritectic, eutectoid and peritectoid reactions in Figures 3.2 and 3.3.

Surprisingly, Figures 3.3 and 3.4 seem to show that temperature and composition can also be varied independently in two-dimensional areas for the two-phase α + liquid, β + liquid and $\alpha + \beta$ regions. This is intentional but misleading. There is an equation that defines the quantities and compositions of the two phases at any composition and temperature. This equation is called the *lever rule*, and is represented on a phase diagram by the construction of a *tie line*.

Figure 3.4 shows the tie line construction for a material of composition c at a temperature T in a two-phase $\alpha + \beta$ region on a eutectic phase diagram. The tie line is a horizontal line passing through the point (c,T), connecting the two single-phase α and β regions on either side of the two-phase $\alpha + \beta$ region, as shown in Figure 3.5. The compositions of the α and β phases in the material are c_α and c_β, and the fractions of the α and β phases in the material, x_α and x_β, are given by the lever rule:

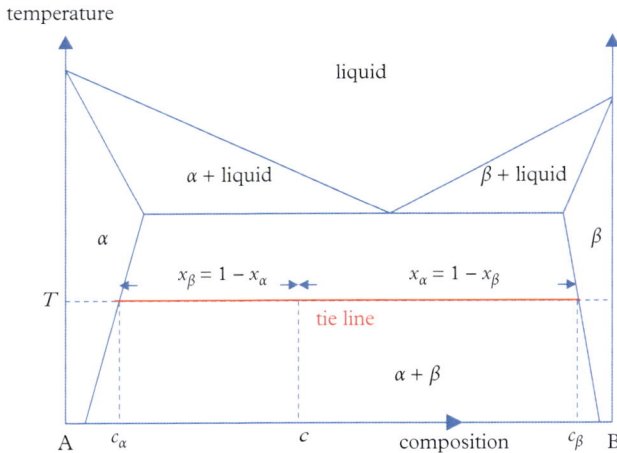

Figure 3.4 *The lever rule for the fractions of α and β in a two-phase $\alpha + \beta$ region*

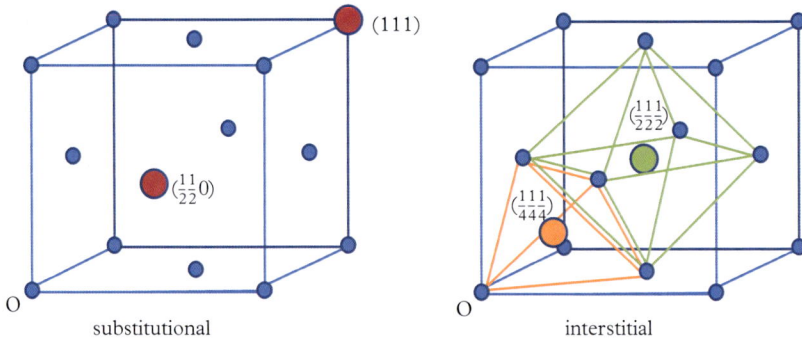

Figure 3.5 *Substitutional (red) and octahedral (green) and tetrahedral (orange) interstitial atoms in a face-centred cubic material*

$$x_\alpha = \frac{c_\beta - c}{c_\beta - c_\alpha}$$

$$x_\beta = \frac{c - c_\alpha}{c_\beta - c_\alpha}.$$

There is a tie line at every temperature in the two-phase region, and the compositions and quantities of the two phases are always given by the lever rule. This means that the temperature and composition cannot be varied independently. Two-phase regions can be thought of as a bundle of horizontal lines at every temperature, rather than a two-dimensional region on the phase diagram.

6   Solid solutions

We are familiar with liquid solutions, e.g. sugar, salt or alcohol dissolved in water. At first, it seems surprising that solid materials can also be solutions, as indicated in Figures 3.3 and 3.4. There are two main types of solid solution, namely *substitutional solid solutions* and *interstitial solid solutions*.

In a substitutional solid solution, the solute atoms or molecules replace or substitute for the solvent atoms or molecules at random throughout the crystal structure. In an interstitial solid solution, the solute atoms or molecules occupy the space between the solvent atoms or molecules, in the *interstices* of the crystal structure.

Figure 3.5 shows examples of substitutional and interstitial solute atoms in an fcc structure. Two (red) substitutional solute atoms have replaced (blue) solvent atoms at the front face centre $\left(\frac{1}{2}\frac{1}{2}0\right)$ and top right back (111) in the unit cell. One (green) interstitial solute atom is occupying an *octahedral interstice*$\left(\frac{1}{2}\frac{1}{2}\frac{1}{2}\right)$, surrounded by an octahedron of (blue) solvent atoms. Another (orange) interstitial solute atom is occupying a *tetrahedral interstice*$\left(\frac{1}{4}\frac{1}{4}\frac{1}{4}\right)$, surrounded by a tetrahedron of (blue) solvent atoms.

7   Compounds

A *compound* is formed when different atoms react, i.e. when *interatomic forces* create *bonds* between the atoms. There are different kinds of compound, depending on the nature of the interatomic forces that cause the reaction and create the atomic bonds. In the rest of this section, we examine one at a time *ionic, covalent, van der Waals* and *electron compounds*, based on, respectively, *ionic, covalent, van der Waals* and *metallic* bonds, as well as the concepts of *stoichiometry* and *line* and *variable compounds*.

Ionic compounds

An ionic bond is created between two atoms when there is an *electron transfer* of one or more electrons between the atoms. A typical example is sodium chloride NaCl, usually known as *common salt*. The overall reaction between sodium Na and chlorine Cl to form NaCl can be written

$$Na + Cl \rightarrow NaCl$$

but the individual reaction steps are

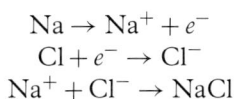

$$Na \rightarrow Na^+ + e^-$$
$$Cl + e^- \rightarrow Cl^-$$
$$Na^+ + Cl^- \rightarrow NaCl$$

i.e. a negatively charged electron e^- is transferred from the sodium atom to the chlorine atom to create a positively charged sodium ion Na^+ and a negatively charged chlorine ion Cl^-. There is an attractive *electrostatic force* between positive and negative charges, so the two ions are attracted electrostatically, creating an ionic bond and forming the ionic compound sodium chloride.

A sodium atom has the electronic structure $1s^2\ 2s^2 2p^6\ 3s^1$. This means that it has two s electrons in its first electron shell $1s^2$, two s electrons and six p electrons in its second electron shell $2s^2 2p^6$, and one s electron in its third electron shell $3s^1$. The sodium atom readily loses its outer s electron to create a stable set of filled first and second electron shells in the positively charged sodium ion Na^+, with an electronic structure $1s^2\ 2s^2\ 2p^6$. Similarly, a chlorine atom has the electronic structure $1s^2\ 2s^2 2p^6\ 3s^2 3p^5$. This means it has two s electrons in its first electron shell $1s^2$, two s electrons and six p electrons in its second electron shell $2s^2 2p^6$, and two s electrons and five p electrons in its third electron shell $3s^2 3p^5$. The chlorine atom readily gains an extra p electron in its outer shell to create a stable set of filled first, second and third electron shells in the negatively charged chlorine ion Cl^-, with an electronic structure $1s^2\ 2s^2 2p^6\ 3s^2 3p^6$.

There are many ionic compounds. *Alkali metal* atoms from the first column of the periodic table all have one outer electron, and *halogen* atoms from the seventh column of the periodic table all have one missing outer electron. They readily, therefore, transfer an electron to create *alkali halides* such as sodium chloride NaCl, potassium chloride KCl

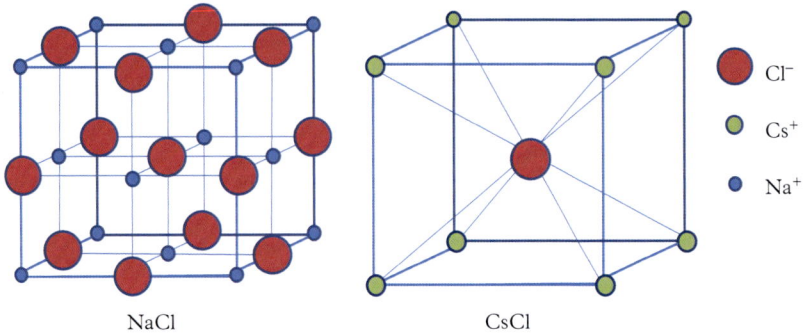

Figure 3.6 *Ionic compounds face-centred cubic sodium chloride NaCl and body-centred cubic caesium chloride CsCl*

and lithium fluoride LiF, which can be written Na^+Cl^-, K^+Cl^- and Li^+F^-. Similarly, *alkaline earth metal* atoms from the second column of the periodic table all have two outer electrons, and *chalcogen* atoms from the sixth column of the periodic table all have two missing outer electrons. They readily, therefore, transfer two electrons to create ionic compounds such as magnesium oxide MgO and calcium oxide CaO, which can be written $Mg^{2+}O^{2-}$ and $Ca^{2+}O^{2-}$. Similarly, an alkaline earth atom can transfer two electrons, one each to two halogen atoms to create ionic compounds such as beryllium chloride $BeCl_2$ and magnesium bromide MgB_2, which can be written $Be^{2+}(Cl^-)_2$ and $Mg^{2+}(Br^-)_2$, and two alkali metal atoms can transfer one electron each to a chalcogen atom to create ionic compounds such as sodium oxide Na_2O and potassium sulphide K_2S, which can be written $(Na^+)_2O^{2-}$ and $(K^+)_2S^{2-}$.

Ionic compounds form crystals in which the ionic bonding extends throughout the material. Typical examples are fcc sodium chloride NaCl and body-centred cubic (bcc) caesium chloride CsCl, as shown in Figure 3.6.

Covalent compounds

A covalent bond is created between two atoms when they share electrons. A typical example is silicon oxide SiO_2, which is often called *silica*, and is an important component of *window glass* and many other ceramic materials. The overall reaction between silicon Si and oxygen O_2 to form silica SiO_2 can be written

$$Si + O_2 \rightarrow SiO_2$$

The molecule SiO_2 has four covalent bonds, two each between the silicon atom and each oxygen atom:

$$O = Si = O$$

Each covalent bond contains two paired electrons, one each from the outer shell of the two bonding atoms. The paired electrons are localised in the region between the silicon and oxygen atoms, which become positively charged as a result. There is an electrostatic attractive force between the localised paired electrons and the positively charged atoms of silicon and oxygen, creating the covalent bonds and forming the covalent compound SiO_2.

A silicon atom has the electronic structure $1s^2\ 2s^2 2p^6\ 3s^2 3p^2$. This means that it has two s electrons in its first electron shell $1s^2$, two s electrons and six p electrons in its second electron shell $2s^2 2p^6$, and two s electrons and two p electrons in its third electron shell $3s^2 3p^2$. With four covalent bonds, it effectively adds four extra electrons from the two oxygen atoms, creating a stable set of filled first, second and third electron shells $1s^2\ 2s^2 2p^6\ 3s^2 3p^6$. Similarly, an oxygen atom has the electronic structure $1s^2\ 2s^2 2p^4$. This means that it has two s electrons in its first electron shell $1s^2$, and two s electrons and four p electrons in its second electron shell $2s^2 2p^4$. With two covalent bonds, it effectively adds two extra electrons from the silicon atom, creating a stable set of filled first and second electron shells $1s^2\ 2s^2 2p^6$.

There are many covalent compounds. Examples include carbon dioxide CO_2, boron nitride BN and acetaldehyde CH_3CHO, which can be written respectively as

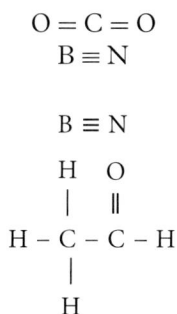

$$O = C = O$$
$$B \equiv N$$

$$B \equiv N$$

$$
\begin{array}{ccc}
H & & O \\
| & & \| \\
H - C & - C & - H \\
| & & \\
H & &
\end{array}
$$

Many solid covalent compounds form crystals in which the covalent bonding extends as a network throughout the crystal. An example is the semiconductor material gallium arsenide GaAs, which exhibits the diamond cubic crystal structure, similar to elemental silicon Si and the diamond form of carbon C. Diamond cubic silicon and gallium arsenide are shown in Figure 3.7. The silicon, gallium and arsenic atoms are tetrahedrally coordinated in an fcc crystal structure.

van der Waals compounds

Many other covalent compounds form crystals in which the covalent-bonded molecules are attracted to each other by van der Waals bonds. A van der Waals bond arises when two atoms or molecules are brought towards each other, and the electronic structure of each atom or molecule creates a small, induced electric charge in the other one. The induced charges are equal and opposite, so the two molecules are attracted electrostatically,

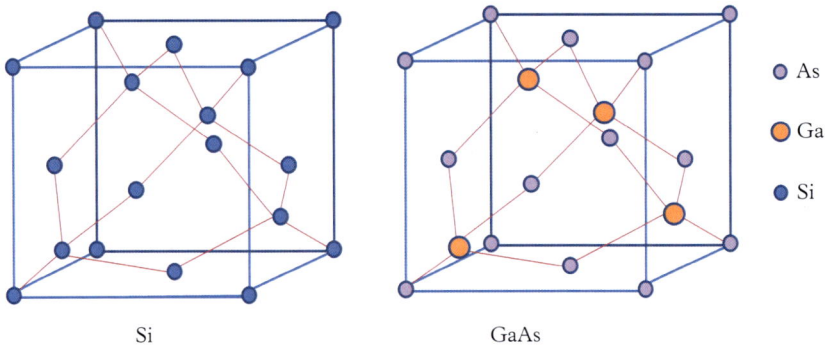

As
Ga
Si

Si   GaAs

Figure 3.7 *Covalent face-centred cubic silicon Si and gallium arsenide GaAs*

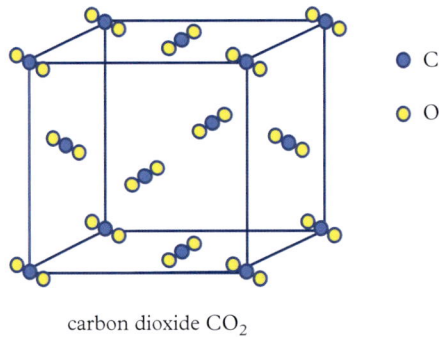

C
O

carbon dioxide CO_2

Figure 3.8 *van der Waals compound carbon dioxide CO_2*

creating a van der Waals bond. van der Waals bonds are weak compared to ionic and covalent bonds, because the induced charges are considerably smaller than the charge on the electron. They are, therefore, less stable as solid crystals or liquids, with much lower melting points and boiling points. Examples of van der Waals compounds are iodine I_2 and carbon dioxide CO_2. Solid carbon dioxide, or *dry ice*, has an fcc arrangement of linear $O=C=O$ molecules bonded together by van der Waals forces, as shown in Figure 3.8. Unreactive elements such as argon Ar form *van der Waals solids* at low temperatures, with van der Waals bonds between their constituent atoms.

A special case is the formation of *hydrogen bonds*, found in solid ice and liquid water H_2O, and in many organic molecules. Local electron transfer produces small positive charges on the hydrogen atoms and correspondingly small negative charges on the rest of the molecule (the oxygen atoms in the case of water). This enhances the normal bonding between the molecules in the solid and liquid forms. Water solidifies as ice in a hexagonal crystal structure, with hydrogen bonds between the H_2O molecules, which explains the beautiful hexagonal shapes of snow crystals.

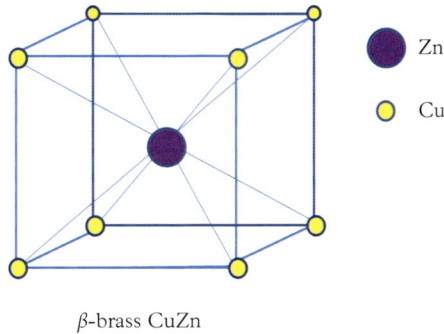

β-brass CuZn

Figure 3.9 *Electron compound β-brass CuZn*

Most polymers consist of long-chain molecules that solidify in either a crystalline array or a tangled amorphous structure, with van der Waals and/or hydrogen bonds between the molecules.

Electron compounds

Metallic elements such as gold, lead, aluminium, copper and iron have one or more loosely bound outer *valence electrons*. In metal solids, the valence electrons become detached and are free to move throughout the material, as so-called *free electrons*, between the remaining fixed ions. A *metallic bond* is created by the electrostatic attraction between the mobile free electrons and the fixed, positively charged ions. *Electron compounds* are formed when different metals combine in an alloy, with valence electrons from the different metals all contributing to the free electrons. Typical examples are β-brass CuZn, which has a bcc structure similar to caesium chloride CsCl, as shown in Figure 3.9, and γ-brass $CuZn_2$, which has a complex cubic crystal structure.

Line and variable compounds

When a compound A_nB_m forms via a simple reaction

$$nA + mB \rightarrow A_nB_m,$$

the *chemical coefficients* n and m are often simple whole numbers and are referred to as the *stoichiometry* of the reaction. The resulting compound is called a *stoichiometric compound*. If the composition of the compound is fixed, i.e. it cannot change its stoichiometry and cannot dissolve other components, it appears as a vertical line on the phase diagram and is called a *line compound*. If the composition of the compound is not fixed, i.e. it can change its stoichiometry and can dissolve other components, it appears on the phase diagram as a two-dimensional single-phase region. Figure 3.10 shows typical phase diagrams with line and variable compounds γ-AB.

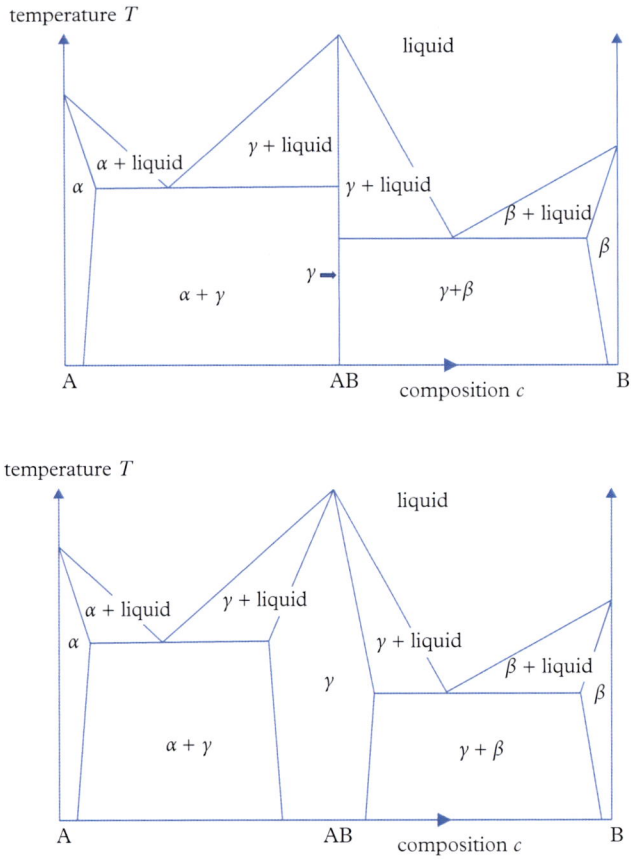

Figure 3.10 *Phase diagrams with line and variable compounds γ -AB*

. . . .

8   REFERENCES

1. Lynde Phelps Wheeler. *Josiah Willard Gibbs: The History of a Great Mind* (Yale University Press, New Haven, 1970), 1.
2. Ibid. 12.
3. Ibid. 21.
4. J. W. Gibbs. 'On the form of the teeth in wheels in spur gearing'. In: L. P. Wheeler, E. O. Waters, and S. Dudley (eds.). *The Early Work of Willard Gibbs* (Schuman, New York, 1947).
5. Wheeler, n 1, 57.
6. See ibid. 73.
7. Ibid. 74.
8. J. W. Gibbs. 'On the equilibrium of heterogeneous substances'. *Transactions of the Connecticut Academy Arts and Sciences* **3** (1875–1876), 108.

9. H. A. Bumstead and R. G. van Name, eds. *The Scientific Papers of J. Willard Gibbs Volumes I and II* (Dover, New York, 1961).
10. Bill Bryson. *A Short History of Nearly Everything* (Broadway Books, Portland, 2003; Transworld, London, 2004), 116.
11. Wheeler, n 1, 84.
12. James Clerk Maxwell. *A Treatise on Electricity and Magnetism* (Clarendon Press, Oxford, 1873).
13. J. W. Gibbs. *Elements of Vector Analysis Arranged for the Use of Students in Physics* (Tuttle, Moorhouse and Taylor, New Haven, 1884).
14. Edwin Bidwell Wilson. *Vector Analysis: A Text-Book for the Use of Students of Mathematics and Physics: Founded upon the Lectures of J. W. Gibbs* (1901, re-printed Dover, New York, 1960 and Nabu Press, Charleston, 2010).
15. Wheeler, n 1, 111.
16. Bumstead and van Name, n 9.
17. See ibid.
18. J. W. Gibbs. *Elementary Principles in Statistical Mechanics* (Dover, New York, 2014).
19. E. B. Wilson. 'Reminiscences of Gibbs by a student and colleague'. *Bulletin of the American Mathematical Society* **37** (1931), 401–416.
20. Wheeler, n 1, 83.
21. H. A. Bumstead. 'Josiah Willard Gibbs'. *American Journal of Science* **16** (1903), 187–202.
22. W. R. Longley and R. G. van Name, eds. *The Collected Works of J. Willard Gibbs*, vol. I, (Yale University Press, New Haven, 1928, 1957), xiii–xxviii.
23. Wheeler, n 1, 186.
24. C. S. Hastings. 'Biographical memoir of Josiah Willard Gibbs 1839–1903'. *National Academy of Sciences Biographical Memoirs* **VI** (1909), 390.
25. Wheeler, n 1, 53, 120.
26. Richard Panek. 'The greatest mind in American history'. *Yale Alumni Magazine* **May/June** (2017). https://yalealumnimagazine.com/articles/4496-josiah-willard-gibbs (accessed 2 January 2020).
27. Asutosh Jogelekar. 'Who's the greatest physicist in history'? *Scientific American* **May** (2013), https://blogs.scientificamerican.com/the-curious-wavefunction/whos-the-greatest-american-physicist-in-history/ (accessed 31 March 2020).
28. Robert A. Millikan. 'Biographical memoir of Albert Abraham Michelson 1852–1931'. *Biographical Memoirs of the National Academy of Sciences of the United States of America* **19** (1938), 121.

· ·

9 BIBLIOGRAPHY

Biographical Memoir of Josiah Willard Gibbs, vol. VI. Charles S. Hastings (National Academy of Sciences, Washington, DC, 1909).
Elementary Principles of Statistical Mechanics. J. W. Gibbs (Dover, New York, 2014).
Josiah Willard Gibbs: The History of a Great Mind. Lynde Phelps Wheeler (Yale University Press, New Haven, 1970).
Phase Equilibria, Phase Diagrams and Phase Transformations: Their Thermodynamic Basis. Mats Hillert (Cambridge University Press, Cambridge, 2007).
The Scientific Papers of J. Willard Gibbs, vols. I and II. H. A. Bumstead and R. G. van Name eds. (Dover, New York, 1961).

4

Boltzmann's Equation

Thermodynamics

1   The laws of thermodynamics

The science of *thermodynamics*, as is fairly obvious from its name, is about how heat
and motion interact. Understanding thermodynamics was essential during the Industrial
Revolution in the 19th century to explain how machinery and engines work, and to
help their development and improvement. For a material, thermodynamics underlies
how its atoms interact and move individually and collectively. There are four *laws of
thermodynamics*, which we examine one at a time.

The zeroth law of thermodynamics

The *zeroth law of thermodynamics* defines the notion of *equilibrium*, when a material
is stable and nothing happens, because all the forces acting on it or within it are in
balance. Temperature and pressure are constant throughout a material at equilibrium.
If the temperature varies, heat flows between different parts of the material and it is not
at equilibrium. If the pressure varies, different parts of the material expand and contract,
and again it is not at equilibrium. Similarly, the temperature and pressure in a material at
equilibrium are equal to those in its surroundings. If the temperatures are different, heat
flows into or out of the material and it is not at equilibrium. If the pressures are different,
the material expands or contracts, and again it is not at equilibrium. The zeroth law
allows us to define and measure the temperature and pressure in a material by putting it
in contact with, and therefore letting it come into equilibrium with, a thermometer or a
pressure gauge.

The first law of thermodynamics

The *first law of thermodynamics* says that the *internal energy* of a material E increases if it
is heated, and decreases if it does work. We can easily see how the notions of heat, work
and motion are closely intertwined. A material is heated or cooled when it is moved from

The Equations of Materials. Brian Cantor. Oxford University Press (2020). © Brian Cantor.
DOI: 10.1093/oso/9780198851875.001.0001

one place to another at a higher or lower temperature. Similarly, it does work when it is moved from one place to another in response to an external applied force, which can be, for instance, gravitational (changing its height), compressional (changing its volume), electrical or magnetic (moving it in an electrical or magnetic field).

For an increase in heat dq and an amount of work done dw, the change in internal energy dE is given by the first law of thermodynamics as

$$dE = dq - dw.$$

This is a definition of the *internal energy* of a material. It is also effectively a statement of the *conservation of energy*. It says that the internal energy of a material changes only when it receives energy in the form of heat dq, or loses energy by moving and, therefore, doing work dw in response to an external applied force. Of course, it can also lose thermal energy if dq is negative, i.e. if it is cooled; and it can also gain energy if dw is negative, i.e. if it moves against rather than with an applied force.

The second law of thermodynamics

Heat or *thermal energy* is different from other forms of energy. Thermodynamics combines the notion of heat and motion in a microscopic as well as a macroscopic way. Heat is absorbed in a material by enhancing the random movements of its component atoms or molecules: predominantly their translations if they are free in a gas; and predominantly their vibrations and rotations if they are constrained in a liquid or solid.

When a material absorbs heat, its component atoms or molecules become more *disordered* because of their increased motion. It turns out that this *disorder* is just as important as the underlying motions themselves. It is convenient, therefore, to define a quantity called the *entropy S*, which measures the *disorder* in a material. The disorder depends on the temperature of the material as well as the amount of heat absorbed. At a high temperature, the atoms or molecules are moving rapidly and at random. They are already fairly disordered, so the extra disorder from being heated is quite limited. At a low temperature, however, the atoms or molecules are not moving so rapidly. They are fairly well ordered, so the extra disorder from being heated is much greater. In other words, the disorder is proportional to the heat absorbed, but inversely proportional to the temperature at which it is absorbed.

For a material, the *second law of thermodynamics* says that its *entropy* increases when it is heated at a particular temperature T:

$$dS = \frac{dq}{T}.$$

Just as the first law of thermodynamics is a definition of the internal energy of a material, the second law is a definition of its entropy.

The third law of thermodynamics

The first and second laws of thermodynamics define the *internal energy* and the *entropy* of a material via changes caused by heating it or moving it. It is natural to take the zero point for internal energy at absolute zero temperature, because this corresponds to no atomic or molecular motion, i.e. no internal thermal energy:

$$E = 0 \text{ at } T = 0.$$

The *third law of thermodynamics* defines the same zero point for entropy:

$$S = 0 \text{ at } T = 0.$$

At absolute zero, therefore, the atoms or molecules of a material have no energy or entropy and are stationary, fixed in place in a perfectly ordered array. This is the definition of a perfect crystal. All materials form a perfect crystal at absolute zero; but at realistic temperatures, the atoms or molecules are in motion: vibrating, rotating and translating, and creating imperfections and disorder in the regularity of the crystal structure. The third law is sometimes called the *Nernst heat theorem*.

Reversibility

The first and second laws of thermodynamics are strictly only accurate for *reversible* changes, i.e. when heat is transferred and work is done very carefully and slowly to make sure that there are no losses. In fact, realistic methods of transferring heat and doing work are never completely efficient, and there are always losses, leading, for instance, to vibration and frictional heating of the surrounding machinery. This inefficiency creates additional disorder, so the entropy increases more rapidly for *irreversible* changes, and a more general version of the second law is

$$dS \geq \frac{dq}{T}.$$

2   Boltzmann's equation

The laws of thermodynamics were developed for macroscopic materials and engines. The concept of atomic and molecular disorder is, however, essentially microscopic. This is made clearer by *Boltzmann's equation*, which defines the relationship between the entropy S and the disorder w, defined as the number of equivalent atomic or molecular structures,

$$S = k \ln w,$$

where k is *Boltzmann's constant*. The probability p of a particular atomic or molecular structure in a material is the inverse of the number of equivalent structures w. If there is only one possible structure, its probability is $p = 1$, the number of equivalent structures is $w = 1$ and the entropy is $S = 0$. On the other hand, when the atoms or molecules are disordered, the probability of each structure is lower, $p < 1$, there are many equivalent structures, $w > 1$, and the entropy is higher, $S > 0$.

When the atoms or molecules of a material vibrate, rotate or translate, the structure of the material varies from moment to moment (i.e. a snapshot of the atomic or molecular positions is different at each moment), the material is accessing a range of equivalent structures, the probability of each one is low, and the entropy of the material is high. Similarly, defects in the crystal structure, such as vacancies, interstitial atoms, solute atoms, dislocations and grain boundaries, can exist at a range of positions in the crystal, the material can again access a range of equivalent structures, the probability of each one is again low, and the entropy of the material is again high. When the material is heated, thermal energy is absorbed by increasing the atomic or molecular vibration, rotation and translation, and by increasing the number of defects, i.e. by increasing the number of equivalent structures, reducing each one's probability and, therefore, increasing the disorder and entropy.

3   Ludwig Boltzmann

Ludwig Boltzmann

In 1906 Ludwig Boltzmann, internationally acclaimed Professor of Theoretical Physics at the University of Vienna, was suffering badly from chronic asthma, poor eyesight, severe depression and neurasthenia. On 16 January 1906, he wrote to his longstanding friend, Franz Brentano, who had retired some years earlier as Professor of Philosophy at the University of Vienna and was now living in Florence:

> Over the Christmas holidays I had a painful and very troublesome illness which kept me chained to my room for almost fourteen days. I have recovered from it, but a great mental depression has remained. How I envy you your constant cheerfulness and satisfaction. You are truly a genuine philosopher. I am already sixty-two years old, but I have won no peace of mind. In such times of melancholy I am downright pessimistic concerning my own philosophy, indeed, even about the very possibility of philosophical knowledge.[1]

Another longstanding friend, Alois Höfler, promoted two years earlier from Lecturer at the University of Vienna to Professor of Pedagogy at the University of Prague, commented about Boltzmann: 'When I visited him during the Easter holidays … he expressed his physical and mental suffering thus: "I would never have believed such an end was possible"'.[2] The Dean of the Philosophy Faculty at the University wrote to the Austrian Minister of Education in May: 'Boltzmann is suffering from severe neurasthenia and according to the doctor in charge must stop every scientific activity … . The Institute for Theoretical Physics as well as the lectures are being taken care of by the Assistant and Privatdozent Dr Meyer'.[3]

Boltzmann's wife, Henriette, persuaded him, albeit with great difficulty, to take a summer trip to try to improve his health. They left Vienna in July with their three daughters, initially to spend several weeks in Schwannberg in Styria. Ludwig, Henriette and their youngest daughter, Elsa, then travelled south to the picturesque and fashionable seaside resort of Duino, on the Adriatic coast just north of Trieste. Duino has a beautiful castle perched on a white rock on the edge of the sea, later made famous by the poet Rainer Maria Rilke, who wrote the *Duino Elegies*,[4] dedicated to the castle's aristocratic owner, Princess Maria von Thurn und Taxis-Hohenlohe. According to a later report in the newspaper *Die Zeit*, Boltzmann was 'upset and nervous, because he wanted to return to Vienna';[5,6] and a Viennese family friend Lili Hahn later reported, 'He was very melancholy for a long time and did not want to send his suit to be cleaned because it would mean a further delay in returning to Vienna'.[7] On 5 September, Henriette and Elsa went swimming, leaving Ludwig in the hotel room. Elsa, of whom her father used to say 'she is the sunshine of my life',[8,9] returned later to find, shockingly, that he had hanged himself with a sash cord from the window. Elsa was just 15 years old.

It seems clear that there were complex causes of Boltzmann's extreme melancholia and manic depression. After all, he had a happy family life and had been phenomenally successful in his career, receiving accolades worldwide. But his scientific work depended absolutely and completely on the belief that matter was made of atoms and molecules, and that their behaviour was determined mechanically, i.e. by the forces acting on them. Both of these fundamental tenets had come under sustained and widespread criticism and even ridicule for decades, particularly in continental Europe. Boltzmann was much exercised and downhearted by these criticisms, which he took personally and which hurt him intensely. He increasingly retreated from physics to philosophy, trying but failing to overcome his critics and detractors.

In 1905, the year before Boltzmann's death, Albert Einstein in a monumental scientific tour de force published four of the most influential papers ever on fundamental physical phenomena, known as the *Annus Mirabilis Papers*,[10–13] explaining: Brownian motion (the random motions of pollen particles suspended in water); the photoelectric effect (the emission of electrons from a surface exposed to light); special relativity (the independence of the laws of physics to the velocity of the frame of reference in which they are measured); and the equivalence of mass and energy (the famous Einstein equation $e = mc^2$). It is not known whether Boltzmann saw or read these papers. But it is heartrendingly ironic that the atomistic view of matter was triumphantly proved by Einstein's theory of Brownian motion, followed by Jean Perrin's experimental verification three years later. By 1908, in other words, Boltzmann's atomistic approach was fully vindicated and established throughout the scientific community. On the other hand, it is doubly ironic that the theory of special relativity demonstrated the interconvertibility of energy and mass via Einstein's famous equation $e = mc^2$, directly contradicting Boltzmann's mechanics, which emphasised the primacy of particles with mass over fields of energy.

Ludwig Eduard Boltzmann was born in Vienna on 20 February 1844. His father, Ludwig Georg Boltzmann was a relatively poor, minor tax official, and his paternal grandfather, Gottfried Ludwig Boltzmann, was a clock and music box maker from Berlin, who had moved to Vienna as a young man. His mother, Maria Pauernfeind was the daughter of a rich Salzburg merchant. Boltzmann was born on the night between Shrove Tuesday and Ash Wednesday, which he often said explained why his moods could swing violently from great happiness to deep depression. He had a younger brother, Albert, who died of pneumonia while at secondary school, and a younger sister, Hedwig. They were all brought up in their mother's Catholic religion, though their father was a Protestant.

The Boltzmann family moved to Wels and then Linz in northern Austria, where the young Ludwig went to the local gymnasium (high school), following his earlier elementary education at home with a private tutor. He was academically gifted, usually top of his class, and particularly keen on maths and science. He took piano lessons from the (later) famous composer Anton Bruckner. Boltzmann ascribed his poor eyesight in later years to long evenings spent studying by candlelight. When he was just 15 years old, his father died of tuberculosis. In 1863, aged 19, he went to study maths and physics at the University of Vienna. The Institute of Physics in Vienna had been founded in 1850 by Christian Doppler, discoverer of the Doppler effect on the frequency of sound waves. It had just appointed a new Director, Josef Stefan, aged only 28, who later became famous for the law named after him, which says that the heat radiated from a material is proportional to the fourth power of its temperature. Stefan became Boltzmann's mentor. Many years later, Boltzmann wrote in Stefan's obituary:

> [T]he first thing he [Stefan] did was to hand me a copy of Maxwell's papers, and since at that time I did not understand a word of English, he also gave me an English grammar; I had received a dictionary from my father.[14]

Boltzmann graduated with his PhD three years later, in 1866, without having to prepare a doctoral thesis, which was not required in Vienna until 1872. In 1867 he gave his inaugural address to become a *Privatdozent* (lecturer or assistant professor) in the Institute; in 1868 he was awarded his *venia legendi* (entitlement to lecture); and in 1869, aged just 25, he was appointed to the Chair of Mathematical Physics at the University of Graz in Styria. He spent four years in Graz, with visiting periods to Heidelberg to work with Robert Bunsen and Leo Königsberger, and to Berlin to work with Gustav Kirchoff and Hermann von Helmholtz. In 1873 he returned to the University of Vienna as Professor of Mathematics, but soon returned to Graz, this time in his preferred position as Professor of Experimental Physics, where he stayed for the next 14 years.

This was a happy and productive period of Boltzmann's life, both personally and professionally. In 1872, the year before leaving Graz for Vienna, he met his future wife, Henriette von Aigentler, who was, according to his biographer Carlo Cerignani, 'a girl with long blonde hair and blue eyes, ten years his junior...living as an orphan in the house of the parents of the composer Wilhelm Kienzl in Stainz, south of Graz'.[15] Henriette was a teacher, but after meeting Ludwig she decided to study maths and began to attend his lectures. The Dean of the Faculty believed, misogynistically, that women should concentrate on cooking and cleaning. He persuaded the faculty to introduce a rule banning female students, but Henriette successfully petitioned the Minister of Public Education to be exempted from this discrimination. Boltzmann continued to correspond with Henriette while he was in Vienna, finally proposing to her in 1875. He believed, rather formally, that a proposal should be written not spoken. His letter to Henriette had a characteristically pragmatic and ironic tone:

> As a mathematician, you will not find that numbers, which dominate the world, are very poetical. Last year my yearly income was 5400 florins. This will be sufficient to support our existence, but if we take into account the huge increase in prices in Vienna, it is not sufficient to offer you many distractions and amusements.[16]

Boltzmann was short and stout, with curly hair and blue eyes, tender-hearted, unable to stop crying whenever he had to leave Henriette, and unable to say no to any requested favour. She called him 'my sweet fat darling'.[17] They married on 17 July 1876 and settled down happily in Graz, gradually building a family of two sons and three daughters over the next few years. Boltzmann loved nature, long walks, ice skating and swimming. He studied botany and was a great collector of butterflies. He was knowledgeable about classical German literature and music, often quoting from his favourite poet, Friedrich Schiller, and playing pieces on the piano from his favourite composer, Ludwig van Beethoven. He liked to write what he called jocular poems. In fact, he was usually jovial and good-humoured, and loved giving parties as well as going to them, often staying up until the late hours. The Boltzmanns bought a farm near Oberkroisbach with wonderful views. He had a faithful Alsatian dog, who would come down from the farm at noon, wait for him outside the Institute, and accompany him to the pub for lunch. At the university he experimented with sound, light, electricity and elasticity, and was fascinated by the technical development of airships and airplanes. His friend, the composer Wilhelm Kienzl, described him as 'the prototype of an unworldly scholar who lived in the realm of the world of science and of pioneering research'.[18] The German physicist Theodor des Coudres said he was possessed of 'inexhaustible friendliness',[19] and his student, the Austrian physicist Lise Meitner, said he was 'kind-hearted, full of belief in ideals and reverence for the wonders of nature'.[20]

In 1872 Boltzmann published a paper entitled, rather modestly, 'Further Studies on the Thermal Equilibrium of Gas Molecules',[21,22] which introduced the famous (extended) *Boltzmann equation*, describing the evolution of the positions and velocities of the atoms or molecules in a gas. This was the first time that probabilistic techniques had been used in theoretical physics. In three dimensions, every atom or molecule is described at any given instant by three position coordinates and three velocities, so an assembly of N atoms or molecules is described by $6N$ variables. The set of all possible values of the $6N$ variables is called the *phase space*, and Boltzmann's equation describes the probability distribution for every possibility, i.e. over all phase space, and how it will evolve with time, for different kinds of atoms and molecules, different interactions between them and different starting configurations. Boltzmann's equation is a generalisation of the previously derived *Maxwell distribution*, which is the probability distribution when the assembly has reached equilibrium. The American science historian Martin Klein wrote that Boltzmann's 'study of Maxwell led him directly to a new approach to the study of gases, one in which the molecular distribution function … played a central part'.[23] Stefan's foresight in handing the young Boltzmann a copy of Maxwell's papers and an English grammar had been prescient indeed.

In the end, Boltzmann's equation relies on the fact that, for a sufficiently large number of atoms or molecules, their average behaviour gives an accurate description of the behaviour of the total assembly. Individual atoms or molecules all have different trajectories, different velocities and different energies, but they are continually interacting and exchanging energy, so the spread of velocities and energies remains stable, and evolves in a coherent and consistent way. Boltzmann's equation is difficult to derive and often difficult to use, not helped by its complexity as an integro-differential equation, or by his writing style, which was long-winded and ponderous. In 1873 Maxwell commented on Boltzmann, 'I have been unable to understand him. He could not understand me on account of my shortness, and his length was and is an equal stumbling block for me'.[24] On the other hand, the young Einstein wrote much later, in 1900, 'Boltzmann is magnificent … . He is a masterly expounder'.[25] It is tempting to comment that, if Maxwell found him difficult, heaven help everyone else (except Einstein). And Boltzmann's equation proved to be the bedrock for our understanding of the behaviour of large collections of atoms and molecules. It is the basis for the *kinetic theory of gases*, and effectively gives a full microscopic underpinning for the laws of thermodynamics as applied to gases. The notion of entropy and disorder arise directly and naturally from the spread of the probability distribution, and their increase with time from the way in which organised arrangements are progressively eroded by atomic or molecular motion.

Thermodynamics had evolved through the 19th century, initially as a set of empirical laws about the interactions of heat, energy and motion. Experimental understanding was driven by the Industrial Revolution, with the development of machines and engines that were transforming the face of civilised society. The German scientist Rudolf Clausius first described the first and second laws of thermodynamics in his famous paper of 1850[26,27] (the zeroth and third laws were developed much later). Clausius' first law was a statement of the conservation of energy, a fairly natural extension of Newtonian mechanics. Clausius' second law was more curious. It was a refinement of the *Carnot principle*, derived by the French engineer Sadi Carnot much earlier, in 1824, which says that there is a maximum efficiency of any engine. Clausius' second law stated that this was because heat cannot pass from a cooler to a hotter body. In 1865 Clausius invented the concept of entropy, and re-defined the second law as saying that entropy is either conserved or increases, but cannot decrease. His compact statement of the two laws was then

The energy of the universe is constant.
The entropy of the universe tends to a maximum.[28,29]

The British novelist and physical chemist C. P. Snow paraphrased these laws as

You can't win
And you can't even break even.[30]

The second law defines an irreversible direction in events, towards ever-increasing disorder. But why can machines or engines not be perfectly efficient? Why can heat not flow 'uphill'? Why can entropy not decrease? Boltzmann's equation gives the answer by linking macroscopic thermodynamics to the probabilistic behaviour of an assembly of atoms or molecules. In principle, all the atoms or molecules of a gas could congregate in one half of a container, leaving the other half completely empty. But this never happens, because of the laws of large numbers. According to Boltzmann's equation, it is indeed possible that all the atoms or molecules move in the same direction at the same time. The probability is not zero, but it is vanishingly, infinitesimally small. If we throw a pack of cards on the floor, they never land as a pack of cards. It is possible, but it never happens. They land as a disordered pile. Disorder and entropy always increase or, at best, remain constant under extremely carefully controlled conditions. For the same reason, heat can only flow in one direction, and machines can never be perfectly efficient. The American engineer Seth Lloyd commented wryly, 'Nothing in life is certain except death, taxes and the second law of thermodynamics'.[31]

While at Vienna and Graz during the 1870s and 1880s, Boltzmann continued to develop his theories of thermodynamics and statistical mechanics, extending them to more complex situations and more complex molecular structures. Most notably, he developed his contentious *H-theorem*, defining the entropy of a non-equilibrium state, and arguing that such an assembly would inevitably exhibit fluctuations moving it towards equilibrium. He also developed the *ergodic thesis*: the assumption that a single assembly will, over time, pass through all possible configurations, i.e. sample all phase space. According to his biographer John Blackmore, he 'climbed out on a limb by stretching his work from equilibrium to non-equilibrium … and advocating the ergodic hypothesis [was] cutting off the limb behind him'.[32] He also worked on many other branches of physics, including electrodynamics, elasticity, rheology, radiation and classical mechanics.

Boltzmann had by now become well known and well respected as a distinguished mathematical, theoretical and experimental physicist. As the 1880s progressed, however, events conspired to disturb, and ultimately disrupt, his happy scientific and family life. First, his mother died in 1885. Second, he was elected Rector of the University of Graz in 1887, an honour but also an onerous duty, which didn't suit his friendly, freewheeling academic style. Pro-German students stole the bust of the Austrian Emperor, and Boltzmann had to take disciplinary action, which led to a general student revolt. He found this painful and stepped down the following year. Third, his son, Ludwig, died in 1889. Fourth, he developed itchy feet, wanting to get a prestigious position in one of the increasingly powerful German universities. The influence of Austria was gradually waning through the second half of the 19th century, defeated in wars by France in 1859 and Prussia in 1866, followed by the unification of Germany in 1871 and the relentless rise of German power. He tried three times to move to Germany, without any real success. When Kirchoff died, he was offered the Chair in Berlin, signed a contract, but then embarrassingly changed his mind at the last minute. Later, he took Chairs of Theoretical Physics, at the University of Munich in 1890 and at the University of Leipzig in 1900, but in neither case did it work out well. Each time he found conditions not to his liking and he rapidly moved back to Vienna.

Finally, and perhaps most importantly, as Boltzmann's thermodynamic theories became better known and understood, they began to attract waves of criticism and challenge. Physicists (with notable exceptions such as Maxwell and Lorentz) did not like his use of probabilistic methods, mathematicians disputed the foundations of his theories and the accuracy of his predictions, and chemists had not fully committed to atomic and molecular theories of matter. Loschmidt's paradox pointed out that fundamental theories of mechanics are time independent, i.e. remain valid when the direction of time is reversed, which is inconsistent with the continual increase of entropy predicted by Boltzmann's equation and the second law. Zermelo's paradox pointed out that an event, even with a small probability, still occurs sooner or later, so Boltzmann's thermodynamics and the inevitable increase of entropy would be invalidated as long as one were sufficiently patient. Boltzmann defended himself against these criticisms. The arguments are deep and complex, but in the end he had the crucial scientific advantage of extensive, not to say overwhelming, agreement with experiment.

But the strongest attack came from closer to home, notably from his colleague Ernst Mach, Professor of Mathematics at the University of Graz and, later, Professor of Physics at Salzburg and Prague, and Professor of History and Philosophy of Science at Vienna. Mach is well known as a physicist for his work on shock waves, inventing the Mach number, the ratio of a body's speed to the speed of sound. Mach had vied with Boltzmann for his first professorship at Graz. As his career developed, Mach became increasingly philosophical. He was a leading exponent of *phenomenalism*, the view that only sensations are real, and there is no objectively verifiable physical world. This is incompatible with a belief in atoms and molecules. Phenomenalists were also close to the scientific *energetics* movement, which developed in the later decades of the 19th century and was supported by leading scientists such as Wilhelm Ostwald and Georg Helm. This also called into question the existence of atoms and molecules, preferring a philosophy of physics based on energy fields, rather than mechanical interactions of atoms and molecules. As the 19th century progressed, it seemed that science might be on the point of explaining everything in the world as a great deterministic machine, and this produced a strong backlash from philosophers and theologians, against Cartesian dualism (the separation of mind and matter), materialism (the existence of an objective physical world), Newtonian mechanics (a clockwork, 'mechanistic' view of the interactions of material objects) and atomic theory (the idea that matter is not fully continuous and consists of assemblies of minute particles or 'atoms'). Mach in particular attacked Boltzmann frequently and fiercely. He famously shouted, after a lecture by Boltzmann in 1897 at the Imperial Academy of Sciences in Vienna, 'I don't believe that atoms exist'![33,34]

Boltzmann's health began to deteriorate and he became increasingly melancholic. He retreated into philosophical arguments, determined to persuade his detractors that they were wrong. This was, of course, doomed to failure, because they were motivated more by theology and emotion than by logic and science. He never wavered from his belief in the reality of atoms and their mechanical interactions, but he felt continually threatened and thought that he was losing the argument. John Blackmore described it as a myth that Boltzmann committed suicide directly because of opposition to atomism or because of the attacks of Ernst Mach. The truth is that we will probably never know exactly why Boltzmann took his life. In the modern world he would almost certainly have been diagnosed as suffering from bipolar disorder (previously known as manic depression). As he grew older, his black moods became worse, almost certainly compounded by a range of factors including, as he saw it, his failure to regain a happy and comfortable academic position, the continual attacks on his scientific integrity and his failure to win the philosophical argument. According to the US National Institute of Mental Health, 'bipolar disorder affects more than 5.7 million adults (about 3% of the population) in the USA each year';[35] and, according to the US National Center for Biotechnology Information, 'at least 25–50% of patients with bipolar disorder attempt suicide at least once'.[36]

It is macabre and strangely fitting that, if Boltzmann had to commit suicide, he did so in Duino, where Rilke much later (in 1912) wrote his famous *Duino Elegies*. The Bohemian Rilke is intensely mystical, 'invoking haunting images of the difficulty of communing with the effable in an age of disbelief, solitude and profound anxiety'.[37] The writer Bibhudutt Dash describes the elegies as

> an ontological torment … an impassioned monologue about coming to terms with human existence … the limitations and insufficiency of the human condition and fractured human consciousness … man's loneliness, the perfection of the angels, life and death.[38]

After Boltzmann's death, Max Planck wrote to Henriette with his commiserations:

> I was deeply shaken by the newspaper report of the sudden passing away of your husband, and I beg to be allowed to give expression to my inner feelings on the tragic end of this rich life .... What must the pitiable sick man have suffered, before he decided to take this final liberating step! And what a blessed life so rich with successes has departed through him. I have always looked up to him as a model that one should strive after even if one cannot reach his standard; for he was ahead of us all .... Never will what he has done for our science be replaced or even approached.[39]

In 1877, when Boltzmann was still working through the ramifications of his famous equation, he mused (according to the science historian Stephen Brush), 'One could even calculate from the relative numbers of the different distributions their probabilities'.[40] Later that year, he wrote a paper that put this idea into mathematical form:

$$S = k \ln w,$$

where 'w is the number of possible molecular configurations corresponding to a given macroscopic state of the system'.[41] This important equation was later described as *Boltzmann's principle* by Einstein, but it is also often (and somewhat confusingly) called *Boltzmann's equation*. It is much shorter and more accessible than his great integro-differential equation, but it encapsulates with beautiful simplicity the core probabilistic truth that Boltzmann discovered about entropy and the second law of thermodynamics. It is inscribed on his gravestone in Vienna.

It may not be too fanciful to quote from the poem Boltzmann wrote about Beethoven, which seems to contain hints of his own sombre end:

> *I never mastered my ways or weeping.*
> *I never won self-possession or keeping*
> *A balanced view of Heaven and Earth.*
> *But now I love angelic mirth. Ha! Ha!*[42]

4   The fundamental equations of thermodynamics

The first fundamental equation

Consider a given amount of a single-component material in the form of a gas (e.g. steam). Work can be done by compressing or expanding the gas with a piston (as in a steam engine), i.e. $dw = PdV$, where P is the pressure and dV is the change in volume caused by compression or expansion. And the second law of thermodynamics can be re-written to define heat supplied to the gas as $dq = TdS$. Inserting these expressions for work and heat into the first law of thermodynamics gives

$$dE = dq - dw = TdS - PdV.$$

This equation combines the first and second laws of thermodynamics, and is called the *first fundamental equation of thermodynamics*. It describes the internal energy of a material in terms of two independent variables, the entropy S and the volume V. This agrees with the Gibbs phase rule, which shows that there are only two independent variables for a single-component, single-phase material such as steam, i.e. when the number of components is $c = 1$ and the number of phases is $p = 1$:

$$p + f = c + 2$$
$$\therefore f = 1 + 2 - 1 = 2.$$

Other thermodynamic functions and fundamental equations

It is convenient to define three new thermodynamic functions, namely enthalpy H, Helmholtz free energy A and Gibbs free energy G:

$$H = E + PV$$
$$A = E - TS$$
$$G = H - TS.$$

Differentiating, and inserting the first fundamental equation of thermodynamics,

$$dH = TdS + VdP$$
$$dA = -PdV - SdT$$
$$dG = VdP - SdT.$$

These are alternative *fundamental equations of thermodynamics*. The different fundamental equations are equivalent, and provide no new information. They just correspond to different natural variables. Which of the functions E, H, A or G, and which of the corresponding fundamental equations are convenient to use depends on the particular material and conditions being considered. Internal energy is particularly suitable when the natural independent variables are the entropy and volume, enthalpy is particularly suitable when the natural independent variables are entropy and pressure, Helmholtz free energy is particularly suitable when the natural variables are volume and temperature, and Gibbs free energy is particularly suitable when the natural independent variables are pressure and temperature.

For materials, the natural variables are usually temperature and pressure. In general we make things out of solid and liquid materials by heating them and stretching or squeezing them, effectively changing their temperature and pressure. It is convenient, therefore, to use Gibbs free energy to understand and explain how the atoms and molecules in a material are affected, both during processing to manufacture useful objects as well as during their subsequent use.

Temperature and pressure are called *intensive variables*, i.e. they are not affected by the amount of material we are considering. As defined by the zeroth law, they are constant throughout a material at equilibrium. Mass, entropy, volume, internal energy, enthalpy and free energy are called *extensive variables*, i.e. they are proportional to the amount of material we are considering.

Other forms of work and energy

We have to include chemical energy terms if the amount of material is not fixed, or if the material is heterogeneous and consists of more than one component or phase, so that chemical reactions or changes of state are possible:

$$dE = TdS - PdV + \Sigma \mu_i dn_i$$
$$dH = TdS + VdP + \Sigma \mu_i dn_i$$
$$dA = -PdV - SdT + \Sigma \mu_i dn_i$$
$$dG = VdP - SdT + \Sigma \mu_i dn_i,$$

where n_i is the number of moles of the ith component and μ_i is called the *chemical potential* of the ith component.

Integrating the first fundamental equation for the internal energy and inserting the result into the definitions of the other thermodynamic functions gives

$$E = TS - PV + \Sigma \mu_i n_i$$
$$H = TS + \Sigma \mu_i n_i$$
$$A = -PV + \Sigma \mu_i n_i$$
$$G = \Sigma \mu_i n_i,$$

which shows (amongst other things) that the chemical potential of each component is the same as its molar Gibbs free energy:

$$G = \Sigma G_i = \Sigma \mu_i n_i$$
$$\therefore \mu_i = \frac{G_i}{n_i}.$$

We can also include many other forms of work in the fundamental equations. For instance, gravitational work is $dw = Fdh$, where F is the applied force and the material changes its height by dh; extensional work is $dw = Fdx$, where F is again the applied force and dx is the longitudinal extension; electrical work is $dw = Qd\mathcal{E}$, where the material has a charge Q and is moved in an electric field, with a change in electric potential $d\mathcal{E}$; magnetic work is $dw = mdH$, where the material has a magnetic moment m and is moved in a magnetic field, with a change in magnetic field dH; and surface work is $dw = \gamma dA$, when a surface is created with surface energy γ and an increase in surface area dA. The fundamental equations become

$$dE = TdS - PdV + \Sigma\mu_i dn_i + \text{other work terms}$$
$$dH = TdS + VdP + \Sigma\mu_i dn_i + \text{other work terms}$$
$$dA = -PdV - SdT + \Sigma\mu_i dn_i + \text{other work terms}$$
$$dG = VdP - SdT + \Sigma\mu_i dn_i + \text{other work terms.}$$

5   Gibbs free energy

The Gibbs free energy G of a material is defined by

$$G = E + PV - TS = H - TS = \Sigma G_i = \Sigma\mu_i n_i$$
$$dG = VdP - SdT + \Sigma\mu_i dn_i.$$

A material is at equilibrium, i.e. no change takes place, when its Gibbs free energy reaches a minimum $dG = 0$. When a material is not at equilibrium, its atoms or molecules will move to reduce the Gibbs free energy until it reaches a minimum. The atoms or molecules are subjected to two basic (and often competing) forces: first, to reduce their internal energy E and, therefore, the enthalpy H; and, second, to increase their disorder, i.e. to increase the entropy S. Both of these contribute to a reduction in the Gibbs free energy G. At any given temperature, the upshot is that equilibrium corresponds to a balance between the atoms or molecules in a material having their lowest energy and highest entropy.

Gibbs free energy is particularly important in describing the behaviour of materials. This is in part because the natural variables for a solid or liquid material are temperature T, pressure P and the amount of the different components n_i, and in part because the chemical potential of each component μ_i, which controls chemical reactions and changes of state, is the same as its molar Gibbs free energy $\mu_i = G_i/n_i$.

Solids and liquids are relatively *incompressible*. This means that, for solids and liquids, PV terms expressing the energy of expansion and contraction are relatively small, so their internal energy and enthalpy are approximately equal, and their Gibbs and Helmholtz free energies are also approximately equal:

$$\text{solids and liquids}: E \approx H; G \approx A.$$

6   Physical properties

Many of the physical properties of a material can be obtained by differentiating the relevant thermodynamic function. We examine this for the *specific heat* and *thermal expansion coefficient*.

Specific heat

When a material is heated, its temperature increases. The more material there is, the more heat is required to raise the temperature by a given amount. In other words, the temperature increase dT is proportional to the amount of heat added dq and inversely proportional to the amount of material or, more accurately, its mass m:

$$dT \propto \frac{dq}{m}$$
$$\therefore dq = mc\,dT,$$

where the constant of proportionality c is called the *specific heat* of the material. In other words, the *specific heat* of a material is the heat required to increase the temperature of unit mass of the material by $1°$. Similarly, the *molar specific heat* is the heat required to increase the temperature of 1 mole of the material by $1°$, and m in the previous equation is now the number of moles of the material.

The temperature rise for a given amount of added heat and, therefore, the specific heat itself depend on whether the heat is added under conditions of constant pressure or constant volume:

$$dq_V = mc_V\,dT$$
$$dq_P = mc_P\,dT,$$

where dq_V and dq_P are the amount of heat added at constant pressure or volume respectively, and c_V and c_P are the corresponding specific heats at constant pressure and volume respectively.

From the first and second laws, the molar specific heats at constant volume and pressure can be written as

$$c_V = \frac{dq_V}{dT} = \left(\frac{dE}{dT}\right)_V = T\left(\frac{dS}{dT}\right)_V$$
$$c_P = \frac{dq_P}{dT} = \left(\frac{dH}{dT}\right)_P = T\left(\frac{dS}{dT}\right)_P.$$

From the fundamental equation of thermodynamics,

$$dG = V\,dP - S\,dT,$$

so the entropy and the specific heat at constant pressure are given by

$$S = -\left(\frac{dG}{dT}\right)_P$$
$$c_P = T\left(\frac{dS}{dT}\right)_P = -T\left(\frac{d^2 G}{dT^2}\right)_P.$$

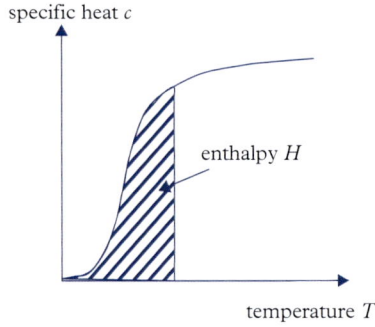

Figure 4.1 *Specific heat c and enthalpy H versus temperature T*

Integrating these equations shows that measurements of the temperature variation of the specific heat of a material can be used to calculate many of its most important thermodynamic properties:

$$H - H^o = \int_{298}^{T} c_P dT$$

$$S - S^o = \int_{298}^{T} \frac{c_P}{T} dT,$$

where H^o and S^o are the *standard enthalpy* and *standard entropy* of the material, i.e. the enthalpy and entropy under *standard conditions*, in a pure state at room temperature (298K) and atmospheric pressure (1 bar or 10^5 Pa). From the definition of Gibbs free energy,

$$G = H - TS$$

$$\therefore G - G^o = \int_{298}^{T} c_P dT - T \int_{298}^{T} \frac{c_P}{T} dT,$$

where G^o is the *standard Gibbs free energy* of the material. Data for standard enthalpies, entropies and Gibbs free energies are given for many materials in chemical tables.

Figure 4.1 shows how the specific heat varies with temperature for a given material and how integration can be used to obtain the enthalpy of the material at any temperature.

Thermal expansion

When a material is heated, it expands. The expansion dV is proportional to its initial volume V_o and the increase in temperature dT:

$$dV = \alpha_V V_o dT,$$

where α_V is called the *volume coefficient of expansion*. Similarly, if the material is in the form of a bar, the increase in length dl is proportional to its initial length l_o and the increase in temperature dT:

$$dl = \alpha_l l_o dT,$$

where α_l is called the *linear coefficient of expansion* and $\alpha_V \approx 3\alpha_l$.

From the fundamental equation of thermodynamics,

$$dG = VdP - SdT,$$

so the volume and the coefficient of expansion are given by

$$V = \left(\frac{dG}{dP}\right)_T$$

$$\alpha_V = \frac{1}{V_o}\left(\frac{dV}{dT}\right)_P = \frac{1}{V_o}\left(\frac{d^2G}{dTdP}\right).$$

7   Chemical reactions

Consider a general chemical reaction,

$$aA + bB \rightleftharpoons cC + dD,$$

where a, b, c and d are the *chemical coefficients*, i.e. the numbers of moles of reactants A and B and products C and D respectively. Depending on temperature, pressure, and reactant and product concentrations, a reaction can usually proceed in either the forward or backward direction, as indicated by the two-way arrow.

The *free energy change of the reaction* ΔG is the change in free energy when the reactants are replaced by the products:

$$\Delta G = G_{\text{products}} - G_{\text{reactants}} = (c\mu_C + d\mu_D) - (a\mu_A + b\mu_B),$$

where μ_i are the *chemical potentials* or *molar free energies* of the reactants and products. The *standard free energy change of the reaction* ΔG^o is the free energy of the reaction when all the reactants and products are at *standard state*, i.e. in a pure state at room temperature (298K) and atmospheric pressure (1 bar or 10^5 Pa):

$$\Delta G^o = G^o_{\text{products}} - G^o_{\text{reactants}} = (c\mu_C^o + d\mu_D^o) - (a\mu_A^o + b\mu_B^o),$$

where μ_i^o are *standard chemical potentials* of the reactants and products.

The chemical potential of the *i*th component μ_i depends on concentration c_i and temperature T according to

$$\mu_i = \mu_i^o + RT \ln c_i,$$

so the free energy change of the reaction can be re-written as

$$\Delta G = \Delta G^o + RT \ln \frac{c_C^c\, c_D^d}{c_A^a\, c_B^b}.$$

Under some conditions, a chemical reaction can be at *equilibrium*. At equilibrium there is no overall free energy change, $G_{\text{products}} = G_{\text{reactants}}$ and $\Delta G = 0$, so

$$\Delta G^o = -RT \ln \frac{c_{Ceq}^c\, c_{Deq}^d}{c_{Aeq}^a\, c_{Beq}^b} = -RT \ln K$$

or, taking anti-logs,

$$K = \frac{c_{Ceq}^c\, c_{Deq}^d}{c_{Aeq}^a\, c_{Beq}^b} = \exp\left(-\frac{\Delta G^o}{RT}\right),$$

where K is called the *equilibrium constant*. Notice that the concentrations in these last two equations are specific values that give equilibrium between the reactants and products, as indicated by labelling them c_{ieq}. This means that the equilibrium constant K is related to the standard free energy change of the reaction ΔG^o, so the balance of the equilibrium concentrations of reactants and products c_{ieq} can be calculated from the difference in their chemical potentials μ_i^o.

8   Solutions

Consider 1 mole of a two-component A–B solid or liquid solution, with a mole fraction x_A of A and $x_B = (1 - x_A)$ of B, as shown in Figure 4.2. The Gibbs free energy of the solution is given by

$$G = x_A \mu_A + x_B \mu_B + \Delta G_{\text{mix}},$$

where μ_A and μ_B are the chemical potentials of A and B respectively, $x_A \mu_A + x_B \mu_B$ is the free energy of the two components before they are mixed to form the solution, and ΔG_{mix} is called the *free energy of mixing*, which is given by

$$\Delta G_{\text{mix}} = \Delta H_{\text{mix}} - T \Delta S_{\text{mix}},$$

where ΔH_{mix} and ΔS_{mix} are the enthalpy of mixing and entropy of mixing respectively.

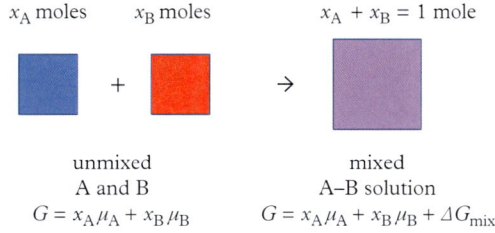

Figure 4.2 *The Gibbs free energy of mixing ΔG_{mix} in an A–B solution*

Ideal solutions

When there is no interaction between the A and B atoms or molecules, the solution is called an *ideal solution*. This is the same as having a random mixture of red and blue billiard balls. There is no energy or enthalpy of interaction between A and B, so the ideal energy and enthalpy of mixing are both zero:

$$\Delta H_{mix}^{ideal} = \Delta E_{mix}^{ideal} = 0.$$

The ideal entropy of mixing is then the disorder associated with a random distribution of x_A moles of A and x_B moles of B on 1 mole of atomic or molecular sites, either disordered sites in a liquid or crystal lattice sites in a solid. This is called the *configurational entropy* of the solution.

The number of ways of placing the first A atom on 1 mole of sites is N, where N is Avogadro's number. The number of ways of placing the second A atom is $N - 1$, but the two atoms could have been placed in reverse order, so there are $N(N-1)/2$ different ways of placing two A atoms. Similarly, the number of different ways of placing three A atoms is $N(N-1)(N-2)/3 \times 2$. If we continue in this way, the number of ways w of placing all A atoms ($x_A N$ of them) is given by

$$w = \frac{N(N-1)(N-2)\dots(N-x_A N+1)}{x_A N(x_A N-1)(x_A N-2)\dots 3 \times 2} = \frac{N!}{(x_A N)!(x_B N)!},$$

because the numerator in the first expression for w is equal to $N!/x_B N)!$. The ideal entropy of mixing can be obtained from Boltzmann's equation:

$$\Delta S_{mix}^{ideal} = k \ln w = k \ln \left(\frac{N!}{(x_A N)!(x_B N)!} \right) = k(\ln N! - \ln (x_A N)! - \ln (x_B N)!).$$

Stirling's approximation for large numbers gives $\ln n! \approx n \ln n - n$, so

$$\Delta S_{mix}^{ideal} = -kN(x_A \ln x_A + x_B \ln x_B) = -R(x_A \ln x_A + x_B \ln x_B),$$

where R is the gas constant. The ideal Gibbs free energy of mixing and the total Gibbs free energy of an ideal solution are, therefore,

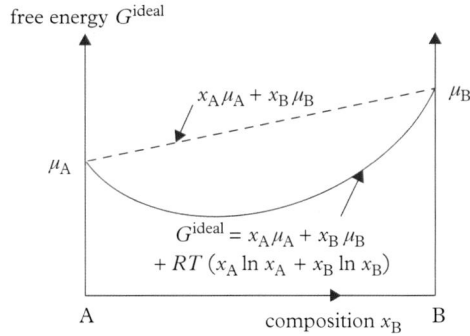

Figure 4.3 *The Gibbs free energy of an ideal solution*

$$\Delta G_{mix}^{ideal} = \Delta H_{mix}^{ideal} - T\Delta S_{mix}^{ideal} = RT\left(x_A \ln x_A + x_B \ln x_B\right)$$

$$G^{ideal} = x_A\mu_A + x_B\mu_B + \Delta G_{mix}^{ideal} = x_A\mu_A + x_B\mu_B + RT\left(x_A \ln x_A + x_B \ln x_B\right).$$

The Gibbs free energy of an ideal solution G^{ideal} is shown as a function of composition in Figure 4.3. Since x_A and x_B are fractions, $\ln x_A$ and $\ln x_B$ are both negative, the ideal entropy of mixing is always positive, and the Gibbs free energy of mixing is always negative, which means that an ideal solution is always preferred over unmixed A and B.

Regular solutions

Most solutions are not ideal and there are positive or negative interactions between A and B atoms or molecules, so the energy and enthalpy of mixing are not zero. In a *regular solution*, the interactions are assumed to be sufficiently small that A and B are still arranged randomly. The entropy of mixing is still the disorder associated with a random distribution of x_A moles of A and x_B moles of B on 1 mole of atomic or molecular sites, either disordered sites in a liquid or crystal lattice sites in a solid, i.e.

$$\Delta S_{mix}^{regular} = \Delta S_{mix}^{ideal} = -R(x_A \ln x_A + x_B \ln x_B).$$

In a mole of solution, the number of A–A, B–B and A–B bonds n_{AA}, n_{BB} and n_{AB} are given respectively for a random distribution by

$$n_{AA} = \frac{1}{2}Nzx_A^2$$

$$n_{BB} = \frac{1}{2}Nzx_B^2$$

$$n_{AB} = Nzx_Ax_B,$$

where z is the coordination number of the atomic or molecular structure.

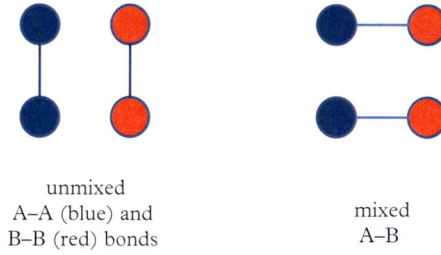

unmixed
A–A (blue) and
B–B (red) bonds

mixed
A–B

Figure 4.4 *In a solution, every two A–B bonds have replaced one A–A bond and one B–B bond*

Comparing the solution to unmixed A and B, every two A–B bonds have replaced one A–A and one B–B bond, as shown in Figure 4.4. Let ϵ_{AA} be the bond energy between two A atoms, ϵ_{BB} be the bond energy between two B atoms, and ϵ_{AB} be the bond energy between one A atom and one B atom. Bond energies of this sort are defined as the decrease in energy as the two atoms are brought together from a long distance apart. The internal energy and enthalpy of mixing are then

$$\Delta H_{mix}^{regular} = \Delta E_{mix}^{regular} = -n_{AB}\,(2\epsilon_{AB} - \epsilon_{AA} - \epsilon_{BB}) = -Nzx_Ax_B\omega,$$

where $\omega = 2\epsilon_{AB} - \epsilon_{AA} - \epsilon_{BB}$ is called the A–B *interaction energy*. When $\omega > 0$, $2\epsilon_{AB} > \epsilon_{AA} + \epsilon_{BB}$, A–B bonds are preferred, and there is a negative energy and enthalpy of mixing, which favours formation of a solution; when $\omega < 0$, $2\epsilon_{AB} < \epsilon_{AA} + \epsilon_{BB}$, A–A and B–B bonds are preferred, and there is a positive energy and enthalpy of mixing, which favours the separation of A and B.

The regular Gibbs free energy of mixing and the total Gibbs free energy of a regular solution are, therefore,

$$\Delta G_{mix}^{regular} = \Delta H_{mix}^{regular} - T\Delta S_{mix}^{regular} = Nzx_Ax_B\omega + RT\,(x_A \ln x_A + x_B \ln x_B)$$
$$G^{regular} = x_A\mu_A + x_B\mu_B + \Delta G_{mix}^{regular}$$
$$= x_A\mu_A + x_B\mu_B + Nzx_Ax_B\omega + RT\,(x_A \ln x_A + x_B \ln x_B).$$

The Gibbs free energy of a regular solution $G^{regular}$ is shown in Figure 4.5 at a high temperature, and in Figures 4.6 and 4.7 at a low temperature for $\omega > 0$ and $\omega < 0$ respectively. At high temperature, the entropy term $RT\,(x_A \ln x_A + x_B \ln x_B)$ dominates. As in an ideal solution, x_A and x_B are fractions, $\ln x_A$ and $\ln x_B$ are both negative, the entropy of mixing is always positive, and the Gibbs free energy of mixing is always negative, which means that at high temperature, a regular solution is always preferred over unmixed A and B. At low temperature, however, the interaction term $Nzx_Ax_B\omega$ dominates. For $\omega > 0$, A–B bonds are preferred and a regular solution is again preferred over unmixed A and B. But for $\omega < 0$, A–A and B–B bonds are preferred, and the solution separates into a two-phase mixture of A-rich α and B-rich β phases with the Gibbs free energy of the mixture given by the common tangent line to the two regular-solution Gibbs free energy minima, as shown in Figure 4.7.

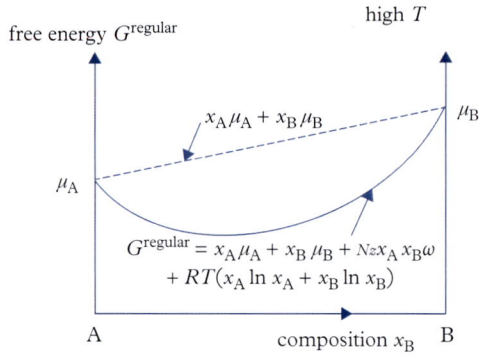

Figure 4.5 *The Gibbs free energy of a regular solution at high temperature*

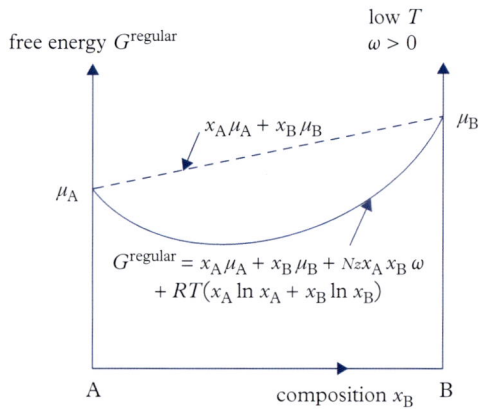

Figure 4.6 *The Gibbs free energy of a regular solution at low temperature for a positive interaction parameter*

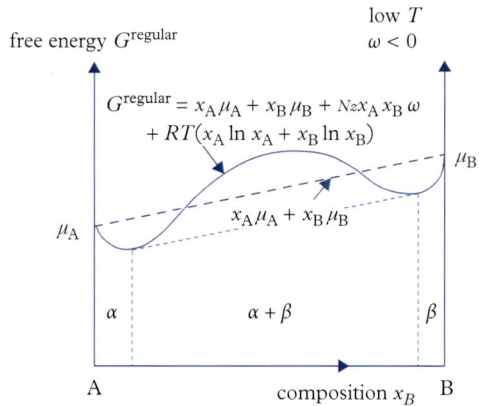

Figure 4.7 *The Gibbs free energy of a regular solution at low temperature for a negative interaction parameter*

....

9 REFERENCES

1. John Blackmore, ed. *Ludwig Boltzmann: His Later Life and Philosophy 1900–1906* (Kluwer-Springer, Dordrecht, 1995), 205.
2. Ibid. 208.
3. Ibid.
4. Rainer Maria Rilke. *Duino Elegies.* David Young, trans. (W. W. Norton, New York, 1978).
5. Blackmore n 1, 209.
6. Carlo Cercignani. *Ludwig Boltzmann: The Man Who Trusted Atoms* (Oxford University Press, New York, 1998), 346.
7. Ibid.
8. Ibid. 226.
9. D. Flamm. 'Ludwig Boltzmann and his influence on science'. *Studies in History and Philosophy of Science* **14** (1983), 255.
10. Albert Einstein. 'On a heuristic view about the creation and conversion of light'. *Annalen der Physik* **17** (1905), 132–148.
11. Albert Einstein. 'Investigations on the theory of Brownian motion'. *Annalen der Physik* **17** (1905), 549–560.
12. Albert Einstein. 'On the electrodynamics of moving bodies'. *Annalen der Physik* **17** (1905), 891–921.
13. Albert Einstein. 'Does the inertia of a body depend on its energy content?' *Annalen der Physik* **18** (1905), 639–641.
14. Cercignani, n 6, 6.
15. See ibid. 11.
16. Ibid. 12.
17. Ibid. 13.
18. E. Broda. *Ludwig Boltzmann: Man, Physicist, Philosopher* (Ox Bow Press, Woodbridge, 1983), 16.
19. Ibid. 17.
20. Ibid.
21. L. Boltzmann. 'Weitere studien uber das warmegleichgewicht unter gasmolekulen' ['Further studies on the thermal equilibrium of gas molecules']. *Sitzungsberichte Akademie der Wissenschaften Vienna* **66**, Pt. II (1872), 275.
22. L. Boltzmann. 'Further studies on the thermal equilibrium of gas molecules'. In: S. G. Brush (ed.). *Kinetic Theory Volume 2: Irreversible Processes* (Pergamon, Oxford, 1965), 88–175.
23. Blackmore, n 1, 11.
24. Cercignani, n 6, 200.
25. See ibid. 216.
26. R. Clausius. 'On the moving force of heat, and the laws regarding the nature of heat itself which are deducible therefrom'. *Annalen der Physik* **79** (1850), 368–397.
27. R. Clausius. 'On the moving force of heat, and the laws regarding the nature of heat itself which are deducible therefrom'. *Philosophy Magazine* **2**, Ser. 4 (1851), 1–21, 102–119.
28. R. Clausius. 'About various forms of the main equations of mechanical heat theory that are convenient for use'. *Annalen der Physik* **201** (1865), 353–400.
29. R. Clausius. *The Mechanical Theory of Heat: With Its Applications to the Steam Engine and to the Physical Properties of Bodies* (John van Voorst, London, 1867).

30. HIGHBROW. 'Laws of thermodynamics', episode #8 of the course 'Scientific laws and theories everyone should know'. (n.d.). https://gohighbrow.com/laws-of-thermodynamics/ (accessed 18 December 2019).
31. Seth Lloyd. 'Going into reverse'. *Nature* **430** (2004), 971.
32. Blackmore, n 1, 15.
33. Palle Yourgrau. *A World without Time: The Forgotten Legacy of Gödel and Einstein* (Allen Lane, London, 2005).
34. Manuel Bächtold. 'Saving Mach's views on atoms'. *Journal of the General Philosophy of Science* **41** (2010), 3.
35. Team Bipolar Lives. 'Bipolar order statistics'. (n.d.). https://www.bipolar-lives.com/bipolar-disorder-statistics.html (accessed 18 December 2019).
36. Danielle M. Novick, Holly A. Swartz, and Ellen Frank. 'Suicide attempts in bipolar I and bipolar II disorder: a review and meta-analysis of the evidence'. *Bipolar Disorder* **12** (2010), 1–9.
37. Wikipedia. 'Rainer Maria Rilke'. (n.d.). https://en.wikipedia.org/wiki/Rainer_Maria_Rilke (accessed 18 December 2019).
38. Bibhudutt Dash. 'In the matrix of the divine: approaches to godhead in Rilke's *Duino Elegies* and Tennyson's *In Memoriam*'. *Language in India* **11** (2011), 355–371.
39. Blackmore, n 1, 210.
40. See ibid. 15.
41. Ibid.
42. Ibid. 224.

· ·

10 BIBLIOGRAPHY

Entropy. J. D. Fast (Macmillan, London, 1970).
Introduction to the Thermodynamics of Materials (*6th edition*). David R. Gaskell and David E. Laughlin (Taylor and Francis, New York, 2018).
Lectures on Gas Theory. Ludwig Boltzmann (Dover, New York, 1964).
Ludwig Boltzmann: Man, Physicist, Philosopher. Engelbert Broda (Ox Bow Press, Woodbridge, 1983).
Ludwig Boltzmann: His Later Life and Philosophy 1900–1906. John Blackmore, ed. (Kluwer-Springer, Dordrecht, 1995).
Ludwig Boltzmann: The Man Who Trusted Atoms. Carlo Cercignani (Oxford University Press, New York, 1998).
Thermodynamics of Materials (*2nd edition*). Richard A. Swalin (Wiley, New York, 1972).

5

The Arrhenius Equation

Reactions

---

1   Chemical reactions

Most materials are manufactured by a *chemical reaction*. For instance, iron and steel are manufactured by reacting iron ore with coke at high temperature in a reaction vessel called a *blast furnace*. Iron ore contains iron oxides, mostly haematite Fe_2O_3 and magnetite Fe_3O_4; coke contains the reducing element carbon C. The oxides are reduced by carbon to metallic iron by the chemical reactions

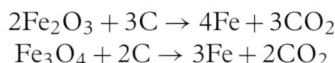

$$2Fe_2O_3 + 3C \rightarrow 4Fe + 3CO_2$$
$$Fe_3O_4 + 2C \rightarrow 3Fe + 2CO_2$$

This process is known as *smelting*. The rates of these reactions depend on the compositions of the iron ore and coke, and the temperature of the blast furnace, which must be controlled carefully to ensure an efficient manufacturing process.

Similar smelting processes are used to manufacture other metallic materials such as copper and zinc from their ores. Copper and zinc ores contain mostly copper and zinc sulphides, CuS and ZnS respectively. The manufacturing process has two stages. The sulphides are initially *roasted* in air to convert them to oxides,

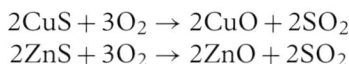

$$2CuS + 3O_2 \rightarrow 2CuO + 2SO_2$$
$$2ZnS + 3O_2 \rightarrow 2ZnO + 2SO_2$$

that are then reduced to metals by smelting with carbon:

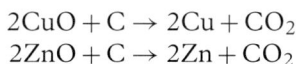

$$2CuO + C \rightarrow 2Cu + CO_2$$
$$2ZnO + C \rightarrow 2Zn + CO_2$$

Once again, the rates of these reactions depend on the compositions of the copper and zinc ores and the furnace temperature, which must, again, be controlled carefully to ensure efficient manufacturing processes.

The Equations of Materials. Brian Cantor. Oxford University Press (2020). © Brian Cantor.
DOI: 10.1093/oso/9780198851875.001.0001

Polymers are usually manufactured from their component *monomer* by a *polymeri-sation reaction*. For instance, polyethylene is manufactured from its *monomer* ethylene C_2H_4, using a *catalyst* such as titanium trichloride $TiCl_3$ or chromic oxide CrO_3 (a catalyst accelerates the reaction while remaining unchanged by it):

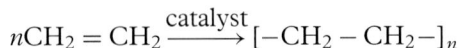

$$nCH_2 = CH_2 \xrightarrow{\text{catalyst}} [-CH_2 - CH_2-]_n$$

Once again, it is important to control the rate and the extent of the reaction to achieve an efficient manufacturing process.

Materials also degrade by chemical reactions. When iron rusts, it is being oxidised by the surrounding air, essentially undoing the reduction reactions with which it was manufactured:

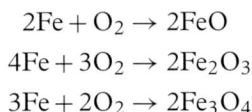

$$2Fe + O_2 \rightarrow 2FeO$$
$$4Fe + 3O_2 \rightarrow 2Fe_2O_3$$
$$3Fe + 2O_2 \rightarrow 2Fe_3O_4$$

To take another example, silicon semiconductor chips also degrade by oxidation in air:

$$Si + O_2 \rightarrow SiO_2$$

If we wish to prevent these unwanted degradation reactions, we must again control the reaction rates and understand how they depend on composition and temperature.

2   Reaction kinetics

Reaction kinetics is the science of the rates at which chemical reactions take place. Consider a general chemical reaction:

$$aA + bB \rightarrow cC + dD$$

A, B, C and D are atomic or molecular species. A and B are the *reactants*, and C and D are the *products*. The integer numbers a, b, c and d are the *chemical coefficients* of the reaction. The chemical reaction scheme indicates that, as the reaction proceeds, whenever a moles of A is consumed, it requires the corresponding consumption of b moles of B, and leads to the corresponding production of c moles of C and d moles of D. Some reactions proceed by a molecules of A and b molecules of B coming together physically to reorganise into c molecules of C and d molecules of D. In many cases, however, reactions proceed by a more complex series of steps.

The *reaction rate r* is defined as the rate at which the reactants are consumed or the rate at which the products are produced,

$$r = -\frac{1}{a}\frac{dc_A}{dt} = -\frac{1}{b}\frac{dc_B}{dt} = \frac{1}{c}\frac{dc_C}{dt} = \frac{1}{d}\frac{dc_D}{dt},$$

where c_i are the concentrations of reactants and products and $i = $ A, B, C or D.

Reaction rates usually vary in a pretty complex way with the concentrations of the reactants. In some cases, however, the reaction rate depends simply on the reactant concentrations according to

$$r \propto c_A^a c_B^b = k c_A^a c_B^b,$$

where the constant of proportionality k is called the *reaction rate constant*. This is linked to the idea that the reaction rate depends on the probability of a molecules of A colliding with b molecules of B, with a probability for each molecule proportional to its concentration.

The *law of mass action* is when the reaction rate is proportional to the concentration of the reacting species. This is often the case in simple reactions such as

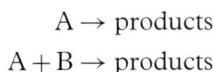

$$A \rightarrow \text{products}$$
$$A + B \rightarrow \text{products}$$

The first is a decomposition reaction with molecules of A decomposing randomly and spontaneously. The second is a reaction by random collision of A and B molecules in a gas mixture or in solution. The corresponding reaction rates are given by the law of mass action as

$$r \propto c_A = k c_A$$
$$r \propto c_A c_B = k c_A c_B.$$

In most cases, however, reactions proceed by a much more complex series of interactions between atomic, molecular, ionic, radical and other species, and the reaction rates depend, correspondingly, in a much more complex way on the reactant, and sometimes product, concentrations. More generally, therefore,

$$r \propto f(c_A, c_B, c_C, c_D) = k f(c_A, c_B, c_C, c_D),$$

where f is a generic function describing the variation of reaction rate with reactant and product concentrations, and k is again the reaction rate constant.

3 The Arrhenius equation

Reaction rates depend on temperature in a much simpler way than their complex dependence on reactant and product concentrations. The reaction rate constant k is

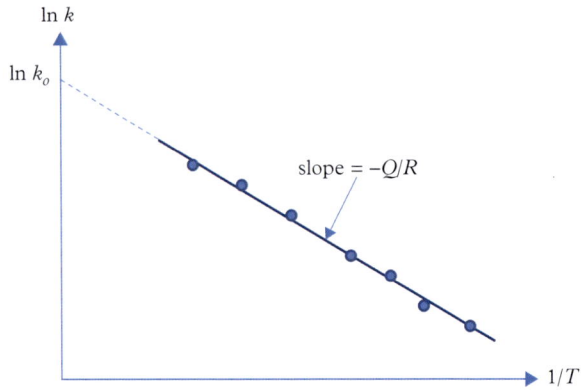

Figure 5.1 *Typical Arrhenius plot for ln chemical reaction rate versus inverse temperature*

almost always found to vary with temperature T according to the *Arrhenius equation*

$$k = k_o \exp\left(-\frac{Q}{RT}\right),$$

where k_o is the *pre-exponential* or *frequency factor*, Q is the *activation energy* and R is the gas constant. Taking logs, the Arrhenius equation can be re-written as

$$\ln k - \ln k_o = -\frac{Q}{RT}.$$

This means that, for any reaction that obeys the Arrhenius equation, plotting the logarithm of the reaction rate constant $\ln k$ against the inverse temperature $1/T$ gives a straight line. This is called an *Arrhenius plot*. The activation energy Q and frequency factor f can then be obtained from the slope and intercept respectively,

$$\text{slope} = -\frac{Q}{R}$$

$$\text{intercept} = \ln k_o,$$

as shown schematically in Figure 5.1.

4   Svante Arrhenius

The 2015 United Nations Climate Change Conference COP 21 was held in Paris from 30 November to 12 December. This led to the Paris Agreement on Climate Change. The following year 174 countries signed the agreement in New York and began their individual ratification processes. The Paris Agreement was the culmination of decades of international diplomacy, aiming to reduce the emission of *greenhouse gases*

Svante Arrhenius

and prevent harmful effects on the world's climate. The centrepiece of the agreement is a goal to limit total *global warming* to less than 2°C compared to pre-industrial levels. Global warming is one of the most important challenges facing humankind in the 21st century. Increasing industrialisation causes ever-greater emission of greenhouse gases, notably carbon dioxide CO_2, which accumulate in the upper atmosphere, increasing the absorption of the sun's rays and therefore raising the earth's temperature. The effects of a temperature rise of only a few degrees are predicted to be quite extreme: melting of the polar ice caps and most of the world's mountain glaciers; rising sea levels, with permanent flooding of many densely populated regions; major changes to crop yields and disruption of food supply; displacement and destruction of many flora and fauna; and increased occurrence of extreme weather events, such as tornados, hurricanes and tsunamis.

The first scientific model linking global warming and climate change to greenhouse gas emissions was developed by Svante Arrhenius. He was a remarkable scientist, only interested in working in novel inter-disciplinary fields. He had three distinct periods in his scientific career, during which he helped start three completely new fields: *physical chemistry*, by developing the theory of *electrolytic dissociation* to explain the behaviour of acid, alkali and salt solutions, for which he won the Nobel Prize in 1903; *earth sciences*, by developing the first theory of greenhouse gas emissions and global warming; and *immunochemistry*, by contributing to the theory of *toxin/anti-toxin* cellular response in the human body.

Svante August Arrhenius was born in February 1859 at Vik near Uppsala in Sweden. At the age of three he taught himself to read and was an arithmetical prodigy. In 1876 he left school and registered at Uppsala University. His professors were Robert Thalén in physics and Per Teodor Cleve in chemistry. After completing his undergraduate degree, he registered for a doctorate, wanting to work at the interface between physics and chemistry. But Thalén was not interested in young students. He had poor laboratory facilities and a terrible reputation for putting off even excellent students so he could concentrate on his own research. Not surprisingly, Arrhenius left Uppsala in 1881 to work instead under the physicist Erik Edlund at the Stockholm Högskola, part of the Swedish Academy of Sciences, and a precursor of the modern Stockholm University. The Högskola was not able to award its own degrees so, although Arrhenius did his research in Stockholm, he stayed registered for his doctorate at Uppsala. Thalén wrote to Edlund: 'Arrhenius . . . has a brilliant mind . . . [but] is terribly lazy, which may be because he is so clumsy'.[1]

In the mid-19th century, chemistry was dominated by the development of organic dyes for the textile mills of the Industrial Revolution. Industrial fortunes were being made, notably in Germany, from synthesising organic compounds as dyes with new and brilliant colours. Universities were establishing departments of organic chemistry to supply trained scientists to this burgeoning industry. But, as late as the 1860s and 1870s, organic chemistry was still based on a relatively rudimentary understanding of valence and chemical bonding to describe the structure of the new organic molecules. John Dalton's atomic theory had been developed in 1803, and Dmitri Mendeleev produced the first periodic table in 1869, but a good theory of chemical bonding was not possible until the discovery of the electron by J. J. Thompson in 1897 and the notion of atomic number in 1913.

Arrhenius was clear from the beginning of his university education that he did not want to be an organic chemist. He wanted to use the methods of physics and apply them to chemical problems. This was a radical agenda. Three scientists working in the late 1870s effectively invented the new discipline of physical chemistry, all working at the geographic fringe, away from the powerhouse of Germanic organic chemistry: Wilhelm Ostwald at Riga Polytechnic in Russia, Jacobus Henricus van't Hoff at the University of Amsterdam in the Netherlands and Arrhenius at the Stockholm Högskola in Sweden.

For his doctoral research, Arrhenius studied the conduction of electricity in *electrolytes*, i.e. solutions of *acids*, *bases* and *salts* dissolved in water. He showed that there were *strong electrolytes*, such as sodium chloride $NaCl$ or sulphuric acid H_2SO_4, with a high conductivity independent of dilution (i.e. the amount of water added to the solution); and *weak electrolytes*, such as acetic acid CH_3COOH or ammonia NH_3, with a low conductivity that increases sharply with dilution. He postulated that electrolyte molecules were either *active* or *inactive*, with only the former able to conduct electricity All molecules are active in a strong electrolyte, so conductivity is high and not affected by dilution, whereas only a few molecules are active in a weak electrolyte, so conductivity is low but increases with dilution. This is, effectively, the accepted modern theory of acids, bases and salts, except that Arrhenius was, at first, not able to identify the nature of the activation process. It was not until 1887, three years after submitting his doctoral thesis, that he completed the theory by describing the activation of a molecule as caused by its dissociation into charged ions, the dissociation reaction obeying Guldberg and Waage's *law of mass action*, and the charged ions producing circular currents, which agreed well with Faraday's laws of *electrolysis*.

In 1884 Arrhenius submitted his thesis. He defended it on May 26, and two days later the science faculty met to determine the outcome. He was awarded the grade *non sine laude approbatur*, which means 'approved not without praise'. This sounds good, but was in fact only just above a pass level. He was criticised for conducting too few poorly controlled experiments, not using sufficiently pure materials, only measuring relative and not absolute dilutions, not measuring solution densities and not controlling the temperature. The underlying problem was that his work fell between the disciplines of physics and chemistry, making it hard to judge by the standards of either. His chemistry examiners thought it was too much physics, and his physics examiners thought it was too much chemistry. The result was a disaster. A low-level pass was insufficient for Arrhenius to be employed as a *docent* (a university teacher), stopping him from pursuing his career as a scientist. And salt was rubbed into his wounds at the graduation ceremony a few days later. After the ceremony, teachers were expected to march up to new graduates to congratulate them, but Thalén and Cleve were too embarrassed to do so and just went home, leaving Arrhenius looking foolish as he stood in line with the other graduates. He never forgave Uppsala and its academics. As much as 40 years later, he was still recounting his humiliation in bitter and graphic detail. He was fond of pointing out that the work for which he received the Nobel Prize had almost been failed by his tutors.

Arrhenius spent the next two years in Uppsala, first caring for his ailing father and then sorting out the estate when he died. On top of not having a job, this hardly made Arrhenius more predisposed to think well of his tutors. He almost decided to give up the idea of a scientific career. Nevertheless, he circulated copies of his thesis to a number of other established scientists, who mostly ignored him, except for Wilhelm Ostwald, who was very enthusiastic and visited him. With Ostwald's support, he applied for and was awarded a special docent position, albeit without a salary or laboratory, and limited only to physical chemistry. He then won a travel grant from the Academy of Sciences that, with his father's legacy, supported him for the next few years, travelling throughout Europe, working with supportive colleagues, notably Ostwald, van't Hoff and Walther Nernst, and gradually cementing his reputation as an up-and-coming scientist. Arrhenius, Ostwald, van't Hoff and Nernst were young men, fond of drinking and carousing. They became known as 'the wild army of the Ionists'.[2]

Finally, Arrhenius returned to Sweden in 1891 to become a physics teacher at the Högskola. In 1895 he was promoted to the position he had always craved, Chair of Physics, and in 1896 he became the Rector. The Högskola was, in many ways, an ideal place for Arrhenius to work. Unlike a university, it was committed to new disciplines, technical subjects and industrial progress; but unlike a German *hochschüle* (high school), there was *lehrfreiheit* (academic freedom) and a liberal approach, notably to gender equality. Arrhenius married his laboratory assistant, Sofia Rudbeck, one of the first generation of Swedish female scientists. Sofia wanted her own career and did not intend to be subordinate, either as an assistant or a companion. She was free-thinking, a theosophist, teetotal and anti-smoking. They were never going to get on well. Two years later they separated, while she was pregnant with Arrhenius' first son, and shortly afterwards they divorced. In 1905 he re-married, to Maria Johansson, with whom he had two daughters and another son.

In 1894 Arrhenius largely abandoned research into physical chemistry to concentrate on what he called 'cosmic physics', which we now call *earth sciences*. He developed his model of global warming, still in use today, using Samuel Pierpoint Langley's infrared observations of the moon to calculate the heat captured by earth's atmosphere, and Stefan's law to calculate how much is radiated away. This showed that 'if the quantity of carbonic acid [i.e. $CO_2 + H_2O$] increases in geometric progression, the augmentation of the temperature will increase nearly in arithmetic progression',[3] i.e. the temperature rise is proportional to the log of the amount of CO_2. He predicted that doubling CO_2 would increase the earth's surface temperature by about 6°C, not far from current estimates of 3 to 5°C. He was, however, much further out with his estimate of the effect of industrial activity. Extrapolating the (then) rates of burning fossil fuels indicated that doubling CO_2 would take about 3,000 years, as Arrhenius said, 'allowing our descendants . . . to live under a warmer sky and in a less harsh environment'.[4] To be fair, levels of industrialisation during the late 19th century were gentle by modern standards, with subsequent population growth and economic development way beyond what might reasonably have been predicted.

In 1900 Arrhenius entered another field, *immunology*, becoming engaged in an extended argument with Paul Ehrlich, a professor of medicine who was developing the theory of *immunisation*. It was already known that, in the presence of a limited amount of a poison (or *toxin*), the body can often produce an *anti-toxin*, which not only counteracts the effect of the poison, but also provides continuing long-term protection. This is called *immunisation*, and it was already being used as the basis for vaccinations against rabies, diphtheria and tetanus. In 1897 Ehrlich proposed his *side-chain* theory of immunisation. Cells in the body have *receptors* on their surface that bind with side chains on the toxin molecules. The body's response is to produce more receptors, some of which are released into the bloodstream as *anti-toxins* (or *antibodies*), thus providing long-term resistance. This is still accepted, in essence, as the basic way in which immunisation works. However, the details are complex and, to explain his observations, Ehrlich went on to propose a series of other chemical molecules: toxoids, toxons, prototoxoids, syntoxoids, epitoxoids, deuterotoxoids, tritoxoids and so on.

Ehrlich met Arrhenius in 1902 and suggested that Arrhenius use the new methods of physical chemistry to try to understand immunological chemical reactions. Arrhenius took up the challenge, rejecting Ehrlich's arbitrary 'zoo of poisons' (as Arrhenius called it),[5] stressing what he saw as the underlying simplicity of toxin/anti-toxin binding, and describing it as a reversible chemical reaction, subject to basic laws of mass action and ionic dissociation. He pooh-poohed Ehrlich's tendency to invent a new compound whenever he needed to explain a new result. He invented the term *immunochemistry* for his book on the subject, which he published in 1907.[6] Ehrlich continued to believe fiercely in the complexity and irreversibility of immunisation processes, and the two scientists became locked for many years in a bitter and often personal debate. In a sense, both were right: Arrhenius was right that toxin/anti-toxin binding is reversible and obeys fundamental chemical laws; Ehrlich was right that the reactions are complex, although his menagerie of chemical agents has now largely been discarded. In the end, Arrhenius' approach was too simplistic. Ehrlich was, practically speaking, more successful, going on to identify important toxins and anti-toxins, developing methods of storing and assessing vaccines, and synthesising the first effective treatment for syphilis.

Ostwald described Arrhenius as 'corpulent with a red face, short moustache and short hair ... more like a student of agriculture than a theoretical chemist with original ideas'.[7] In fact he was a truly brilliant scientist in at least two respects: his ability to detect exciting new inter-disciplinary fields where scientific progress could be made, and his intuitive gift for discerning the underlying scientific theory from limited experimental data. By the late 19th century, he had helped initiate three major new fields, achieved a strong scientific reputation and was working happily in the relative backwater of the Stockholm Högskola. But he was about to be hit by one of the most important socio-political events affecting science and scientists in the past two centuries.

Alfred Nobel was born in Stockholm in 1833. He became interested in explosives through his father, who ran a business making armaments in St. Petersburg. He went on to invent dynamite, gelignite, ballistite (a rocket propellant), detonators, blasting caps and many other explosive devices, taking out a total of 355 patents. When one of his brothers died while visiting Cannes, a local newspaper wrote an obituary of Alfred by mistake, saying: *'Le marchand de la mort est mort'* ['The merchant of death is dead'] and continuing, 'Alfred Nobel, who became rich by finding ways to kill more people faster than ever before, died yesterday'.[8] Nobel was upset and, having no wife or children, resolved to leave his wealth to charity. By the time of his death in 1896, he owned 90 armaments factories and was worth about 30 million Swedish kroner (about half a billion US dollars in today's terms). His will surprised and upset his family (not surprisingly). It put almost all his wealth into a trust to fund the Nobel Prizes in Physics, Chemistry, Medicine, Literature and Peace.

Importantly for Arrhenius, the Nobel Prizes were to be administered by the Swedish Academy of Sciences. The sleepy and relatively undistinguished backwater of the Stockholm Högskola was suddenly at the centre of the international scientific stage. Arrhenius had just become Rector and was determined to make the most of this opportunity. Nobel's will was poorly drafted, and it took four years to complete the litigation with his unhappy family and set up the Nobel Foundation itself. Arrhenius was heavily involved, and badly conflicted: he was friendly with the estate's lawyer; he was consulted on the nomination and assessment procedures; he stood to gain professionally as Rector; he bid for funds from the estate to set up a Nobel Institute at the Högskola; and he was likely to be a candidate for a prize. In the end, he became a key member of the Nobel Committees for both Physics and Chemistry, and the first Director of the Nobel Institute, where he remained until his death in 1927. He did his best to use these positions to support scientists whom he regarded as his friends, and to frustrate those whom he perceived as his enemies.

Arrhenius was certainly not averse to making scientific enemies. Perhaps because of his early treatment, he viewed scientific research in military terms, advising students, for instance, that 'you have to win the battle . . . do you have allies or are you alone?'[9] He worked hard to ensure the chemistry prize for Ostwald, who had given him critical support in the aftermath of his disastrous doctoral thesis, and he tried to prevent awards to Nernst, Mendeleev and Ehrlich, with whom he had fallen out, effectively delaying their recognition for several years. The first Nobel Prize in Chemistry went to van't Hoff in 1901, the second to Edmond Fischer in 1902, and the third, finally, to Arrhenius in 1903. He described it as *die Krönung* (the coronation). In fact, he had wanted the physics prize as revenge over Thalén and the Uppsala physicists, but it became clear that he would not get physics and would have to settle for chemistry. In his Nobel biography and speech, he again took the opportunity to rail against the poor grading of his doctoral thesis. According to his biographer Patrick Coffey, 'At the Nobel banquet, Arrhenius brushed aside an apology by Cleve and made it clear he owed nothing to his Uppsala professors'.[10]

The decade before the First World War saw Arrhenius enjoying sustained and established success and influence, both as a scientist and through the Nobel Committees and Institute. Living in neutral Sweden meant that, during the war, he could help scientists on both sides, which he did. In 1924 he suffered a stroke, from which he never fully recovered. He died in 1927 after an attack of acute intestinal catarrh and was buried in the town cemetery in Uppsala. Elisabeth Crawford, in her biography of Arrhenius, analyses his citations. Three papers account for almost three quarters of all his citations. The largest number, almost a third, refer to his paper in *Zeitschrift für Physicalische Chemie* in 1889,[11] describing the Arrhenius equation for the exponential temperature variation of the rate constants of chemical reactions. Interestingly, the equation was given first by van't Hoff, but Arrhenius claimed that van't Hoff's derivation was not valid. The other two most cited publications are the 1891 paper in *Philosophical Magazine*[12] containing his model for the greenhouse effect, and the 1887 *Zeitschrift für Physicalische Chemie* paper[13] on his Nobel Prize–winning work on ionic dissociation.

5   Reaction equilibria

Consider a general chemical reaction:

$$a\mathrm{A} + b\mathrm{B} \rightleftharpoons c\mathrm{C} + d\mathrm{D}$$

Depending on temperature, pressure, and reactant and product concentrations, a reaction can usually proceed in either the forward or backward direction, and is indicated by a two-way arrow.

The *Gibbs free energy change of the reaction* ΔG is the change in Gibbs free energy when the reactants are replaced by the products:

$$\Delta G = G_{\mathrm{products}} - G_{\mathrm{reactants}} = (cG_{\mathrm{C}} + dG_{\mathrm{D}}) - (aG_{\mathrm{A}} + bG_{\mathrm{B}}),$$

where G_i are the Gibbs free energies per mole (or *molar free energies*) of the reactants and products. The *standard Gibbs free energy change of the reaction* ΔG^o is the Gibbs free energy of the reaction when all the reactants and products are at *standard state*, i.e. in a pure state at atmospheric pressure and room temperature (usually defined, more accurately, as composition $c_i = 1$, pressure $P = 10^5$ Pa and temperature $T = 25\,^{\circ}C$):

$$\Delta G^o = G^o_{\text{products}} - G^o_{\text{reactants}} = \left(cG^o_C + dG^o_D\right) - \left(aG^o_A + bG^o_B\right),$$

where G^o_i are *standard molar free energies* of the reactants and products.

The Gibbs free energy of the *i*th component G_i depends on concentration c_i and temperature T according to

$$G_i = G^o_i + RT \ln c_i,$$

so the Gibbs free energy change of the reaction can be rewritten as

$$\Delta G = \Delta G^o + RT \ln \frac{c^c_C c^d_D}{c^a_A c^b_B}.$$

Under some conditions, a chemical reaction can be at *equilibrium*. At equilibrium there is no overall Gibbs free energy change, $G_{\text{products}} = G_{\text{reactants}}$ and $\Delta G = 0$, so

$$\Delta G^o = -RT \ln \frac{c^c_{Ceq} c^d_{Deq}}{c^a_{Aeq} c^b_{Beq}} = -RT \ln K,$$

or, taking anti-logs,

$$K = \frac{c^c_{Ceq} c^d_{Deq}}{c^a_{Aeq} c^b_{Beq}} = \exp\left(-\frac{\Delta G^o}{RT}\right),$$

where K is called the *equilibrium constant*. Notice that the concentrations in these last two equations are specific values that give equilibrium between the reactants and products, as indicated by labelling them c_{ieq}. This means that the reaction rate constant K is related to the standard Gibbs free energy change of the reaction ΔG^o, so the balance of the equilibrium concentrations of reactants and products c_{ieq} can be calculated from the difference in their standard free energies G^o_i.

When a chemical reaction is at equilibrium, it doesn't mean that nothing is happening. The forward and backward reactions are in balance, equal and opposite, so there is no overall change with time. The reaction is in a *dynamic equilibrium*. At different places and at different times, reactant molecules A and B are reacting to form product molecules C and D, and at other places and other times, product molecules C and D are reacting to form reactant molecules A and B. Averaged over time and place, the two effects are equal and opposite so there is no overall change. In other words, the forward and backward reactions rates are equal:

$$r_f = r_b$$

$$\therefore k_f c_{Aeq}^a c_{Beq}^b = k_b c_{Ceq}^c c_{Deq}^d$$

$$\therefore K = \frac{k_f}{k_b} = \frac{c_{Ceq}^c c_{Deq}^d}{c_{Aeq}^a c_{Beq}^b} = \exp\left(-\frac{\Delta G^o}{RT}\right) = \frac{k_{fo} \exp\left(-\frac{Q_f}{RT}\right)}{k_{bo} \exp\left(-\frac{Q_b}{RT}\right)}.$$

The pre-exponential factors are approximately equal, $k_{fo} = k_{bo}$, so taking logs,

$$\Delta G^o = Q_f - Q_b.$$

This is an important result, which relates the thermodynamics of the reaction (i.e. its standard Gibbs free energy change ΔG^o) to the kinetics of the reaction (i.e. the forward and backward reactions rates k_f and k_b) and their activation energies Q_f and Q_b.

6   Transition states

Figure 5.2 shows how the Gibbs free energy changes for a general reaction

$$aA + bB \rightleftharpoons cC + dD$$

at standard state. The energies of the reactants and products are $G^o_{reactants}$ and $G^o_{products}$, with the standard Gibbs free energy change of the reaction $\Delta G^o = G^o_{products} - G^o_{reactants}$. The activated state $A_a B_b\star$ is a complex of a molecules of A and b molecules of B midway between the start and endpoints of the reaction. The asterisk indicates that the activated state has excess Gibbs free energy $G^o_{activated\ state}$. The reaction can be written

$$aA + bB \rightleftharpoons A_a B_b^* \rightleftharpoons cC + dD$$

The activated state is, equivalently, $C_c D_d\star$, a complex of c molecules of C and d molecules of D.

The forward reaction proceeds when the reactant molecules come together and acquire the forward activation energy Q_f. In doing so, they reach the activated state, which decomposes rapidly with decreasing energy to form the products. The backward reaction proceeds similarly when the product molecules come together and acquire the backward activation energy Q_b. In doing so, they also reach the activated state, which decomposes rapidly with decreasing energy to form the reactants.

energy

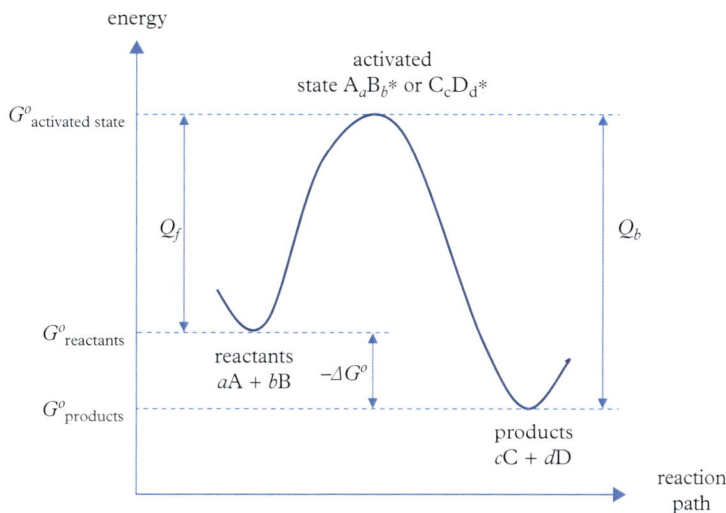

Figure 5.2 *The energy change for a general reaction* $aA + bB \rightleftharpoons cC + dD$ *at standard state*

Figure 5.2 shows clearly the relationship between the thermodynamics and kinetics of the reaction, as represented by the relationship between the standard Gibbs free energy change of the reaction ΔG^o and the two activation energies Q_f and Q_b:

$$\Delta G^o = Q_f - Q_b.$$

7 Molecularity and order

The *molecularity* of a chemical reaction is the number of molecules that are involved. Generic examples of *unimolecular*, *bimolecular* and *trimolecular* reactions are as follows:

$$\text{unimolecular}:\ A \rightarrow \text{product}$$
$$\text{bimolecular}:\ 2A \rightarrow \text{products}$$
$$\text{bimolecular}:\ A + B \rightarrow \text{products}$$
$$\text{trimolecular}:\ 2A + B \rightarrow \text{products}$$
$$\text{trimolecular}:\ A + B + C \rightarrow \text{products}$$

The *order* of a chemical reaction is the sum of the exponents of the concentration terms in the rate equation. Generic examples of *first-order*, *second-order* and *third-order* reactions have reaction rates as follows:

$$\text{first order}: r \propto c_A = kc_A$$
$$\text{second order}: r \propto c_A^2 = kc_A^2$$
$$\text{second order}: r \propto c_A c_B = kc_A c_B$$
$$\text{third order}: r \propto c_A^2 c_B = kc_A^2 c_B$$
$$\text{third order}: r \propto c_A c_B c_C = kc_A c_B c_C.$$

Unimolecular reactions are sometimes first order, but not always; bimolecular reactions are sometimes second order, but not always; and so on. Similarly, first-order reactions are sometimes unimolecular, but not always; second-order reactions are sometimes bimolecular, but not always; and so on.

The evolution of a chemical reaction can be obtained by integrating its rate equation. For instance, consider a unimolecular, first-order decomposition reaction:

$$A \rightarrow \text{products}$$
$$r = -\frac{dc_A}{dt} = kc_A.$$

Integrating gives the concentration of A at any time t during the course of the reaction:

$$\int_{c_A^o}^{c_A} \frac{dc_A}{c_A} = -\int_0^t k \, dt$$
$$\therefore \ln\left(\frac{c_A}{c_A^o}\right) = -kt$$
$$\therefore c_A = c_A^o \exp(-kt),$$

where c_A^o is the initial concentration at time $t = 0$. In other words, the concentration of A decreases exponentially with time. The *half life* of a decomposition reaction $\tau = (\ln 2)/k$ is the time it takes for the concentration to drop to half the initial concentration $c_A = \frac{1}{2}c_A^o$.

As another example, consider a unimolecular second-order reaction:

$$2A \rightarrow \text{products}$$
$$r = -\frac{dc_A}{dt} = kc_A^2.$$

Integrating again gives the concentration of A at any time t during the course of the reaction:

$$\int_{c_A^o}^{c_A} \frac{dc_A}{c_A^2} = -\int_0^t k \, dt$$
$$\therefore \left(\frac{2}{c_A^o} - \frac{2}{c_A}\right) = -kt$$
$$\therefore c_A - c_A^o = \frac{1}{2}c_A c_A^o kt.$$

The evolution of other reactions can, similarly, be obtained by integrating the relevant rate equation, although it is often necessary to use numerical rather than analytical integration methods. This procedure is also used in reverse: the evolution of the concentration of the reactants and products with time are measured to deduce the rate equation and, therefore, the reaction mechanism. Except in simple cases, this is not straightforward.

8 Collision theory

Consider a billiard ball A of radius r_A fired at a velocity v towards a second stationary billiard ball B of radius r_B, as shown in Figure 5.3. The balls collide if the trajectory of A passes within a distance $r_A + r_B$ of B. Consider a circle drawn with its centre coincident with the centre of B, normal to the trajectory of A, and with a radius $r_c = r_A + r_B$. This circle defines the area within which collision can take place. It is called the *collision cross section* σ, and is given by

$$\sigma = \pi r_c^2 = \pi (r_A + r_B)^2.$$

Its radius r_c is the distance of nearest approach for collision not to take place.

Now consider firing A at a number n of stationary B balls in a region of area a. The probability of a collision p is given by the number of balls n times the collision cross section σ divided by the total area a:

$$p = \frac{n\sigma}{a}.$$

This probability can be determined by firing the A ball many times and measuring the fraction of collisions. The collision cross section is obtained by inverting the previous equation:

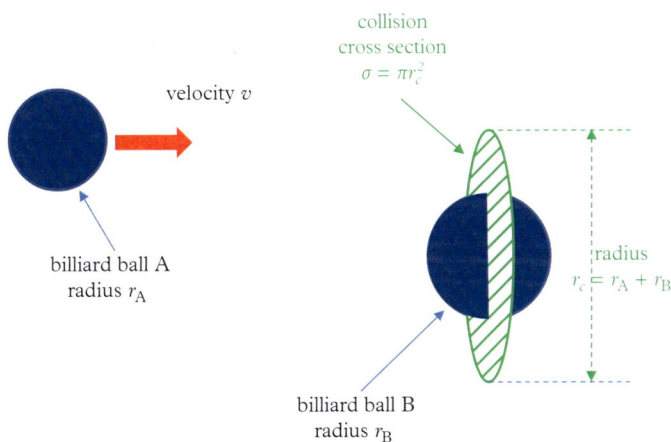

Figure 5.3 *Collision cross section*

$$\sigma = \frac{ap}{n}.$$

Now consider a simple chemical reaction between two molecules A and B:

$$A + B \rightarrow \text{products}$$

If the A and B molecules are like small billiard balls, a chemical reaction can only take place when they collide, i.e. when they approach each other within the collision cross section σ. In general, however, molecules are not like billiard balls, and they interact with attractive and repulsive forces. By analogy with the collision cross section, we can define a *reaction cross section* σ_r as the area within which a reaction can take place:

$$\sigma_r = \pi r_r^2 > \pi (r_A + r_B)^2,$$

where r_r is the distance of nearest approach for the reaction not to take place. The reaction cross section is larger than the collision cross section because of the attractive forces between reacting molecules.

The reaction rate r and reaction rate constant k can, therefore, be written

$$r = \sigma_r v c_A c_B$$
$$k = \sigma_r v,$$

where v is the average relative velocity of the A and B atoms or molecules.

9 Complex reactions

A few examples of complex reactions are described briefly in this section.

Consecutive reactions are when the product of one reaction becomes the reactant of a second one:

$$A \xrightarrow{k_1} B \xrightarrow{k_2} C$$

with reaction rate constants k_1 and k_2 for the first and second reactions respectively.

Parallel reactions are when a reaction or decomposition takes place by two different mechanisms, i.e. following two different pathways, in parallel:

$$A \xrightarrow{k_1} B$$
$$A \xrightarrow{k_2} C$$

with reaction rate constants k_1 and k_2 in this case for the two parallel reactions.

Chain reactions are when the reaction regenerates the reactant, which then leads to further reaction:

$$A \to B + A$$

so, once the reaction begins, it is self-sustaining.

Catalytic reactions are when a catalyst is added to speed up the reaction, without being affected itself:

$$A + catalyst \to products + catalyst$$

Catalysts often work by acting as a template to which the reactants can bind. The activation energy is reduced, and the reaction rate is sharply increased.

$$Q_{catalysed} \ll Q_{uncatalysed}.$$

10 REFERENCES

1. Elisabeth Crawford. *Arrhenius: From Ionic Theory to the Greenhouse Effect* (Science History Publications, Canton, 1997), 14.
2. Patrick Coffey. *Cathedrals of Science: The Personalities and Rivalries That Made Modern Chemistry* (Oxford University Press, Oxford, 2008), 23.
3. Crawford, n 1, 152.
4. See ibid. 154.
5. Ibid. 201, 34.
6. Svante Arrhenius. *Immunochemistry: The Application of the Principles of Physical Chemistry to the Study of the Biological Antibodies* (Macmillan, New York, 1907).
7. Coffey, n 2, 17.
8. The Editors of Encyclopaedia Britannica. 'Alfred Nobel'. 2012. https://www.britannica.com/biography/Alfred-Nobel (accessed 2 January 2020).
9. Crawford, n 1, 17.
10. Coffey, n 2, 33.
11. Svante Arrhenius. 'Über die reaktiongeschwindigkeit bei der inversion von rohrzucker durch säuren' ['On the reaction rate in the inversion of cane sugar by acids']. *Zeitschrift für Physicalische Chemie* **4** (1889), 226–248.
12. Svante Arrhenius. 'On the influence of carbonic acid in the air upon the temperature on the ground'. *Philosophical Magazine* **31** (1891), 415–418.
13. Svante Arrhenius. 'Über die dissociation der in wasser gelösten stoffe' ['On the dissociation of dissolved substances in water']. *Zeitschrift für Physicalische Chemie* **1** (1887), 631–648.

....

11 BIBLIOGRAPHY

Arrhenius: From Ionic Theory to the Greenhouse Effect. Elisabeth Crawford (Science History Publications, Canton, 1997).

Cathedrals of Science: The Personalities and Rivalries That Made Modern Chemistry. Patrick Coffey (Oxford University Press, Oxford, 2008).

Chemical Kinetics. Keith J. Laidler (Pearson, London, 1997).

Physical Chemistry from Ostwald to Pauling: The Making of a Science in America. John W. Servos (Princeton University Press, Princeton, 1990).

Reaction Kinetics. Michael J. Pilling and Paul W. Seakins (Oxford University Press, Oxford, 1995).

6

The Gibbs-Thomson Equation

Surfaces

1 Surface tension and surface energy

Every child enjoys dipping a wire circle into soapy water to create a soapy film, and then blowing into the film to make soap bubbles that then float away in the wind. How do the soap bubbles survive? Figure 6.1 shows a section through a soap bubble of radius r. There is a pressure inside each bubble created by blowing into it that is balanced by a force in the surface of the surrounding soapy film. The *internal pressure p* is a force per unit area, which acts over the cross sectional area of the bubble πr^2. The *surface tension γ* is a force per unit length, which acts everywhere along the perimeter of the film $2\pi r$. Balancing forces gives

$$\pi r^2 p = 2\pi r\gamma$$
$$\therefore p = \tfrac{2\gamma}{r}.$$

The internal pressure inside any bubble is inversely proportional to its radius, and the constant of proportionality is twice the surface tension in the bubble's surface film. This equation is called *Young's equation* or sometimes the *Young-Laplace equation*.

The internal pressure is obviously caused by blowing air into the bubble. But what causes the balancing surface tension? Atoms or molecules at a surface or an interface between any two phases, such as the surface between the soapy water film and the surrounding air, have a higher energy than atoms or molecules in the bulk of the two phases on either side of the surface. This is because they are neither one thing nor the other, not fully embedded in either phase, and not, therefore, wholly part of either of them. The interatomic or intermolecular structure and bonding on one side of the surface is characteristic of one phase; on the other side it is characteristic of the other phase; and along the surface it is intermediate between the two.

The number of atoms or molecules at the surface with higher energy is proportional to the surface area, so the total excess energy of the surface is also, therefore, proportional

The Equations of Materials. Brian Cantor. Oxford University Press (2020). © Brian Cantor.
DOI: 10.1093/oso/9780198851875.001.0001

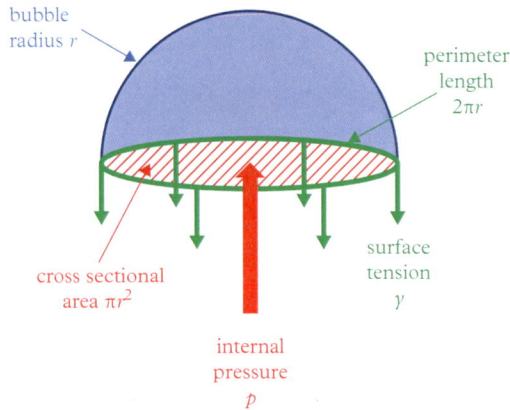

Figure 6.1 *Internal pressure and surface tension in a bubble*

to the surface area. We say that the surface has an excess *surface energy* γ per unit area, which can often manifest itself as a force or *surface tension* γ per unit length, as shown, for instance, for the soap bubble in Figure 6.1. Surface energy and surface tension are essentially the same thing, both arising from the higher energy of the surface atoms or molecules. Physical effects caused by surfaces and their associated surface energy and surface tension are referred to as *capillarity* effects.

2   The Gibbs-Thomson equation

All surfaces have a *surface energy*, i.e. an excess energy per unit area of surface, because of the intermediate nature of the local structure and bonding of atoms or molecules at the surface. The Gibbs free energy G of any material containing two phases and a surface between them is, therefore, given at a pressure P and a temperature T by

$$dG = VdP - SdT + \gamma dA,$$

where V is the volume, S is the entropy and A is the area of the surface between the two phases. Fairly obviously, a two-phase material will try to minimise the surface area between the phases because atoms or molecules at the surface have higher energy than atoms or molecules in either of the two phases. This can be seen more formally by minimising the Gibbs free energy at any given temperature and pressure:

$$\left(dG_{P,T}\right)_{\min} = \gamma \, dA_{\min}.$$

There are, therefore, two possibilities for equilibrium in a two-phase material consisting of phases α and β, as shown in Figure 6.2. One possibility is that the two phases have the same free energy, $G_\alpha = G_\beta$, and they are separated by a flat surface that minimises

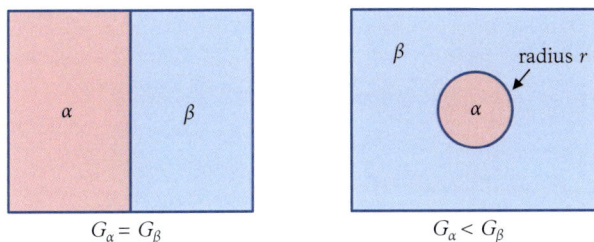

Figure 6.2 *Equilibrium in a two-phase material with flat and spherical surfaces*

the surface area between them. The other possibility is that one phase has a lower free energy, $G_\alpha < G_\beta$, and is, therefore, in the form of a particle contained within a matrix of the other phase. Just as in the soap bubble, the lower free energy in the particle creates an outward pressure that is balanced by the surface energy of the interface between the two phases:

$$\Delta G_V = \frac{2\gamma}{r},$$

where $\Delta G_V = \Delta G/v_m = (G_\beta - G_\alpha)/v_m$ is the free-energy difference between the two phases per unit volume, v_m is the molar volume, and r is the radius of the α particle, which is spherical to minimise the surface area. This is called the *Gibbs-Thomson equation*.

3   J. J. Thomson and William Thomson (Lord Kelvin)

One of the peculiar things about the Gibbs-Thomson equation is the confusion and uncertainty about how it acquired its name. It is clear that the name Gibbs refers to Josiah Willard Gibbs, the American scientist who laid the mathematical foundations for thermodynamics and statistical mechanics. But, there has been some debate about whether the name Thomson refers to William Thomson (later, Lord Kelvin) after whom the absolute temperature scale is named, or J. J. Thomson, who discovered the electron. In any case, and more curiously, the equation wasn't derived directly by any of these three scientists, not Gibbs nor either of the two Thomsons, though all three played an important part in our understanding of the thermodynamics of surfaces. And there are other quite closely related equations, also describing the behaviour of curved surfaces: the Young-Laplace equation, the Kelvin equation, the Ostwald-Freundlich equation, and so on. The name *Gibbs-Thomson equation* began to be used in 1910, referring to the equation for the adsorption of solutes at surfaces. Its more general modern form was derived independently in the 1920s by Friedrich Meissner, a student at the University of Göttingen of the Estonian physical chemist Gustav Tammann, and the Austrian physicist Ernst Rie in his dissertation at the University of Vienna.

Gibbs is discussed in Chapter 3; the biographies of both Thomsons are presented in this chapter.

J. J. Thomson

J. J. Thomson

On the afternoon of 4 September 1882, the American inventor and businessman Thomas Edison was standing in the office of one of his investors, J. Pierpoint Morgan, of the bankers Drexel, Morgan & Co in downtown New York. At about three o'clock in the afternoon, Edison's chief electrical engineer, John W. Lieb, threw a switch to send electrical power from the newly constructed Pearl Street Power Station to a distribution network supplying about one square mile of offices and businesses in the First District, adjacent to Wall Street and the East River in Lower Manhattan. Simultaneously, Edison screwed in his newly invented *incandescent light bulbs*, which burst into light in the banker's office. Similar lights were turned on in the same way in many other buildings throughout the district. It was the beginning of the world's first commercial *electric power station*, delivering electricity to 400 lights at the offices of 90 customers, including Drexel, Morgan & Co., the New York Times, the New York Herald and the famous Sweet's Restaurant, patronised many years earlier by Thomas Jefferson and Aaron Burr. The whole district lit up. Never since it was first used was the biblical phrase 'let there be light'[1] so appropriate. According to an article that appeared in the *New York Times* the following day:

> Yesterday for the first time the Times building was illuminated by electricity . ... At 7 o'clock, the 27 lamps in the editorial rooms and the 25 lamps in the counting rooms

made those departments as bright as day. ... The light was soft, mellow and grateful to the eye, and it seemed almost like writing by daylight. ... The electric lamps were thoroughly tested last evening by men who have battered their eyes by years of nightwork ... and the decision was decisively in favour of the Edison electric lamp.[2–4]

Edison held more than 1,000 patents and was already famous for contributing to inventing and commercialising the phonograph, motion picture camera, car battery and telegraph, ultimately creating major industries of sound recording, film and telecommunications. By 1870, arc lights were in use for street lighting, but these operated at high voltage, far too bright, unstable and dangerous to be used safely indoors. To solve this problem, Edison invented the low-voltage carbon-filament light bulb, which he first switched on and shone continuously for more than 13 hours at 110 volts at his research lab in Menlo Park, New Jersey, in October 1879. Edison had a genius for promotion, and in December he arranged a public demonstration of his new light bulbs, which was so popular that the Pennsylvania Railroad Company had to put on special trains for the throngs of visitors who attended. Later, in 1882, he displayed publicly the first Christmas tree with electric lights in the home of his associate Edward Johnson; and in 1884 he put on the first electric light street parade down Fifth Avenue. After trying a small prototype electricity distribution system at Holborn Viaduct in London, he purchased 255 and 257 Pearl Street in New York, won a franchise from the New York City Board of Aldermen to install 80,000 feet (24,000 m) of underground cables throughout the First District, developed the 27-ton (24,500-kg), 100-kW Jumbo constant-voltage dynamo (named after a circus elephant owned by the showman P. T. Barnum) and strengthened the Pearl Street floors for the installation of six of the Jumbo dynamos driven by Armington & Sims reciprocating steam engines powered by four coal boilers. It was a heroic engineering project and it was wildly successful. Everyone wanted light. By November, 90 customers had grown to 946 in 107 buildings in Manhattan, and by the end of the decade, similar Edison systems were licensed for installation in other parts of New York and in cities throughout North America, Europe, South America and Japan. Edison is reputed to have said, 'We will make electricity so cheap, only the rich will use candles'.[5] Pearl Street continued to supply electricity to the First District until 1890, with only one break of three hours. In January 1890, the Pearl Street Power Station was damaged by a major fire. The electricity supply was restored in 11 days and the Power Station continued operating until 1895, when it was finally dismantled, Edison sold the buildings and they were later demolished. The site is now a parking lot. Only one of the Jumbo dynamos survives, and it is on display in the Henry Ford museum in Dearborn, Michigan.

Electricity was, of course, known to the ancient Egyptians and Greeks, who used electric catfish and electric rays to shock people as a treatment (probably not very successfully) for headache, arthritis and gout. The first known study of electricity was conducted by the Greek philosopher Thales of Miletus, who demonstrated the magnetic-like behaviour of static electricity created by rubbing the gemstone amber, which then attracts small particles of dust and leaves. In fact, the words *electron* and *electricity* come from the Greek word for amber. Benjamin Franklin, one of the founding fathers of the United States, made extensive studies of electricity in the mid-18th century. He famously (and somewhat dangerously) showed that lightning in the sky is an electric discharge, by flying a wet kite with a trailing wire in the middle of a thunderstorm, to short out the electricity to the ground. Franklin had the good sense to stand on a dry insulator to avoid getting an electric shock. The Swedish physicist Georg Richmann was not so careful and not so fortunate, killing himself by electrocution when he repeated the experiment a few months later. In 1791, the Italian physicist Luigi Galvani showed that the brain communicates with muscles in the body using electric signals carried by *nerve cells* or *neurons*; another Italian physicist, Alessandro Volta, built the first *battery* or *voltaic pile* in 1800 from alternating copper and zinc plates separated by sheets of paper soaked in brine; the English chemist Sir Humphry Davy invented the *electric arc light* in 1808, with an electric discharge between two carbon electrodes in air; and another English scientist, Michael Faraday, invented the *electric motor* in 1821 and the *electric generator* or *dynamo* in 1832, both based on Faraday's laws of *induction*, the first using the interaction between an electric current and a magnet to create a rotating motion, and the second, reversing the effect, using the interaction between a rotating motion and a magnet to create an electric current. The beautiful underlying theory of electromagnetism was published by the Scottish physicist James Clerk Maxwell in 1865.[6]

By 2017, annual world electricity production had grown to approximately 25,000 TWh (terawatt hours) or 10^{20} J, produced at typical voltages of 10 to 100 kV. Electrical energy is voltage times charge ($E = VQ$), so total world production of charge in 2017 was about 10^{16} coulombs. Because each electron has a charge of 1.6×10^{-19} coulombs, this corresponds to a total world production of electrons in 2017 of about 10^{35}. This is an enormous number of electrons, which we use to power our home comforts such as fridges, ovens, TVs and computers, and our industrial machines such as lathes, blast furnaces and robots. For comparison, there are estimated to be about 10^{28} atoms in a person, i.e. about 10^{35} atoms in the population of Britain, and 10^{50} atoms in the world. As the American TV star and comedian George Gobel said, 'If it weren't for electricity, we'd all be watching television by candlelight'.[7] The humble electron, of which we use so many, was, of course, discovered by the British scientist J. J. Thomson (his full name was Joseph John Thomson, but he was always known as J. J.).

J. J. Thomson was born on 18 December 1856 in Cheetham Hill, a suburb in North Manchester. His father, Joseph James Thomson, was an antiquarian bookseller and publisher who ran a bookshop founded by J. J.'s great-grandfather. His mother, Emma Swindells, came from a local textile family, a branch of the Vernon family, who owned a well-known cotton-spinning mill and company in Manchester. He had one brother, Frederick Vernon Thomson, who was two years younger than him. The family was middle class, close, happy and loving, devout Anglicans, not poor but not very wealthy. J. J.'s early education was in private schools. He was shy but determined. He wanted to become an apprentice engineer at a local firm of engine-makers, but was too young, having finished school precociously, aged just 14. Instead, therefore, he went to Owens College in the centre of Manchester, commenting later that this was the lucky chance that shaped his life. At about this time, a cousin asked him what he wanted to do when he grew up, and he replied 'original research',[8] to which the cousin responded, 'Don't be such a little prig, Joe'.[9] Owens College was the precursor of the Victoria

University of Manchester, one of Britain's first civic universities, and had been founded in 1846 as a non-religious educational establishment with a legacy of £96,000 from the local businessman John Owens, described starkly by the historian William Whyte as 'a parsimonious, work-obsessed, easily offended bachelor, who gave little to charity, and was disagreeable with deep pockets but few friends'.[10] J. J. studied engineering physics, maths and chemistry. There was a great scientific tradition in liberal, free-thinking Manchester, following on from John Dalton, who discovered atomic theory, and James Joule, who discovered conservation of energy and gave his name to the standard unit of energy. J. J.'s father introduced him once to Joule, and commented afterwards, 'Some day you will be proud to say you have met that gentleman'.[11] Sadly, his father died two years after J. J. started at Owens College, which left the family in straightened circumstances. The family made sacrifices to support his studies, selling the bookshop and moving to a smaller terraced house. He couldn't now afford to be apprenticed to a local engineering firm, so he stayed at Owens and completed his engineering and physics courses with distinction, winning scholarships and prizes, which helped financially, and writing a paper entitled 'Experiments on Contact Electricity between Non-conductors'.[12]

In 1876, aged 19, Thomson went to study maths at Trinity College, Cambridge, and in 1880 graduated as *second wrangler* (second-best student). He was to stay living in Cambridge for the rest of his life, the best part of 65 years, until his death in 1940 at the grand old age of 83. According to his grandson David, he never lost his Mancunian accent, his northern sense of humour or his liking for Lancashire hotpot and Manchester jam pudding. For the first 13 years at Cambridge, he was unmarried, as an undergraduate, then College Fellow, Lecturer and Professor, mostly living in rooms in Trinity College. His early work was mostly theoretical, initially using analytical dynamics to investigate the interconversion between different forms of energy, effectively discovering *Le Châtelier's principle* independently from and almost simultaneously with the great French chemist Henri Louis Le Châtelier. Le Châtelier's principle states that a system always reacts to oppose any change, a kind of scientific version of *sod's law* or *Murphy's law*. For instance, the stiffness of a spring usually increases with decreasing temperature, so stretching it causes a drop in temperature, making it stiffer and thus more difficult to stretch any more. Similarly, magnetising a bar of iron increases its length, so stretching it reduces its magnetisation, again opposing the stretching effect.

In 1881 J. J. completed his thesis and was awarded a College Fellowship; in 1882 he won the Adams Prize for his theory of *vortex atoms*; the same year he was appointed as a College Lecturer; in 1883 he published his first experimental paper on *electrical discharge* in gases,[13] work that would ultimately lead to the discovery of the electron; and the same year he was appointed as a University Lecturer. Rapid promotion meant that, for the first time, he had sufficient income to be financially secure. And more was to come. Remarkably, in 1884, the following year, at the callow age of 27 and just eight years after arriving in Cambridge, J. J. was elected as a Fellow of the Royal Society and, a few months later, as the third Cavendish Professor of Experimental Physics and Head of the Cavendish Laboratory, succeeding the first holder, the great James Clerk Maxwell, who founded the Cavendish more than a decade earlier, and the second holder, the equally great Lord Rayleigh, who had been in post for the previous five years. J. J.

was selected ahead of better known and more experienced candidates: the German-born British physicist Arthur Schuster, who invented the concept of antimatter and was one of his teachers in Manchester; the Irish fluid dynamicist Osborne Reynolds, who gave his name to the Reynolds number describing the transition between laminar and turbulent flow, and who was also one of J. J.'s teachers in Manchester; Richard Glazebrook, an experienced lecturer at the Cavendish, who had been Rayleigh's student, was Rayleigh's choice, and was widely thought to be the favourite; and the Irish physicist Joseph Larmor, J. J.'s friend and classmate at Cambridge, who had beaten him to be *first wrangler* in 1880, and was already a Professor of Natural Philosophy at Queen's College in Galway. Thomson himself was surprised to be successful, and wrote, 'I felt like a fisherman who with light tackle had casually cast a line in an unlikely spot and hooked a fish much too heavy for him to land'.[14] A senior Cambridge academic commented tartly, 'Matters have come to a pretty pass when they elect mere boys Professors'.[15] But it was an inspired choice. J. J. was to stay as Head of the Cavendish for 35 years, turning it into arguably the pre-eminent physics laboratory in the world. The science writer J. G. Crowther wrote in his history of the Cavendish, 'Clerk Maxwell brought the genius of the Scottish Celtic lairds and Rayleigh that of the English landed aristocracy. J. J. Thomson added to this the potent spirit of the 19th century British middle class … shrewdness, industry [and] common sense'.[16]

In the mid-19th century, experimental science at Cambridge had lagged well behind the rest of Britain, with much better facilities in Glasgow, London and Manchester. The wealthy and independent Cambridge Colleges were (often) in a power struggle with an impoverished central University and did not want to waste their money on expensive laboratories. Moreover, there was a somewhat snooty view of mere 'practical sciences'. For instance, the mathematician and Fellow of St John's College and Fellow of the Royal Society Isaac Todhunter refused Maxwell's offer to see an experiment on refraction of light, on the grounds that, 'I have been teaching it all my life, and I do not want to have my ideas upset'.[17] In the end, the Chancellor of the University, William Cavendish, the 7th Duke of Devonshire, gave the University £6,300 to found the Chair of Experimental Physics in 1871 and build the associated laboratories, completed in 1874, both named after his scientific ancestor Henry Cavendish, who had discovered hydrogen in 1766. Thomson proved to be the ideal successor to Maxwell and Rayleigh. His theoretical background and commitment to Cambridge earned the respect of the snobbish College dons, and his pragmatic approach, good sense and hard work helped him to build the funding and reputation of the Cavendish. He was popular as a member of national scientific committees, as an editor of journals and as a spokesman for British science, leading ultimately to his knighthood in 1908, the Order of Merit from the King in 1912 and his election as Master of Trinity in 1917, a position he retained until his death.

As the 28-year-old Cavendish Professor, with a handsome salary of £500 per annum, he was one of Cambridge's most eligible bachelors. In 1888 he met Rose Paget, daughter of Sir George Paget, the Regius Professor of Physic (i.e. Medicine), one of the grand old men of Cambridge, who had lived there since the 1830s and had just been knighted, a rare honour for an academic in those days. Rose was born in 1860, one of 10 Paget children, the elder of twin girls. According to Thomson's grandson David, 'Rose and her twin sister Violet … remained devoted to each other all their lives, though the two were almost opposite in character. Rose was intellectual, precise and calmly organised, Violet was scarcely educated, extrovert and a passionate gossip'.[18] Rose herself was passionate about science and joined the Cavendish as a researcher in 1888. J. J. and Rose were engaged the following year and married on 2 January 1890. Their correspondence during this period shows a touching transition from formal discussion of teaching matters, as befits a professor and his student, to increasing informality, passion and love. At the key point, when J. J. had just proposed that they become engaged, Rose wrote, 'I have been thinking unceasingly on all that happened this morning and the perplexity becomes more distressing the more I think of it. … Whatever happens I know only this that I cannot leave Papa'.[19] To which, J. J. replies soothingly, 'I am sorry to find from your letter that you have been exerting your unrivalled powers of making yourself miserable so successfully. … I must see your father this afternoon and ask his consent to our marriage'.[20] As David comments, 'Evidently the meeting went well and Sir George persuaded Rose that her scruples on his behalf were unnecessary. After all he still had his wife to run the house as well as two other unmarried daughters'.[21] J. J. and Rose remained happily married for the next 50 years, with two children: George Paget Thomson, born in 1892, like his father a noted physicist and Nobel Laureate, famous for discovering the wave properties of the electron as evidenced by electron diffraction; and a daughter, Joan Paget Thomson, born a decade later in 1903. Initially, they moved into a house in Scroope Terrace just south of the centre of Cambridge, later into a larger house on the *backs* near the river Cam and opposite the Colleges, and, finally, following his appointment as Master, into the Master's Lodge at Trinity.

After becoming Cavendish Professor of Experimental Physics, Thomson began to work intensively on issues associated with the discharge of electricity through gases, and he continued to do so for the rest of his life. This was not, when he started, a popular field of study, in part because of the lack of any theoretical understanding, in part because of the wide variety of observed effects, and in part because of the instability of electrical discharges, making it difficult to conduct controlled experiments. *Gaseous discharge* is seen when an electric potential is applied across two electrodes (a cathode and an anode) in a sealed glass vessel filled with gas at low pressure. Nothing happens at low potential; but, above a threshold voltage, there is an electrical discharge, current passes between the electrodes and the gas breaks down, emitting a fluorescent glow or spark. The light intensity, its colour and its distribution all vary widely depending on the nature of the gas, the gas pressure, the applied electric potential, the shape of the glass tube and the nature and shape of the electrodes. According to Thomson, 'Gaseous discharge is pre-eminent for the beauty and variety of the experiments and for the importance of its results on electrical theories'.[22]

Thomson was manually clumsy, which explains why he concentrated previously on theoretical work. Now, as Head of the Cavendish, he could afford to pay an assistant to do experiments while he watched. He developed an extraordinary ability to see what was happening in a discharge tube, diagnose its behaviour and interpret it theoretically. His research student E. V. Appleton said, 'J. J. though quite innocent of manipulative skill … was … unique in his ability to conceive some new experimental method or some way of overcoming practical difficulties'.[23] Thomson was at first fortunate to be able to appoint D. S. Sinclair as Cavendish Chief Instrument Maker and Technician. Sinclair was an excellent glassblower, which was essential for the manufacture of sealed discharge tubes. When Sinclair left in 1886 to set up his own business, J. J.'s experiments ground to a halt, until Sinclair was replaced by Ebenezer Everett, poached from the Cambridge chemistry labs, with his salary paid for by Thomson personally. The relationship between professor and technician is shown beautifully by an interchange between Thomson and Everett described by J. J.'s biographers Davis and Falconer:

> J. J. (eye on microscope to watch the experiment): 'Put the magnet on', followed by a click as Everett closed a large switch. J. J: 'Put the magnet on'.
> Everett: 'It is on'.
> J. J (eye still on microscope): 'No it isn't. Put it on'.
> Everett: 'It is on'.
> A moment later J.J. called for a compass needle [a 10-inch long, heavy, steel magnet used in lectures], which Everett fetched. J. J. took it and approached the electromagnet. When about a foot away the needle was strongly attracted, swung round and flew off its pivot, crashing into the bulb, which burst with a loud report. Everett was glowing with triumph and J. J. looking at the wreck with an air of dejection said [somewhat shamefacedly], 'Hm. It was on'.[24]

When Everett died in 1933, J. J. wrote in his obituary in *Nature*: 'He made all the apparatus I used in my experiments for the more than forty years in which he acted as my assistant. I owe more than I can express to his skill and the zeal which he threw into his work.'[25]

According to J. J.'s initial theory in the mid-1880s, the electrical field in a gas discharge disturbed the ether, destabilising chemical bonds between component atoms in the molecules of gas, leading to their dissociation and electrical transport in a manner similar to electrolysis in a liquid solution. This was half right, but also wrong in a number of ways: there was, of course, no ether; Thomson conceived of atoms, confusingly, as vortices in the ether; and gases are not like liquid solutions and do not contain pre-existing cations and anions (positively and negatively charged species). Not surprisingly, it took a long time to unravel complex experimental observations, given the difficulty of obtaining reproducible results, the lack of quantitation and the poverty of initial theoretical understanding. By the mid-1890s, progress was being made. Thomson was placing less emphasis on the importance of the ether, had discarded the notion of atoms as vortices, and was concentrating on the separation and motion of small electric and mass charges as the key phenomena in his gas discharges. This gradual change of perspective, built up after arduous experimentation, set him up for the breakthrough and discovery of the electron in 1897.

One of the phenomena associated with gas discharge tubes is the production of *cathode rays*, first reported in 1858 by the German physicist Julius Plücker, who observed a fluorescent patch on the wall of the glass tube, created at right angles to the cathode and independent of the anode shape or position. There was considerable debate about the nature of these rays. It wasn't clear whether they were waves or a stream of particles. Moreover, the British chemist and pioneer of the vacuum tube William Crookes claimed they could be deflected by a magnetic field, but this was disputed by the German physicist Heinrich Hertz, who claimed they couldn't. In 1895, Wilhelm Röntgen at the University of Würzburg discovered *X-rays*, generated from a discharge tube when cathode rays struck the anode, and demonstrated that these X-rays could pass through a human body, leaving an internal image of the body on a photographic plate. Within days, X-ray mania gripped the world, with enormous public interest, more than ever before in a scientific discovery. Röntgen subsequently received the first-ever Nobel Prize for Physics in 1901.

There was redoubled interest in the nature of the precursor cathode rays. In 1897 Thomson made the critical breakthrough, first during a Friday evening lecture at the Royal Institution in London on 30 April 1897, which was written up and published in *The Electrician*,[26] and then re-published in fuller detail six months later in the *Philosophical Magazine*.[27] He showed that cathode rays were indeed deflected by electric and magnetic fields, but only at low pressures, which had misled previous investigators. The cathode rays were deflected in just the way expected for a stream of very small and light, negatively charged particles. He further measured their charge by collecting them in a metallic tube; he measured their size by calculating the distance (about half a centimetre) required to absorb them in atmospheric air and he measured their energy from their impact on a thermal junction (a junction between two dissimilar metal wires). Remarkably, this all showed that cathode rays were composed of particles about a thousand times smaller than atoms or molecules, all with the same charge independent of the cathode material from which they were generated. Thomson commented in his paper in the *Philosophical Magazine*:

> We have in the cathode rays matter in a new state, a state in which the subdivision of matter is carried very much further than in the ordinary gaseous state: a state in which all matter – that is matter derived from different sources such as hydrogen, oxygen etc. – is of one and the same kind, this matter being the substance from which all the chemical elements are built up.[28]

This was Thomson's discovery of the *electron* as the fundamental unit of electricity, and a fundamental particle of all matter. Thomson called the cathode ray particles *corpuscles*, but scientists later reverted to the use of an earlier term, *electron*, first used by the Irish physicist George Johnston Stoney.

Thomson's theory that electrons were sub-atomic particles, universal constituents of all materials, took a few years to be fully accepted, strongly supported by the discovery of radioactive decay by French physicist Henri Becquerel and its further investigation by Marie and Pierre Curie, demonstrating that atoms of different materials could indeed be broken into smaller pieces. Thomson published his book *Conduction of Electricity through Gases* in 1903,[29] and received the Nobel Prize for Physics for his work on electric discharges in 1906. He went on to propose the so-called plum pudding model of the atom, consisting of hundreds and thousands of electrons embedded and spinning in a sphere of positive electric charge. The mass of the atom was all in the electrons, so there had to be very large numbers of them (this proved later to be incorrect). Different numbers of electrons explained the different chemical valency of different atoms, and their spinning instability explained the possibility of radioactive decay (these proved later to be essentially correct). However, Thomson's plum pudding model was overturned by Rutherford's experiments in 1911, which proved conclusively that the positive charge was in a small, central atomic nucleus, leading to the Bohr model of a central, positively charged nucleus surrounded by relatively few orbiting electrons.

In the meantime, J. J. began to study *anode rays* (also called *positive rays*), which streamed backwards through a hole cut in the cathode, and had been first discovered by the German physicist Eugen Goldstein. These were beams of ions, i.e. the positively charged particles left behind after the outer or valence electrons were stripped off the gas atoms in the discharge tube and emitted as cathode rays. Effectively, the gas discharge was the ionisation of the gas atoms, breaking them into a mixture of electrons and ions, and producing equal and opposite beams of cathode and anode rays. The first anode rays to be identified were hydrogen ions H^+ or *protons*, like electrons, also a key constituent of all matter. And J. J. gradually identified ions of many other gas atoms, in all cases measuring their charge and mass in similar ways to those used for the electron. He continued to be involved in arguments about the nature of electromagnetic waves, light and X-rays, and the nature of radioactive decay into different nuclear species of ions and isotopes. He remained a firm believer in the existence of an all-permeating ether within which all these rays and particles moved, an idea that was gradually being discarded by physicists with increasing understanding of relativity and quantum mechanics.

At the outbreak of the First World War, most of the students at Cambridge left to fight in France, the labs were occupied by billeted soldiers and the workshops were devoted to making war equipment. In July 1915, the government set up an Admiralty Board for Invention and Research to oversee the use of science in the naval war effort. It was chaired by the First Sea Lord, Admiral Jackie Fisher, and Thomson was one of the three scientific members, the others being Sir Charles Parsons, inventor of the steam turbine, and the applied chemist, Sir George Beilby, Chair of the Technical College in Glasgow. It oversaw diverse scientific developments such as the use of helium in airships, lighting for aerodrome runways, the prevention of erosion and corrosion of metal ships' hulls, and anti-submarine listening devices. It also vetted ideas put forward by the public, which were not always viable. Thomson wrote later:

> I have just written one letter ... to a charwoman who was upset by a bad smell and thinks it might be bottled up and used against the Germans.... Some of the schemes are wild enough, one was to train large numbers of cormorants to peck the mortar from between bricks and let them loose near Essen so they might peck the mortar from Krupp's chimneys and bring them down.[30]

Thomson was elected President of the Royal Society in 1915, appointed as Master of Trinity College in 1917, resigned as Cavendish Professor in 1919 to be succeeded by Rutherford, became a member of the UK University Grants Commission in 1919 and also in that year helped set up the government's new Department of Scientific and Industrial Research. He continued to do scientific research in Cambridge and influence science policy in Britain for the next 20 years. He died on 30 August 1940, aged 83, after suffering from senile dementia. He was cremated and his ashes were buried in the nave of Westminster Abbey near the graves of Newton, Darwin, Kelvin and Rutherford. According to his biographers Davis and Falconer:

> Thomson's greatest achievement is generally held to be his identification of cathode rays as electrons, which opened up the whole field of subatomic physics to experimental investigation. But his influence was far wider. His work established the transition from nineteenth- to twentieth-century physics. He established one of the world's great research schools [the Cavendish].[31]

William Thomson (Lord Kelvin)

William Thomson (Lord Kelvin)

William Thomson was born on 26 June 1824 in Belfast. His father was James Thomson, born on 13 November 1786 near Ballynahinch in County Down in Ireland, son of a smallholder farming family originally of Scottish extraction. James was gifted intellectually and was elected in 1815 to be Professor of Maths at the Royal Belfast Academical Institution (an independent grammar school known as the *Inst*), teaching mathematics, natural philosophy and classics. In 1817 James married Margaret Gardener, the daughter of a Glasgow merchant, and they had seven children: four boys (including William) and three girls, the youngest of which died in infancy.

The late 18th and early 19th centuries were dominated by the great French mathematical physicists, Joseph-Louis Lagrange, Pierre-Simon Laplace, Adrien-Marie Legendre, Siméon Denis Poisson, Jean-Baptiste Joseph Fourier and Augustin-Jean Fresnel, who were putting scientific maths on a modern analytical basis, replacing the more complex and less convenient geometric methods introduced almost a century and a half earlier by Isaac Newton. James Thomson was an early adopter of these new mathematical methods, which were strongly resisted in England, and his growing reputation for innovation led to him being offered and accepting in 1832 the Chair of Maths at Glasgow University, where he stayed until his death on 12 January 1849. Thomson is one of the most common clan names in Scotland: remarkably, at the time of his appointment, as one of his biographers Andrew Grey writes, '[T]he leaders of that illustrious corporation [Glasgow University] had become bound to [professors with] the name Thomson … [which] filled one half of the chairs in the University'[32]

William and his brothers were educated in their early years at home by their father. According to James' contemporary John Pringle Nichol, Professor of Astronomy at Glasgow, 'He was a stern disciplinarian, and did not relax his discipline when he applied it to his children'.[33] In 1834, at an age of just 10 years, William began to take scientific classes at the University—natural philosophy with William Meikleham, astronomy with John Nichol, physics with David Thomson and chemistry with Thomas Thomson (neither direct relatives)—winning first prizes in astronomy, maths, natural philosophy, Latin and the humanities. Under his father's influence, young William was already reading many of the great 18th-century mathematical treatises, most notably Laplace's *Mécanique Céleste*[34] and Lagrange's *Mécanique Analytique*.[35] William was also adept at languages, spending substantial time in France and Germany. In the summer of 1840, James Thomson took his two eldest sons (William and his elder brother, also James) on a tour of Germany, ostensibly to study the culture and language, but William was concentrating on reading Fourier's great book *La Théorie Analytique de la Chaleur* (*The Analytical Theory of Heat*),[36] and consequently writing his first scientific paper, 'On Fourier's Expansions of Functions in Trigonometric Series', published the following year under the pseudonym P. Q. R. in the *Cambridge Mathematical Journal*.[37] *Fourier analysis* is the description and approximation of arbitrary mathematical functions by a series of sine or cosine waves (called *Fourier series*) and is particularly convenient for solving the differential equations that govern heat flow (as well as many electrical, magnetic, optical and other physical phenomena). Fourier's arguments were relatively informal and therefore were not, at first, fully accepted in Britain, with claims they contained analytical errors, but William demonstrated in his paper that this was not the case. William was well aware of his intellectual abilities; he liked to style himself 'William Thomson BATAIAP (Bachelor of Arts to all intents and purposes)'.[38]

William's father was sufficiently well-off to pay for William's further education, and sufficiently persuasive in introductory letters to arrange for him to be accepted and accommodated for this purpose at St Peter's College (Peterhouse), Cambridge, in October 1841 to study maths with the famous tutor William Hopkins (who also taught, amongst others, George Stokes and James Clerk Maxwell). William was incredibly precocious, publishing a total of 16 papers, mostly on the analysis of heat flow, but some on other topics such as electricity and gravitation, between 1841 and 1845, while he was still an undergraduate student and still not yet 21 years old. This is an amazing output for someone who was nominally a student. When he finally took his *tripos* (final) exams, one of his two examiners commented to the other, 'You and I are just about fit to mend his pens'.[39] And when they later investigated why two students had given almost exactly the same answer to a particular question, one said he had seen this answer previously in a paper in the *Cambridge Mathematical Journal*, and the other (William, of course) said he had written the paper. William graduated in 1845 as second wrangler and won the Smith Prize for original research. During his undergraduate years, William was not just an outstanding mathematician and scientist, he was also successful as an athlete, winning the Colquhoun Sculls for rowing, and as a musician, playing the cornet and French horn and becoming President of the Cambridge University Musical Society. He developed what was to become a lifelong friendship with another great mathematical physicist, George Stokes, who had graduated a few years earlier as senior wrangler and who later took Newton's old chair as Lucasian Professor of Maths at Cambridge.

On graduating in 1845, William was elected to a Fellowship at Peterhouse, but the following year he was appointed to be the Professor of Natural Philosophy back at Glasgow University, succeeding his old tutors, William Meikleham (who died in 1846) and David Thomson (who had temporarily taken over Meikleham's duties while Meikleham was ill between 1842 and 1846), not very long after being one of their students. The professorship was conferred by election:

> The Faculty having deliberated on the respective qualifications of the gentlemen who have announced themselves as candidates for this chair, and the vote having been taken, it was carried unanimously in favour of Mr William Thomson BA, Fellow of St Peter's College, Cambridge, and formerly a student of this University, who is accordingly declared to be duly elected. ... The Faculty hereby prescribe Mr Thomson an essay on the subject 'De caloris distributione per terrae corpus' ['The Temperature Distribution in the Body of the Earth'].[40]

It was a standard convention for new professors to deliver an inaugural essay and lecture in Latin on some suitable topic before being inducted into their position at the University. It was not surprising that William wanted his inaugural topic to be on a heat-related topic, given his early study of this subject, and the issue of the heating and cooling of the earth was to become and remain one of his lifelong interests. He was given by the University an opening endowment of £100 and, in the following few years, additional sums amounting to a total of several hundred pounds to build up gradually what was then called *philosophical apparatus*.[41] This was unusually generous, and was a consequence of the complete inadequacy of physical and chemical equipment in the University at that time. Clearly William was pleased by the support he received. He was to remain in post at Glasgow for more than 50 years, until his retirement, aged 75, in 1899.

The 19th century was a heroic period for classical physics. Many of the world's physical phenomena were succumbing to detailed and insightful explanation using a mix of innovative experimental investigation and creative mathematical theory-building. It was a period when physicists were developing successful theories of light (wave theory), heat (thermodynamics) and electromagnetism (field theory). The mechanical structure of the world seemed to be gradually giving up its secrets, without yet any hint of the existential crisis and breakdown in classical scientific understanding that was to arrive early in the next century in the twin form of quantum mechanics and relativity.

William Thomson was to play a significant role in many of these developments. At Cambridge and then Glasgow, William was much exercised over many years in experimental studies, discussions and theoretical developments aimed at understanding heat flow and the generation of work and energy, leading to the emergence of the complex discipline of thermodynamics. He communicated and collaborated extensively with James Joule, the Salford-born businessman and engineer. Joule was the son of a brewing family, tutored by the Manchester-based scientist John Dalton (Salford is adjacent to the city of Manchester, just across the River Irwell), who was famous for the early introduction of modern atomic theory. Joule became manager of his family's brewing business and an active and able amateur scientist. William worked hard to reconcile Joule's experiments demonstrating the interconvertibility of heat and work, with the earlier observations of heat conservation by the French engineer Sadi Carnot. This led, in 1851, to one of the major pieces of work in Thomson's career, 'On the Dynamical Theory of Heat', published in 1851 in the *Transactions of the Royal Society of Edinburgh*,[42] and then re-published in 1852 in the *Philosophical Magazine*.[43] Effectively, these papers give an early version of the first and second laws of thermodynamics. Joule's interconvertibility of heat and work was within an overall principle of *conservation of energy*, essentially the first law of thermodynamics: 'When equal quantities of mechanical effect are produced by any means whatever from purely thermal sources, or lost in purely thermal effects, equal quantities of heat are put out of existence or are generated'.[44]

But Thomson also stressed the importance of *dissipation of energy*, i.e. the difficulty of fully recovering mechanical energy because of frictional effects (essentially the second law of thermodynamics): 'It is impossible, by means of any inanimate material agency, to derive mechanical effect from any portion of matter by cooling it below the temperature of the coldest of the surrounding objects'.[45]

He proposed an *absolute temperature* scale in which

> a unit of heat descending from a body A at the temperature T° of this scale, to a body B at the temperature $(T-1)°$, would give out the same mechanical effect [work], whatever be the number T. ... [Such a scale] would be quite independent of the physical properties of any specific substance.[46]

This defines the modern unit of temperature, the *Kelvin* or *K*, and, together with the concept of *absolute zero*, proposed previously by the French physicist Guillaume Amontons, the *Kelvin temperature scale* (all, fairly obviously, named after him).

In the 1850s and 1860s, Thomson used thermodynamics to try and estimate the ages of the sun and the earth. His first estimate for the age of the earth was between 20 and

400 million years old, with a wide range because of uncertainties in the nature of the rocks in the earth's constitution, and unknown parameters such as melting point, thermal conductivity and specific heat. In later years he refined this range to between 20 and 40 million years old, but he was out by a long way and was moving in the wrong direction. Our current best estimates for the ages of the sun and the earth are approximately 4.6 and 4.5 billion years respectively. Unfortunately, Thomson had arrived at wildly incorrect answers because of ignoring the (then-unknown) effects of thermonuclear reactions in the sun and radioactive heating in the earth. Even more unfortunately, his incorrect estimates led him into fierce arguments with early proponents of Darwinian evolution such as T. H. Huxley, because he claimed that the earth was too young and there was insufficient time for evolution to take place. He continued to rail against geologists, biologists and physicists on this topic, never accepting a longer lived earth, until his death in 1907.

Thomson wrote important papers on electricity and magnetism: linking the two effects, demonstrating similarities between the flow of electricity and heat, suggesting wave-like behaviour in each case, and proposing that magnetism was caused by microscopic rotations. These ideas were influential, stimulating the ground-breaking subsequent work on electromagnetic theory of his younger contemporary, James Clerk Maxwell. Thomson also made major advances in hydrodynamics, fluid flow, electrolysis, electrical distribution and telecommunications. He developed a vortex theory of atomic structure, which was also influential throughout much of the 19th century, though it later proved to be quite incorrect. Thomson and Peter Guthrie Tait, his opposite number as Professor of Natural Philosophy at Edinburgh University, co-wrote a widely read textbook on mechanics, kinematics and dynamics entitled *Treatise on Natural Philosophy* that was first published in 1867, with a second edition in 1883,[47] and a more detailed exposition entitled *Elements of Natural Philosophy* that was first published in 1872.[48]

Thomson was an outstanding engineer, innovator and businessman as well as a scientist. He invented electrometers and current balances for measuring electrical charge and current; he invented methods of deep-sea depth sounding, tide prediction and adjusting compasses to correct for the iron construction of ships; he was the first President of the Electrotechnical Commission for world electrical standards; and he chaired the International Commission to design the Niagara Falls hydrodynamic power station. He was not, however, always right: as well as championing vortex atoms and a 20- to 40-million-year-old earth, he refused to believe in X-rays or the possibility of heavier-than-air planes. Like most 19th-century scientists, he believed firmly in the need for an ether to carry light waves (amongst other things), and he is reputed to have said that nothing new remained to be discovered in physics.

Perhaps most notably in engineering, he was influential in the development of the transatlantic telegraph. In the mid-19th century, direct communication between Europe and America became a possibility through laying copper cables on the floor of the Atlantic. Thomson was the key consultant on the project. He performed critical calculations on the required structure and dimensions of the copper conductor and its insulating sheath, and he went on the cable-laying ship voyages to help and advise. William Cooke and Samuel Wheatstone had first demonstrated practical telegraphic communication using copper cables laid along a rail track, partly underground and partly on raised posts, initially in 1837 over a short distance between Euston and Camden Town in London (1.5 miles), and then in 1839 over longer distances between Paddington and West Drayton (13 miles), subsequently extended to Slough (18 miles). The telegraph was wildly successful and was publicly exhibited at Paddington as a marvel of science, transmitting information at 280,000 miles per second (7,500 km/s), with the public charged an admission price of one shilling (£0.05) per person to view communication taking place. One observer commented, 'After each word came a sign from Slough signifying "I understand", coming certainly in less than one second'.[49] A few years later, Wheatstone advised the Railway Committee of the British House of Commons on the successful laying of the first submarine telegraph cable in 1850 between Dover and Calais.

Beginning in 1851, Cyrus West Field and Frederick Newton Gisborne set up the New York, Newfoundland and London Telegraph Company and then the Atlantic Telegraph Company to lay a cable across the Atlantic, with Charles Tilson Bright as Chief Engineer, Wildman Whitehouse as Chief Electrician, and William Thomson as Scientific Advisor. Funds for the project were provided by the US and British governments, Field himself (a quarter of the total) and shares sold in Atlantic Telegraph. The subsidy bill only just passed, by a single vote, in the US Senate, because of opposition from protectionist senators. The project began in earnest in 1854 and, in 1857, William Thomson sailed on the *HMS Agamemnon*, a converted warship loaned by the British government. The *Agamemnon* and the *USS Niagara*, also a converted warship loaned by the US government, were attempting to lay a cable from Telegraph Field on Valentia Island, just south of Ballycarbery Castle in County Cork on the southwest coast of Ireland, to the romantically named township of Heart's Content, in Trinity Bay on the Bay de Verde peninsula near the easternmost point of Newfoundland. The *Agamemnon* laid cable westward from Ireland; the *Niagara* laid cable eastward from Canada; and the two cables were to be spliced together in the mid Atlantic. The project was called off when the *Agamemnon* cable broke after 380 miles. The following summer, the two ships tried again, and again the cable broke, this time after making the splice, while lowering it into the water.

The Board of Atlantic Telegraph wanted to abort the project, but Field argued for a third attempt, backed by Thomson, who claimed that the technical problems could be solved. Without much enthusiasm, the Board agreed and, at the third attempt on 5 August 1858, the cable was successfully laid, spliced and connected. Test messages were sent on 10 August, with the first official message on 16 August reading: 'Directors of Atlantic Telegraph Company, Great Britain, to Directors in America: Europe and America are united by telegraph. Glory to God in the highest: on earth peace, good will towards men'.[50,51] This was followed by a second message from Queen Victoria to President James Buchanan, hoping for 'an additional link between the nations whose friendship is founded on their common interest and reciprocal esteem'.[52] To which he replied,

It is a triumph more glorious, because far more useful to mankind, than was ever won by
conqueror on the field of battle. May the Atlantic telegraph, under the blessing of Heaven,
prove to be a bond of perpetual peace and friendship between the kindred nations, and
an instrument destined by Divine Providence to diffuse religion, civilization, liberty, and
law throughout the world.[53]

Clearly there were great hopes for the future. The next day there was a 100-gun salute
in New York, there were flags in the streets, church bells were rung, and at night the city
was illuminated; and at the beginning of September, there was a daytime parade and
an evening torchlight procession, which caused a fire in the Town Hall. Unfortunately,
after a few days the cable insulation began to crack up and fail, with communication
gradually deteriorating, getting slower and slower until it finally ceased. In the end, the
cable was never put into service commercially. Some commentators claimed that the
whole project was a hoax, and a public inquiry into associated stock market speculation
concluded that the Chief Electrician Wildman Whitehouse was to blame. The company
was criticised for employing a Chief Electrician with only limited technical qualifications.
It took another six years before the Telegraph Construction and Maintenance Company
(Telcon, later part of BICC) was formed in 1864 to lay a new cable and successfully re-
initiate transatlantic telegraphic communication. By the end of the 19th century, British,
American, French and German cables criss-crossed the Atlantic in a complex web of
telecommunications between Europe and North America. Thomson took part in laying
the French North Atlantic cable in 1869, and the Western and Brazilian and Platino-
Brazilian cables across the South Atlantic between Portugal and Brazil and along the
Brazilian coast in 1873. In the end, laying telegraph cables made Thomson rich and
famous, and led to him being knighted along with others for their contributions to
telecommunications.

The original Atlantic Telegraph cable consisted of seven copper wires, each weighing
26 kg/km covered with three coats of insulating gutta-percha (a rubber-like plastic-setting
resin made from the pulp of Malaysian trees), wound with tarred hemp and sheathed
in a helix with 18 strands each of seven iron wires, weighing a total of almost a ton
(907 kg) per nautical mile (1.9 km). Whitehouse had disagreed with Thomson about the
best kind of cable to use, believing erroneously that conduction would just be through
the surface of the cable. Thomson advised that current would be carried throughout its
volume, and the cable conductivity and cross section should be maximised to enhance
signal carrying capacity (what we now call *bandwidth*). He also argued for maximum
flexibility to prevent fracture during cable-laying on the seabed. Thomson was ignored
and Whitehouse prevailed, leading to the problems of cable failure encountered by the
Agamemnon, first during laying and then during use.

In 1852, Thomson married his childhood sweetheart Margaret Crum, daughter of
the Scottish chemist and businessman Walter Crum. Sadly, her health broke down on
their honeymoon and she continued to be in some distress and ill health for the next 18
years, until her early death in 1870. Thomson slipped on some ice in 1860, fracturing his
leg badly and causing him to limp for the rest of his life. Nevertheless, he remained an
avid yachtsman, addicted to seafaring, and, shortly after his wife's death, he used some

of the money made from cable patents to purchase a 126-ton (114-Mg) schooner, the *Lalla Rookh*. Thomson wrote to Tait to say, 'My desk at the NPL [Natural Philosophical Laboratory] and the LR [Lalla Rookh] are the only places in the world for which I am fit'.[54] In 1871 he invited Tait, Maxwell, the German physicist Hermann von Helmholtz and the Irish physicist John Tyndall (famous for discovering the *greenhouse effect*) to go cruising with him on the *Lalla Rookh* through the beautiful Western Isles of Scotland. All but Helmholtz demurred, with Tait commenting that he would rather be playing golf. Helmholtz wrote to his wife: '[Golf is] a kind of ball game, which is played on the green sward with great vehemence by every male visitor.... W. Thomson must be now just as much absorbed in yachting as Mr Tait is in golfing'.[55]

Some years later, Thomson was helping to lay the Brazilian South Atlantic cable while sailing onboard the *Hooper* when the cable developed faults and the ship was forced to make a two-week stopover in Madeira. There, he met and became friends with Charles Blandy, a member of one of the founding families of the Madeira wine trade, and his three daughters. He was smitten with Blandy's second daughter, Fanny. The following year, on 2 May 1874, he set sail for Madeira on the *Lalla Rookh* and signalled to the Blandy residence as he approached the harbour, to ask Fanny, 'Will you marry me'?[56] She signalled back yes and they were married shortly thereafter in the British Consular Chapel on 24 June 1874. Returning to Scotland, he bought land and built a very fine new family home called *Netherhall* at Largs, about 25 miles (40 km) west of Glasgow, baronial in style, complete with peacocks supplied by Maxwell. According to Thomson's biographer, Mark McCartney,

> Fanny was a very different wife to Margaret. Margaret, almost permanently ill, languishing at home, writing poetry about sadness and death, was replaced by Fanny, who enjoyed sailing and loved to entertain and see the house filled with guests. And filled with guests it often was.[57]

Despite the considerable difference in their character and health, Thomson had no children by either marriage.

Thomson became one of the most famous scientists of the 19th century and received awards and prizes throughout his career. He was elected as a Fellow of the Royal Society of Edinburgh in 1847, was awarded its Keith Medal in 1861 to 1863, and was its President from 1873 to 1878 and from 1886 to 1890. He was elected as a Fellow of the Royal Society of London in 1851, was awarded its Copley Medal in 1883, and was its President from 1890 to 1895. He was President of the British Association for the Advancement of Science in 1871, giving a presidential address on the kinetic theory of gases and the associated sizes of atoms; and in 1884 covered similar topics in his famous Baltimore Lectures. In 1892 he was elevated by Queen Victoria to a peerage, becoming Baron Kelvin of Largs, named after the River Kelvin, a small tributary of the River Clyde, which runs close to Glasgow University. Later that decade he was made a member of the Order of Merit in Britain, a Knight of the Ordre Pour le Mérite in Prussia and a Grand Officier de la Légion d'Honneur in France. In 1896 Glasgow University and the City of Glasgow celebrated Kelvin's Jubilee as Professor of Natural Philosophy with a convention attended by scholars, kings and princes from all round the world, and a memorial volume of the proceedings that was published in the journal *Nature*.[58]

After retiring in 1899, Kelvin continued in a state of excellent health: his only significant illness was a tendency to suffer from occasional but painful facial neuralgia. In 1907 his health seemed fine as usual. He presided as Chancellor at the installation of the Chancellor of the Exchequer Herbert Asquith (soon to become Prime Minister) as Lord Rector of the University; he attended the funeral of the Vice-Chancellor, Principal Story; he presided in April over the graduation ceremonies and the public opening of the new Philosophy Institute and Medical Buildings by the Prince of Wales; and he conferred degrees of Doctor of Laws on the Prince and Princess. He and Lady Kelvin left in July, as they did most years, to spend the summer in Aix-les-Bains in the French Alps, returning to Netherhall early in September. They planned to visit Belfast to open the new scientific buildings at Queen's College, but cancelled when Lady Kelvin fell ill. Fanny's illness persisted for several weeks, causing William much anxiety, and perhaps the worry over Lady Kelvin played a role, but he caught a chill in late November, deteriorated rapidly, and died on 18 December. He was buried five days later, next to Newton and Darwin in Westminster Abbey, with a service attended by dignitaries from all round the world, including representatives of King Edward and the Prince of Wales. According to Andrew Grey, 'There he sleeps well who toiled during a long life for the cause of natural knowledge, and served nobly as a hero of peace, his country and the world'.[59]

4   Nucleation of solidification

Figure 6.3 shows liquid and solid free energies G_l and G_s versus temperature T for a pure material:

$$G_l = H_l - TS_l$$
$$G_s = H_s - TS_s,$$

where H_l, H_s, S_l and S_s are the liquid and solid enthalpies and entropies respectively. In general, $H_l > H_s$ and $S_l > S_s$ because atoms and molecules have higher energy and are more disordered in the liquid state.

The liquid and solid free energies are equal at the *melting point* T_m:

$$G_l = G_s$$
$$\therefore H_l - T_m S_l = H_s - T_m S_s$$
$$\therefore L = T_m \Delta S; \text{ and } \Delta S = \frac{L}{T_m},$$

where $L = H_l - H_s$ is the *latent heat of solidification* and $\Delta S = S_l - S_s$ is the entropy change on solidification.

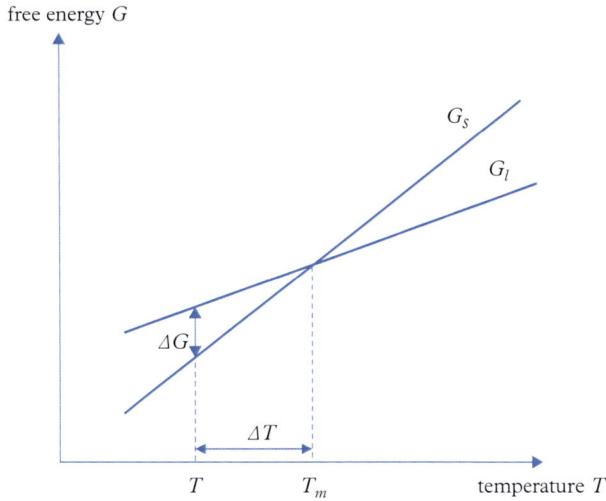

Figure 6.3 *Driving force ΔG for solidification at an undercooling ΔT below the melting point*

For a liquid material to solidify, it must be cooled below its melting point so that the free energy of the solid is lower than that of the liquid, and there is a *driving force for solidification ΔG*. At any temperature T below T_m, i.e. at an *undercooling $\Delta T = T_m - T$*, the *driving force for solidification ΔG* is given by

$$\Delta G = G_l - G_s = (H_l - TS_l) - (H_s - TS_s)$$
$$= L - T\Delta S = \frac{L\Delta T}{T_m}.$$

In other words, the driving force for solidification increases linearly with increasing undercooling below the melting point, as can be seen in Figure 6.3.

Consider a small particle of solid that forms in the liquid at an undercooling ΔT, as shown in Figure 6.4. The solid has lower free energy than the liquid, as shown in Figure 6.3, and the Gibbs-Thomson equation shows that the solid particle is in equilibrium with the liquid when

$$\Delta G_V = \frac{(G_l - G_s)}{v_m} = \frac{L_V \Delta T}{T_m} = \frac{2\gamma}{r}$$
$$\therefore \ \Delta T = T_m - T = \frac{2\gamma T_m}{L_V r}$$
$$\therefore T = T_m - \frac{2\gamma T_m}{L_V r} = T_m \left(1 - \frac{2\gamma}{L_V r} \right),$$

where L_V is the latent heat per unit volume. Effectively, the melting point, which is the point of equilibrium between solid and liquid, varies with the size of the solid particle. This can be seen more clearly by defining the temperature $T = T_m(r)$ as the melting

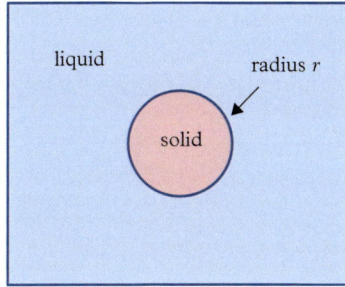

Figure 6.4 *Nucleation of solidification*

point for a particle of radius r, and the normal melting point $T_m = T_m(\infty)$ as the melting point for a particle of infinite radius $r = \infty$:

$$T_m(r) = T_m(\infty)\left(1 - \frac{2\gamma}{L_V r}\right).$$

The melting point decreases inversely with decreasing particle size, with the constant of proportionality equal to twice the surface energy divided by the latent heat per unit volume. This is another common form of the Gibbs-Thomson equation. The change of melting point with particle radius is called a *capillarity effect*, i.e. an effect caused by surface energy.

Solidification takes place in two stages: *nucleation*, which is the initial formation of small particles of solid within the undercooled liquid; and *growth*, which is the increase in size of the solid particles to consume the liquid. Consider an initial *nucleus* or small particle of solid that forms in the liquid, as shown in Figure 6.4. On the one hand, the solid–liquid interface has excess energy, which acts as a barrier to the solidification process. On the other hand, the solid has a lower Gibbs free energy than the liquid, which acts to drive the solidification process. As noted previously, the Gibbs-Thomson equation shows that the nucleus is in equilibrium with the surrounding liquid when the difference in free energy is given by

$$\Delta G_V = \frac{(G_l - G_s)}{v_m} = \frac{L_V \Delta T}{T_m} = \frac{2\gamma}{r} = \frac{2\gamma}{r^*},$$

where $r = r^*$ is called the *critical nucleus size*. Particles with $r = r^*$ are called *critical nuclei* and they are in equilibrium with the surrounding liquid, with the driving force for solidification balanced exactly by the surface energy. Smaller particles with $r < r^*$ are called *sub-critical nuclei* and they decay away because the surface energy barrier is bigger than the driving force for solidification. Larger particles with $r > r^*$ are called *super-critical nuclei* and they can grow into the liquid because the driving force for solidification is bigger than the surface energy barrier. The critical nucleus size is

$$r^* = \frac{2\gamma}{\Delta G_V} = \frac{2\gamma T_m}{L_V \Delta T}.$$

In other words, the critical nucleus size decreases inversely with increasing undercooling below the melting point.

Similar capillarity effects are seen with any phase transformation that takes place in a material with increasing or decreasing temperature, e.g. melting, boiling, condensation or change of crystal structure. In each case, the equilibrium temperature varies with particle size, and the transformation requires a nucleation stage to overcome the surface energy barrier.

5   Nucleation of precipitation

Figure 6.5 shows a eutectic phase diagram in a two-component A–B material. A material of composition c is cooled from a high temperature T_1 in the single-phase α region above the α solvus to a lower temperature T_2 in the two-phase $\alpha + \beta$ region below the α solvus to form a material consisting of β particles in an α matrix. The difference $\Delta c = c - c_e$ between the composition c and the equilibrium composition c_e at the α solvus is called the *supersaturation*, which drives the precipitation process at temperature T_2.

Figure 6.6 shows the variation of the Gibbs free energy with composition for α and β at the lower temperature T_2. The variation of α free energy with composition c for dilute solutions, i.e. when c is not too large, is approximately

$$G_\alpha(c) = G_\alpha^o - RT \ln c,$$

where G_α^o is the free energy of pure A. On cooling to T_2, the material is initially still single-phase α with free energy G_α; but, after some time, β particles precipitate and the

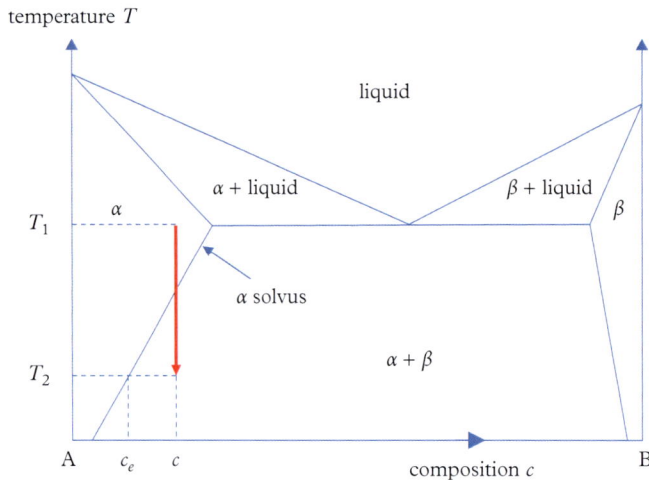

Figure 6.5 *Precipitation of β from α*

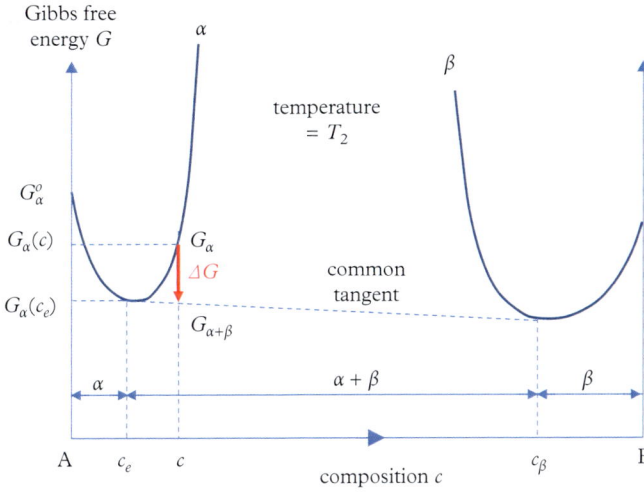

Figure 6.6 *The Gibbs free energy–composition curves for* α *and* β

free energy reduces to $G_{\alpha+\beta}$. At a supersaturation Δc, the difference between G_α and $G_{\alpha+\beta}$ is the *driving force for precipitation* ΔG, which for a dilute solution is given by

$$\Delta G = G_\alpha - G_{\alpha+\beta} \approx G_\alpha(c) - G_\alpha(c_e) \approx \left(G_\alpha^o - RT \ln c\right) - \left(G_\alpha^o - RT \ln c_e\right)$$

$$= RT \ln \left(\frac{c}{c_e}\right) = RT \ln \left(1 + \frac{\Delta c}{c_e}\right) \approx RT \frac{\Delta c}{c_e},$$

where $G_\alpha(c)$ and $G_\alpha(c_e)$ are the free energies of α at compositions c and c_e, and we have used the Taylor series approximation $\ln(1+x) = x - \frac{x^2}{2} + \frac{x^3}{3} - \cdots \approx x$ for small x. In other words, the driving force for precipitation increases approximately linearly with supersaturation above the equilibrium composition at the α solvus.

Figure 6.7 shows a typical β particle precipitated in the α matrix. The Gibbs–Thomson equation shows that the β particle is in equilibrium with the α matrix when

$$\Delta G_V = \frac{G_\alpha - G_{\alpha+\beta}}{v_m} = RT \frac{\Delta c}{c_e} = \frac{2\gamma}{r}$$

$$\therefore \Delta c = c - c_e = \frac{2\gamma c_e}{RTr}$$

$$\therefore c = c_e + \frac{2\gamma c_e}{RTr} = c_e \left(1 + \frac{2\gamma}{RTr}\right).$$

Effectively, the α solvus composition, which is the point of equilibrium between α and $\alpha + \beta$, varies with the size of the solid particle. This can be seen more clearly by defining the composition $c = c_e(r)$ as the solvus for a particle of radius r, and the normal solvus $c_e = c_e(\infty)$ as the solvus for a particle of infinite radius $r = \infty$:

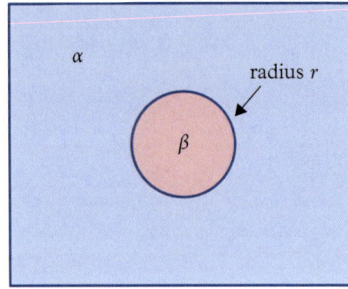

Figure 6.7 *β particle in an α matrix*

$$c_e(r) = c_e(\infty)\left(1 + \frac{2\gamma}{RTr}\right).$$

The solvus composition increases inversely with decreasing particle size, with the constant of proportionality equal to twice the surface energy divided by the gas constant and the temperature. This is another common form of the Gibbs-Thomson equation, and another example of a capillarity effect.

Precipitation, like solidification, takes place in two stages: *nucleation*, which is the initial formation of small β particles within the supersaturated α matrix; and *growth*, which is the increase in size of the β particles to complete the precipitation process. Consider an initial *nucleus* or small particle of β that forms in the α matrix, as shown in Figure 6.7. On the one hand, the $\alpha - \beta$ interface has excess energy, which acts as a barrier to the precipitation process. On the other hand, the β precipitate has a lower Gibbs free energy than the supersaturated α matrix, which acts to drive the precipitation process. As noted previously, the Gibbs-Thomson equation shows that the nucleus is in equilibrium with the surrounding matrix when the difference in free energy is given by

$$\Delta G_V = \frac{(G_\alpha - G_{\alpha+\beta})}{v_m} = RT\frac{\Delta c}{c_e} = \frac{2\gamma}{r} = \frac{2\gamma}{r^*},$$

where $r = r^*$ is again called the *critical nucleus size*. Particles with $r = r^*$ are *critical nuclei* and they are in equilibrium with the surrounding matrix, with the driving force for precipitation balanced exactly by the surface energy. Smaller particles with $r < r^*$ are *sub-critical nuclei* and they decay away because the surface energy barrier is bigger than the driving force for precipitation. Larger particles with $r > r^*$ are *super-critical nuclei* and they can grow into the matrix because the driving force for precipitation is bigger than the surface energy barrier. The critical nucleus size is

$$r^* = \frac{2\gamma}{\Delta G_V} = \frac{2\gamma c_e}{RT\Delta c}.$$

In other words, the critical nucleus size decreases inversely with increasing supersaturation above the α solvus.

The variation of solvus with supersaturation during precipitation is clearly similar to the depression of melting point with undercooling during solidification, and the variation of critical nucleus size with supersaturation during precipitation is similar to the variation of critical nucleus size with undercooling during solidification.

6   Ostwald ripening

During precipitation in a two-component A–B material as shown in Figure 6.7, many β particles nucleate and grow at different times until the supersaturation is exhausted and precipitation is complete. The material then consists of a collection of differently sized β particles in an α matrix, as shown in Figure 6.8. When precipitation is complete, the Gibbs-Thomson equation gives the equilibrium composition in the matrix surrounding each particle:

$$c = c_e \left(1 + \frac{2\gamma}{RTr} \right).$$

The matrix composition surrounding each particle depends on its size, with a higher concentration of B atoms or molecules surrounding the smaller particles. The resulting concentration gradients cause diffusion of B atoms or molecules from the smaller particles to the larger particles. Larger particles grow in size, fed by inward diffusion of B, and smaller particles reduce in size until they disappear, gradually denuded by outward diffusion of B. This process is called *Ostwald ripening* of the particles and is another example of a capillarity effect. In many materials, it is important to have a fine dispersion of particles to develop optimal properties, so it is important to minimise Ostwald ripening both during and after the precipitation process.

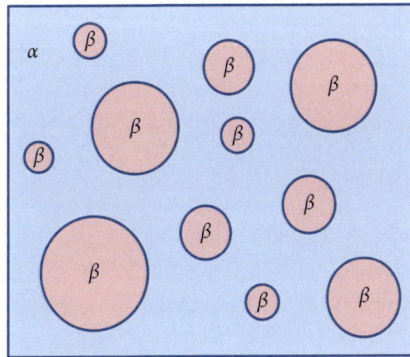

Figure 6.8 *Collection of β particles in an α matrix*

7 Crystals

The equilibrium shape of a particle of one phase embedded in another one is only spherical when the surface energy is independent of orientation, so that minimising the Gibbs free energy at constant temperature and pressure is equivalent to minimising the surface area. When the α particle or the surrounding β matrix are crystalline, however, the surface energy depends on orientation, often quite strongly, because different crystal planes have different atomic or molecular structures. Instead of being spherical, an equilibrium crystalline particle is, therefore, bounded by sharp, flat facets corresponding to particular crystalline surfaces with low-energy atomic or molecular structures. This is why, for instance, salt and sugar are collections of small crystals rather than spherical powder particles, and why solid precipitates, vapour deposits and rock formations are also often collections of small crystals or, sometimes, quite large-faceted crystals such as gemstones.

The equilibrium shape of any material, crystalline or otherwise, is determined by the *Gibbs-Wulff theorem*. To allow for varying surface energy with orientation, the Gibbs-Wulff theorem says that the minimum Gibbs free energy for a particle embedded in a matrix is obtained not simply by minimising the surface area, but by minimising the sum of all surface energy–area combinations for all surfaces or facets in the equilibrium shape:

$$\left(dG_{P,T}\right)_{\min} = \left(\sum_{\text{all facets } i} \gamma_i dA_i\right)_{\min}.$$

The equilibrium shape can be found by using the *Wulff construction*. The first step is to construct a γ-plot of the material $\gamma(n)$. This is a graph of the variation of surface energy with orientation, in the form of a polar plot of γ versus surface normal n, as shown in two dimensions in Figure 6.9. For a liquid or a non-crystalline solid, the γ-plot is spherical, because the surface energy is independent of orientation. For a crystal, however, the γ-plot is shaped like a raspberry, with cusps of low surface energy at each simple crystal plane with a low-energy atomic or molecular structure. The second step is

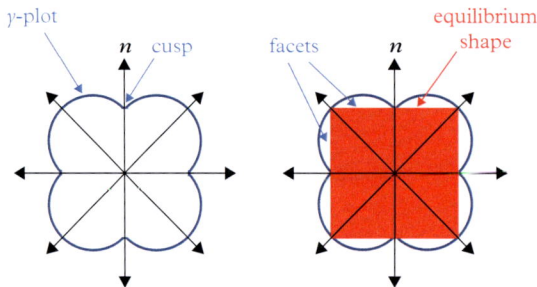

Figure 6.9 *The Wulff construction*

to draw planes perpendicular to every normal n at the surface of the γ-plot. The resulting inner envelope of these planes is the equilibrium shape of the crystal, as shown in Figure 6.9. For a liquid or non-crystalline solid this is again spherical, but for a crystal it is a faceted particle, as shown in Figure 6.9.

8   Segregation and adsorption

Surfaces have an excess surface energy because of their atomic or molecular structure, which is intermediate between the structures of the two phases on either side of the surface. The surface energy can often be reduced by inserting impurity atoms or molecules into the surface structure. This process is called *adsorption* when it refers to gaseous impurities being attracted to the free surface of a solid or liquid material; it is called *segregation* when it refers to impurities within a solid material being attracted to an internal surface between two phases or to a grain boundary. *Physisorption* is when the impurity atoms or molecules occupy surface sites with no special change in the bonding at the surface. *Chemisorption* is when the impurity atoms or molecules form new covalent or ionic bonds with the surface. The reverse process, when impurities are released from a surface, is called *desorption* or *desegregation*.

The extent of adsorption or segregation can be measured in different ways. The *surface excess* is defined as the difference between the amount of impurity at the surface and the amount that would be present if there was no adsorption. The fraction adsorbed θ is defined as the fraction of surface sites occupied by impurity atoms or molecules.

The variation of adsorption or segregation with pressure and composition at a given temperature is called the *adsorption isotherm*. One of the most well-known isotherms is the *Langmuir adsorption isotherm*, which describes the adsorption or segregation process as a chemical reaction between impurity atoms or molecules and surface sites. For adsorption of gaseous impurities at the free surface of a solid or liquid material, this can be written as

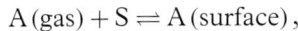

$$A\,(\text{gas}) + S \rightleftharpoons A\,(\text{surface}),$$

where A is an impurity atom or molecule, S is a surface site, and the forward and backward reactions represent adsorption and desorption respectively. The equilibrium constant for this reaction K is given by

$$K = \frac{k_d}{k_a} = \frac{c_{A(\text{surface})}}{c_{A(\text{gas})}c_S}.$$

where k_a and k_d are reaction rate constants for the adsorption and desorption reactions respectively, and $c_{A(\text{surface})}$, $c_{A(\text{gas})}$ and c_S are concentrations of adsorbed impurity, gaseous impurity and surface sites respectively. The amount of adsorbed impurity $c_{A(\text{surface})}$ is proportional to the fraction adsorbed θ, the amount of gaseous impurity is proportional to the pressure p, and the number of surface sites available for adsorption is proportional to $(1 - \theta)$, so

$$K = \frac{\theta}{p\,(1-\theta)}$$

or

$$\theta = \frac{Kp}{1+Kp}.$$

This is the *Langmuir adsorption isotherm*. With low impurity levels in the gas, i.e. for low values of pressure p, $1 + Kp \approx 1$, so the fraction adsorbed is $\theta \approx Kp$ and it increases linearly with increasing pressure. With high impurity levels in the gas, i.e. for high values of pressure p, $1 + Kp \approx Kp$, so the fraction adsorbed is $\theta \approx 1$ and the surface is completely saturated with impurity.

9 REFERENCES

1. The Bible, Genesis 1:3.
2. 'Edison's electric light: the Time building illuminated by electricity'. *New York Times*, 5 September 1882.
3. Randall Stross. *The Wizard of Menlo Park: How Thomas Alva Edison Invented the Modern World* (Crown Publishers, Random House, New York, 2007), 133.
4. Marie-Claire Le Reun. 'Thomas Edison: "the greatest adventure of my life"'. n.d. https://www.leclindoeilpetillant.com/en/2012/02/26/thomas-edison-la-plus-grande-aventure-de-ma-vie/ (accessed 11 March 2020).
5. Chris J. Magoc. *Chronology of Americans and the Environment* (ABC-CLIO, Santa Barbara, 2011), 46.
6. James Clerk Maxwell. 'A dynamical theory of the electromagnetic field'. *Philosophical Transactions of the Royal Society of London* 155 (1865), 459–512.
7. 'George Gobel quote'. *Cue*, 6 November 1954. https://libquotes.com/george-gobel/quote/lbx8f4l (accessed 11 March 2020).
8. E. A. Davies and I. J. Falconer. *J. J. Thomson and the Discovery of the Electron* (CRC Press, Taylor and Francis, Boca Raton, 1997), xiii.
9. Ibid.
10. William Whyte. *Redbrick: A Social and Architectural History of Britain's Civic Universities* (Oxford University Press, Oxford, 2015), 71.
11. Davies and Falconer, n 8, xiii.
12. J. J. Thomson. 'Experiments on contact electricity between non-conductors'. *Philosophical Magazine* 3 (1877), 389.
13. J.J. Thomson. 'On a theory of the electric discharge in gases'. *Philosophical Magazine* 15 (1883), 427–434.
14. Davies and Falconer, n 8, 45.
15. George Paget Thomson. *J. J. Thomson and the Cavendish Laboratory* (Nelson, London, 1964), 73.
16. Davies and Falconer, n 8, xvii.
17. James G. Crowther. *The Cavendish Laboratory 1874–1974* (Macmillan, London, 1974), 9.
18. Davies and Falconer, n 8, xvii.

19. See ibid. xix.
20. Ibid. xx.
21. Ibid.
22. Ibid. 47.
23. Ibid. 48.
24. Ibid. 56.
25. Ibid.
26. J. J. Thomson. 'Cathode rays (Friday evening meeting of the Royal Institution, 30 April 1897)'. *The Electrician* **39** (1897), 104.
27. J. J. Thomson. 'Cathode rays'. *Philosophical Magazine* **44** (1897), 303–326.
28. G. L. Squires. *J. J. Thomson Autobiography* (Department of Physics, University of Cambridge, 1997), https://www.phy.cam.ac.uk/history/electron/autobiography (accessed 3 February 2020), n.p.
29. J. J. Thomson. *Conduction of Electricity through Gases* (Cambridge University Press, Cambridge, 1906; reprinted Wexford College Press, Wexford, 2005 and Cambridge University Press, Cambridge, 2013.
30. Davies and Falconer, n 8, 209.
31. See ibid. 211.
32. Andrew Grey. *Lord Kelvin: An Account of his Scientific Life and Work* (Createspace Publishers, North Charleston, 2014), 2.
33. Ibid. 3.
34. Pierre Simon Laplace. *Mécanique Céleste*, tomes I–V [*Celestial Mechanics*, vols. I–V] (L'Imprimerie de Crapelet, Paris, 1798–1825).
35. Joseph Louis Lagrange. *Mécanique Analytique* (Courcier, Paris, 1811; re-issued by Cambridge University Press, Cambridge, 2009).
36. Jean-Baptiste Joseph Fourier. *La Théorie Analytique de la Chaleur* [*The Analytical Theory of Heat*] (Firmin Didot, Paris, 1822).
37. P. Q. R. 'On Fourier's expansions of functions in trigonometric series'. *Cambridge Mathematical Journal* **2** (1841), 258–262.
38. Raymond Flood, Mark McCartney, and Andrew Whitaker. *Kelvin: Life, Labours and Legacy* (Oxford University Press, Oxford, 2008), 4.
39. Grey, n 32, 8.
40. See ibid. 19.
41. Ibid. 20.
42. William Thomson. 'On the dynamical theory of heat, with numerical results deduced from Mr Joule's equivalent of a thermal unit, and M. Regnault's observations on steam'. *Transactions of the Royal Society of Edinburgh* **XX**, Pt. II (1851), 261–268, 289–298.
43. William Thomson. 'On the dynamical theory of heat, with numerical results deduced from Mr Joule's equivalent of a thermal unit, and M. Regnault's observations on steam'. *Philosophical Magazine* **4** (1852), 8–21.
44. David B. Wilson. *Kelvin and Stokes: A Comparative Study in Victorian Physics* (Adam Hilger, Bristol, 1987), 4.
45. William Thomson. *Mathematical and Physical Papers*, vol. 1 (Cambridge University Press, Cambridge, 1882; reprinted 2011), 179.
46. Don S. Lemon. *Thermodynamic Weirdness: From Fahrenheit to Clausius* (MIT Press, Cambridge, 2019), 64.

47. W. Thomson and P. G. Tait. *Treatise on Natural Philosophy* (Clarendon Press, Oxford, 1867); 2nd ed., vols. 1 and 2 (Cambridge University Press, Cambridge, 1878 and 1883; re-issued 2009).
48. W. Thomson and P. G. Tait. *Elements of Natural Philosophy* (Cambridge University Press, Cambridge, 1872; re-issued 2010).
49. Gertrude Sullivan. *A Family Chronicle* (John Murray, London, 1908), 216–217.
50. Harry Granick. *Underneath New York* (Fordham University Press, New York, 1991), 115.
51. Joint Committee Appointed by the Lords of the Committee of Privy Council for Trade and the Atlantic Telegraph Company to Inquire Into the Construction of Submarine Telegraph Cables. *Manipulation of the Atlantic Telegraph Line* (Eyre and Spottiswoode, London, 1861), 230–232.
52. Jesse Ames Spencer. 'The queen's message and the president's reply', Chapter 10. In: *History of the United States: From the Earliest Period to the Administration of President Johnson* (Johnson, Fry, New York, 1856), 542.
53. Ibid.
54. Flood, McCartney, and Whitaker, n 38, 17.
55. Ibid.
56. Ibid. 18.
57. Ibid.
58. A. Gray. 'Lord Kelvin's jubilee'. *Nature* 54 (1896), 173–181.
59. Grey, n 32, 89.

··

10   BIBLIOGRAPHY

Interfaces in Crystalline Materials. A. P. Sutton and R. W. Baluffi (Oxford University Press, New York, 1995).
J. J. Thomson and the Cavendish Laboratory. George Paget Thomson (Thomas Nelson and Sons, London, 1964).
J. J. Thomson and the Discovery of the Electron. E. A. Davis and I. J. Falconer (Taylor and Francis, London, 1997).
Kelvin: Life, Labours and Legacy. Raymond Flood, Mark McCartney, and Andrew Whitaker (Oxford University Press, Oxford, 2008).
Lord Kelvin: An Account of His Scientific Life and Work. Andrew Grey (Createspace Publishers, North Charleston, 2014).
Thermodynamics of Surfaces and Interfaces. Gerald H. Meier (Cambridge University Press, Cambridge, 2014).
The Scientific Papers of J. Willard Gibbs, vols. I and II. J. W. Gibbs (Dover, New York, 1961).

7

Fick's Laws

Diffusion

1 Fick's first law

Materials can exist in three phases: solid, liquid or gas. At high temperatures, all materials are gaseous. The constituent atoms or molecules have enough energy to be fully mobile. They move rapidly and independently, spending most of their time travelling unconstrained between collisions with other atoms or molecules or the container wall. Because of this motion, they expand to fill any container. At lower temperatures, materials become liquid or solid, which are called *condensed phases*. The constituent atoms or molecules have less energy and become packed closely together, constrained by interatomic forces that prevent them from moving rapidly and independently. They have a fixed density and volume independent of the container. At intermediate temperatures, materials are liquid, with the constituent atoms or molecules constrained by interatomic forces but still able to move collectively, a process called *fluid flow*. And at low temperatures, materials are solid, with the constituent atoms or molecules constrained completely by interatomic forces, so they are fixed in place and are, more or less, unable to move, apart from vibrating about their lattice point.

Surprisingly, however, we find that atomic movement can still take place within a solid material, albeit fairly slowly and mainly at higher temperatures. This process is called *atomic diffusion* or just *diffusion*. Consider a simple experiment in which a thin foil of solid material is sealed inside a glass tube with a gas at high pressure P_1 on one side and low pressure P_2 on the other side, as shown in Figure 7.1. We find that the gas atoms are able to diffuse through the solid material. If the glass tube is sealed with a plug at each end, the high pressure P_1 decreases continuously and the low pressure P_2 increases continuously until they become equal. If the pressures P_1 and P_2 are maintained by pumps, there is a continuous flow of gas atoms through the solid material.

While diffusion is taking place, the concentration of gas atoms c is found to be linear with distance x through the thickness of the solid, as shown in Figure 7.2. The rate at which the gas atoms move dn/dt is called the *diffusion rate* and is found to be proportional to the cross sectional area A of the solid and the concentration gradient through its thickness dc/dx:

The Equations of Materials. Brian Cantor. Oxford University Press (2020). © Brian Cantor.
DOI: 10.1093/oso/9780198851875.001.0001

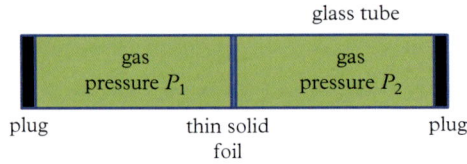

Figure 7.1 *Gas diffusion through a thin foil of solid material*

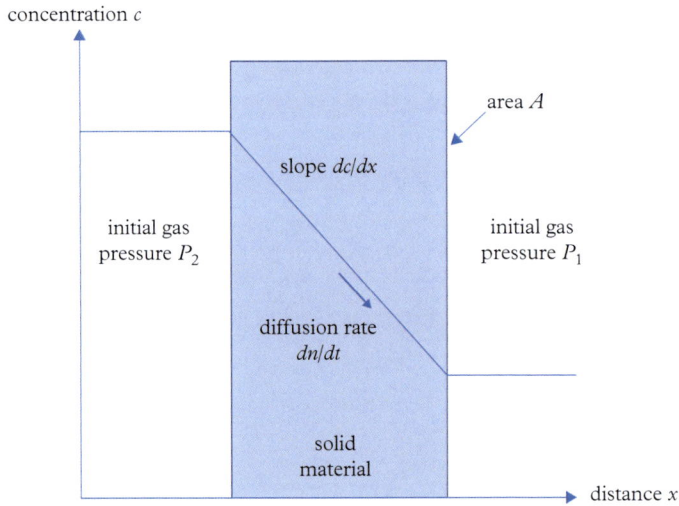

Figure 7.2 *Fick's first law: the diffusion rate dn/dt is proportional to the cross sectional area A and the concentration gradient dc/dx*

$$\frac{dn}{dt} = -DA\frac{dc}{dx},$$

where the constant of proportionality is called the *diffusivity* or *diffusion coefficient D*. This equation is *Fick's first law*. It can also be written in terms of the *flux* $\mathcal{J}$, which is the diffusion rate per unit area:

$$\mathcal{J} = \frac{1}{A}\frac{dn}{dt} = -D\frac{dc}{dx}.$$

The diffusion rate dn/dt measures the number of gas atoms n moving per second, and the concentration gradient dc/dx measures the variation with distance of the concentration c, i.e. the number of gas atoms n per unit volume. This means that the diffusion coefficient D has units of (length) squared per time, i.e. metres squared per second.

2 Fick's second law

Fick's first law describes the diffusion rate or flux as a function of the concentration gradient for a linear concentration profile, i.e. when the concentration gradient is constant, as shown in Figure 7.2. Fick's second law generalises diffusion behaviour to cases in which the concentration profile is non-linear and the concentration gradient is not constant. Fick's second law gives the change in concentration with time dc/dt as a function of the variation in concentration gradient d^2c/dx^2,

$$\frac{dc}{dt} = D\frac{d^2c}{dx^2},$$

which is a well-known differential equation called the *diffusion equation*.

Fick's second law can be derived from Fick's first law by considering two planes X and Y at positions x and $x + dx$, i.e. separated by a small distance dx, in a solid material that has a non-linear concentration profile and, therefore, a varying concentration gradient, as shown in Figure 7.3. The diffusion rates at the two planes are different because of the non-linear concentration profile, and are given by Fick's first law:

$$\left(\frac{dn}{dt}\right)_x = -DA\left(\frac{dc}{dx}\right)_x$$

$$\left(\frac{dn}{dt}\right)_{x+dx} = -DA\left(\frac{dc}{dx}\right)_{x+dx} = -DA\left\{\left(\frac{dc}{dx}\right)_x + \left(\frac{d^2c}{dx^2}\right)dx\right\}.$$

Because of the difference in diffusion rates, there is an accumulation of atoms in the volume $dv = Adx$ between the two planes:

$$Adx\frac{dc}{dt} = \left(\frac{dn}{dt}\right)_x - \left(\frac{dn}{dt}\right)_{x+dx} = DA\left(\frac{d^2c}{dx^2}\right)dx,$$

and dividing by the volume $dv = Adx$ leads to Fick's second law:

$$\frac{dc}{dt} = D\left(\frac{d^2c}{dx^2}\right).$$

When the concentration varies in more than one dimension, there is a concentration field $c(x,y,z)$, and Fick's first and second laws can be re-written as

$$\boldsymbol{J} = -D\nabla\mathbf{c}$$

$$\frac{\partial c}{\partial t} = D\nabla^2 c,$$

where $\boldsymbol{J}$ is the flux vector, ∇ is the gradient operator, so $\nabla \mathbf{c} = \frac{\partial c}{\partial x}\overrightarrow{\boldsymbol{x}} + \frac{\partial c}{\partial y}\overrightarrow{\boldsymbol{y}} + \frac{\partial c}{\partial z}\overrightarrow{\boldsymbol{z}}$ is the three-dimensional concentration gradient vector where $\overrightarrow{\boldsymbol{x}}$, $\overrightarrow{\boldsymbol{y}}$ and $\overrightarrow{\boldsymbol{z}}$ are unit vectors in

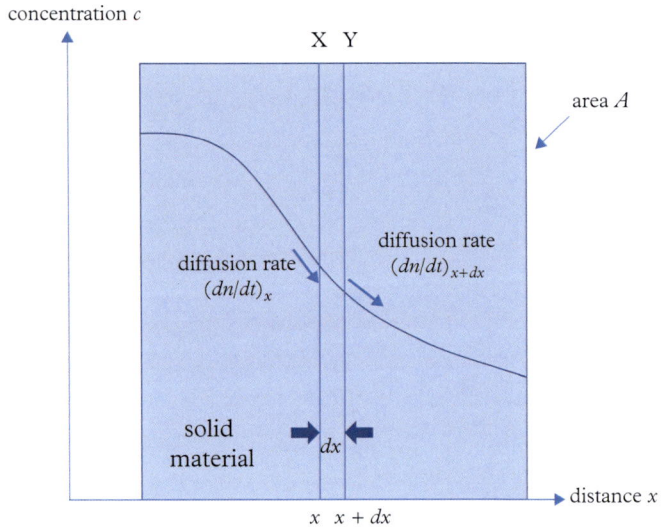

Figure 7.3 *Fick's second law for a non-linear concentration profile*

the x, y and z directions, and ∇^2 is the Laplacian operator, so $\nabla^2 c = \frac{\partial^2 c}{\partial x^2} + \frac{\partial^2 c}{\partial y^2} + \frac{\partial^2 c}{\partial z^2}$ is the scalar three-dimensional curvature in the concentration field.

3 Adolf Fick

Adolf Fick

Adolf Eugen Fick was enormously influential in the fields of physiology, anatomy, medicine, physics and chemistry, yet he has been described as 'the forgotten genius'[1] and, compared to other similarly influential scientific figures, there is very little biographical material available about his life and times. In many ways, he is quite a mystery man. He was born on 3 September 1829 in the city of Kassel, at that time the principal city and capital of the Grand Duchy of Hesse, later to be incorporated, following annexation by Prussia in 1867 and the subsequent Franco-Prussian war of 1870 to 1871, as one of the Länder (states) in the North German Federation, the precursor of modern Germany. Amongst other things, Kassel was, at that time, the home of the Brothers Grimm, Jacob and Wilhelm, where they spent the first 30 years of the 19th century employed as court librarians, and in their spare time collected and published their famous folk stories, fairy tales and legends. Adolf's father was Friedrich Fick, who worked as a municipal architect. Adolf was the youngest of nine children, and his brothers and sisters were all successful in different ways, in fields such as anatomy, chemistry and law.

Adolf's early schooling was at the local gymnasium (high school). In 1847 he began studying maths and physics at the nearby University of Marburg, where two of his brothers already occupied senior academic positions, Ludwig as Professor of Anatomy and Heinrich as Professor of Law. Heinrich persuaded Adolf to study maths and physics, arguing, as it turned out correctly, that they would be invaluable for undertaking medical research, which Adolf wanted to do. In 1849 Adolf spent time at the University of Berlin to attend lectures by the surgeon Bernard von Langenbeck, who invented the house staff model for surgeons and is known as the 'father of modern [surgical] training programmes'[2], the neurologist Mauritz von Romberg, who invented the *Romberg test* for loss of motor coordination, the naturalist and medic Johann Schönlein, who discovered Schönlein's disease (vasculitis) and the ringworm parasite, and the physicist and physician Hermann von Helmholz, who worked on electromagnetism, fluid flow and vortices, and invented the ophthalmoscope and the theory of colour vision. Adolf completed his doctorate in 1851 as an assistant to his older brother, Franz Ludwig Fick, who was Professor of Anatomy at Marburg, with a doctoral thesis entitled 'Tractatur de Errore Optico Quodam Asymetria Bulbi Oculi Effecto' ('Treatment of Some Visual Errors Caused by Eyeball Asymmetry'),[3] a subject he selected because of his own severely astigmatic eyesight.

After receiving his doctorate, Adolf worked briefly with his brother as a prosector, a job that involved teaching and studying as well as dissecting dead bodies for anatomical examination. Six months later he moved to the University of Zürich, working again as a prosector, this time with the Professor of Anatomy and Physiology, Carl Ludwig, a previous close colleague, who had himself recently been appointed from a position as prosector and special Professor at Marburg. Ludwig was a charismatic and influential figure. Adolf found this a stimulating environment and he prospered. He was appointed Associate Professor at Zürich in 1855, when Ludwig left for a professorship at the University of Vienna, and was later promoted to be a full Professor of Physiology at Zürich in 1862. While in Zürich, Adolf married and had two sons, one who later became an anatomist and the other who later became a lawyer. In 1868 he moved to succeed Alfred von Bezold as Professor of Physiology at the prestigious University of Würzburg, where he stayed for the next 31 years, becoming Dean of Medicine and then Rector of the University. He retired in 1899 and died from apoplexy shortly after his 70th birthday in 1901 in Flanders. In 1929 his sons founded the *Adolf Fick fund*, which gives a prize every year for an outstanding contribution to physiology.

Adolf Fick spent his life and career at the heart of one of the most important and exciting socio-political developments in modern European history, namely Bismarck's post-Napoléonic unification of Germany, and the associated flowering of German science, engineering, medicine and culture. The end of the 18th century and the beginning of the 19th century were dominated across Europe by the French Revolution in 1789, the fall of the *ancien regime*, the harsh reaction, and the subsequent Napoléonic wars. Napoléon Bonaparte rose to conquer almost all of Europe, but then faltered and retreated, leading ultimately to his final defeat at the Battle of Waterloo in 1815 and his exile to the tiny island of St Helena, 2,000 miles west of Africa in the South Atlantic. The Holy Roman Empire had existed in the centre of Europe as a multi-ethnic alliance of up to 500 separate states, territories and princedoms ever since the crowning of Charlemagne as Emperor in 800, but was effectively destroyed when Emperor Francis II was defeated by Napoléon at the Battle of Austerlitz in 1806. German nationalism flowered under the Napoléonic Empire; but, after Napoléon's defeat, the Congress of Vienna established separate and overlapping spheres of influence for the four great powers: Britain, France, Russia and Austria. Most of the German-speaking states were put into a loose German confederation, under Austria's leadership, with a federal *Diet* or *Bundestag* (assembly or parliament) that met in Frankfurt am Mein (close to Fick's birthplace in Kassel). This was an uneasy alliance of war-ravaged states that ignored the rising power of Prussia, exemplified by the important role of the Prussian cavalry in defeating Napoléon at Waterloo. It staggered on until it was overturned by the revolutions across Europe of 1848, when, amongst other political develop-ments, leadership of the German federation transferred to the Prussian King Frederick William IV. Johann Droysen, a member of the Frankfurt parliament commented, 'the whole German question is a simple alternative between Prussia and Austria . . . [it] is a question of power, and the Prussian monarchy is now wholly German, while that of Austria cannot be'.[4] Frederick William IV suffered a stroke in 1857, and his brother Wilhelm was made Prince Regent, and then King Wilhelm I in 1860. He appointed Otto von Bismarck as his Minister-President. Droysen was, however, wrong. The alliance was not so simple. Prussian leadership alone was insufficient to resolve the fundamentally fragmented nature of the disparate German states. As the *New York Times* commented in 1860:

> There is, in political geography, no Germany proper to speak of. There are Kingdoms and Grand Duchies and Duchies and Principalities, all inhabited by Germans, and each separately ruled by an independent sovereign.... Yet there is a natural undercurrent tending to a national feeling and toward a union of the Germans into one great nation.[5]

Bismarck worked hard to create such a nation. In his famous 'blood and iron' speech in 1862, he recognised that welding the German peoples into an effective nation under the leadership of Prussia would probably require common enemies against which to react and bond:

> The position of Prussia in Germany will not be determined by its liberalism but by its power.... Not through speeches and majority decisions will the great questions of the day be decided – that was the great mistake of 1848 and 1849 – but by iron and blood.[6]

This proved to be an accurate analysis. After a succession of wars—the second war of Schleswig between Denmark and Prussia in 1864, the Austro-Prussian war of 1866 and finally the Franco-Prussian war of 1870—Bismarck finally established Germany as a full nation-state, rather than a loose confederation, at the Treaty of Versailles in 1871.

These political changes in the second half of the 19th century facilitated the German industrial revolution and the growth of German science and technology. Before 1850, Germany lagged well behind France and Britain economically, but by the end of the century it was a world leader in manufacturing and industrialisation based on excellent science and engineering, alongside Britain and the United States. This led to the dominance of the German engineering, automotive and chemicals industries in the 20th century, and indirectly to the two terrible World Wars. Adolf Fick and his mentor, Carl Ludwig, played enormous leadership roles in the development of the fields of physiology and anatomy and associated applications to medical practice, rejecting the then commonly accepted notion of special biological forces and (unknown) 'vital forces' in favour of much more solidly based and scientifically sound physical and chemical explanations.

Adolf's time at Zürich and Würzburg was very productive scientifically. He invented, amongst other things, an aneroid manometer for measuring blood pressure, a machine to measure the working power of a muscle, a plethysmograph for measuring the change in volume of an arm or leg, a myotonograph for studying skeletal muscles, and a pendulum myograph to measure the force produced when a muscle contracts. He developed the *Imbert-Fick law*, relating deformation of the cornea to intra-ocular pressure, improving the accuracy of applanation–tonometry for the diagnosis of glaucoma. He wrote a classic textbook on medical physics, as well as other books on physiology, anatomy and blood circulation. Most notably, he developed *Fick's principle* for the circulation of blood from the heart, which says that cardiac output is given by the difference in oxygen content between inspired and expired air divided by the difference in oxygen content between the left and right ventricles. As he commented at the time, 'It is astonishing that no one has arrived at [this] method by which the amount of blood ejected by ... the heart can be determined'.[7] In his *Handbook of Physiology*,[8] the American biologist William F. Hamilton says of Fick's principle,

> [It] carries a very simple message that is self evident once it is grasped ... it is a turning point in the development of quantitative measurement of blood flow, and from

its central idea have come many and various techniques that have given us the soundest measurements of the output of the heart and the flow of blood through the organs.[9]

Fick's principle is difficult to prove because it is difficult to measure blood flow and oxygen content directly and accurately. Three years before Fick died, Nathan Zuntz and Oscar Hageman, as part of their monumental work on the metabolism of horses, passed a tube through the jugular vein of a horse to obtain its right ventral blood and verify Fick's principle. This was not a feasible procedure to use with humans, and Fick's principle was not fully verified in humans until 1930, with the advent of cardiac catheterisation as developed by Werner Forssman, André Frédéric Cournand and Dickinson W. Richards, for which they received the Nobel Prize for Medicine in 1956.

In 1858, Fick's brother Franz Ludwig died young, aged just 45, shortly after his wife had also died young. Adolf was godfather to his nephew, Franz's six-year-old son, with a very similar name, Adolf Gaston Eugen Fick, who was now left an orphan, so Adolf took him into his family to adopt and raise him. His nephew was much influenced by Adolf, and also studied medicine at Würzburg, Zürich, Marburg and Freiburg. He later became famous for inventing and manufacturing the first contact lenses, made from heavy brown glass, which he tested on rabbits before using them on himself and then other patients.

Adolf Fick's most important work was his discovery of the laws controlling *diffusion* of gases through membranes and tissues in the body. In the late 1840s and early 1850s, the English chemist Thomas Graham had published extensively on the subject of diffusion. *Graham's law* says that the speed at which a gas diffuses varies inversely with its molecular weight. He also showed that diffusion increases with temperature, but he failed to identify the underlying mechanism. The French mathematician and engineer Siméon-Denis Poisson suggested that diffusion was caused by capillarity (i.e. surface effects) and the French physicist Henri Becquerel suggested it was caused by electrical effects, but these theories were abandoned when Carl Ludwig showed it was caused by random molecular motion. Fick analysed these results, and compared molecular diffusion to the conduction of heat as described by *Fourier's law*.

Jean-Baptiste Joseph Fourier was born in Auxerre. His father was a tailor, but he was orphaned at the age of nine. Through the good offices of the Bishop of Auxerre, he was taken in, looked after and educated by the Benedictine Order of the Convent of St Mark. He became a military lecturer and played a prominent part in the French Revolution, serving on his local revolutionary committee and being imprisoned briefly during the Terror. He was appointed Professor at the École Normale, before succeeding Joseph-Louis Lagrange as Professor at the École Polytechnique. He accompanied Napoléon as a scientific advisor on his Egyptian expedition in 1798, was appointed by Napoléon to be secretary of the Institut d'Egypt, and then returned to France as Prefect of the Department of Isère, based in Grenoble. While in Grenoble, Fourier began to study the propagation of heat, publishing in 1822 his famous book *Théorie Analytique de la Chaleur* (*The Analytical Theory of Heat*).[10] He is most famous for the development of Fourier series in which, with some restrictions, any mathematical function can be represented by an infinite series of sines or cosines. The physical applications of Fourier series are extremely important and wide ranging in fields such as heat flow, electrical engineering, acoustics, optics, signal processing, image processing and quantum mechanics, but the technical details are quite complex and have exercised mathematicians for many years, finally being put on a fully firm basis by Lejeune Dirichlet and Bernhard Rieman. Fourier also invented the idea of dimensional analysis of engineering equations, and developed *Fourier's law*, which says that the rate of heat flow through a material is proportional to its cross sectional area and the imposed temperature gradient. In 1822 he became Permanent Secretary of the French Academy of Sciences. He had suffered for many years from aneurism (bleeding) of the heart and, sadly, on 4 May 1830, he fell down the stairs, aggravating his heart condition, and died in his bed 12 days later.

In analysing diffusion, Fick concluded that 'the volume of gas flow per unit time moving across a tissue sheet is directly proportional to the area of the sheet and the difference in pressures between the two sides but inversely proportional to tissue thickness'.[11] This is the same as the modern version of Fick's law, which says that the diffusion rate is proportional to the cross sectional area and the concentration gradient. It is also clearly analogous to *Fourier's law* for heat flow (published previously in 1822), as well as *Ohm's law* for the flow of electricity (also published previously, in 1827). Fick published his work twice in 1855, in German in the *Annalen der Physik*[12] and in English in the *Philosophical Magazine*[13], because it applied to gas diffusion in both physiological and physical contexts. It is remarkable that a scientist who studied medicine, gained a medical doctorate and committed all his working life to investigating anatomy and physiology had such a major impact on the physical sciences by identifying correctly the laws that govern the motion of atoms and molecules in solid and liquid materials. The physicist H. G. V. Tyrrell wrote in 1964,

> Fick's ideas have proved fruitful for more than a century.... Yet his paper on diffusion is usually misquoted: his name does not appear in general works of reference such as the Encyclopedia Britannica ... and the details of his career are unknown to most physical scientists.[14]

4   Thin- and thick-film solutions

There are many different cases of complex concentration profiles, with no general solution to Fick's second law to describe their evolution. However, there are two important cases for which Fick's second law can be solved analytically to obtain the evolution of the concentration profile with time. These are called the *thin-film solution* and *thick-film solution*.

Thin-film solution

Consider an initial thin layer of material B deposited on the surface of a bar of material A, as shown in Figure 7.4. If the bar is heated, the B atoms diffuse into it and the resulting concentration profile evolves with time, as shown in Figure 7.4. The concentration profile $c(x,t)$ as a function of distance x and time t is obtained by solving Fick's second law with two boundary conditions:

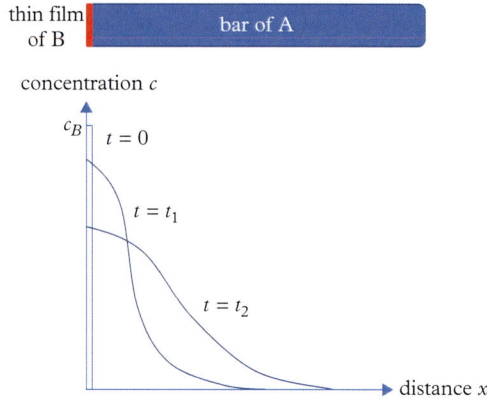

Figure 7.4 *Thin-film solution of Fick's second law*

1. Initially there are no B atoms in A:

$$c(x,0) = 0.$$

2. The number of deposited B atoms is conserved, i.e. the total number of B atoms N_B remains constant at all times:

$$\int_0^\infty c(x,t)\,dx = N_B.$$

The resulting thin-film solution of Fick's second law is

$$c(x,t) = \frac{N_B}{\sqrt{\pi Dt}} \exp\left(\frac{-x^2}{4Dt}\right).$$

After a time t, in other words, the concentration c decays exponentially with distance x, with a surface concentration $c_s = c(0,t) = N_B/\sqrt{\pi Dt}$ and a characteristic decay length $\lambda = 2\sqrt{Dt}$. The thin-film solution can, therefore, be re-written as

$$c(x,t) = c_s \exp\left(-x/\lambda\right)^2.$$

Thick-film solution

Consider an initial thick layer of material B deposited on the surface of a bar of material A, as shown in Figure 7.5. If the bar is heated, the B atoms again diffuse into it and the resulting concentration profile evolves with time, as also shown in Figure 7.5. The

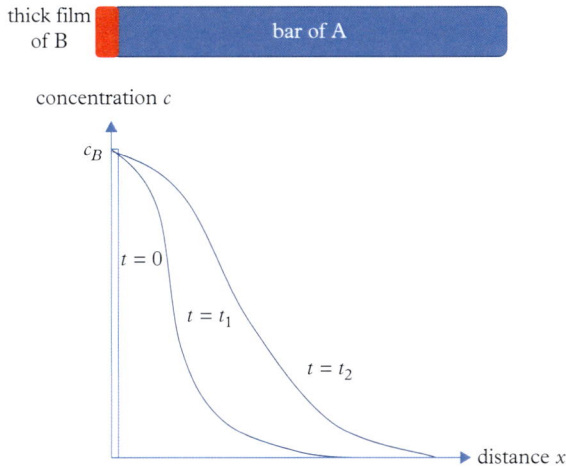

Figure 7.5 *Thick-film solution of Fick's second law*

concentration profile $c(x,t)$ is again obtained by solving Fick's second law with two boundary conditions:

1. Initially there are no B atoms in A:

$$c(x,0) = 0.$$

2. The thick film maintains a pure B surface concentration c_B at all times:

$$c(0,t) = c_B.$$

The resulting thick-film solution of Fick's second law is

$$c(x,t) = c_B \left\{ 1 - \text{erf} \left(\frac{x}{2\sqrt{Dt}} \right) \right\}$$

where the error function erf is defined as

$$\text{erf } x = \frac{2}{\sqrt{\pi}} \int_0^x \exp\left(-u^2\right) du$$

and has values of $\text{erf}(0) = 0, \text{erf}(\infty) = 1$ and $\text{erf}(-\infty) = -1$. Similar diffusion behaviour is obtained, and the thick-film solution applies, when the surface of the bar of material A is exposed to a continuous source of pure B atoms in the gas phase.

Welded bars

Consider two bars of materials A and B welded together, as shown in Figure 7.6. If the bars are heated, B atoms diffuse into A, and A atoms diffuse into B. This is called *interdiffusion*. The concentration profile $c(x,t)$ (c is, as noted previously, taken as the concentration of B atoms) is again obtained by solving Fick's second law with two boundary conditions:

1. Initially there are no B atoms in A, and no A atoms in B:

$$c(x,0) = 0 \text{ for } x > 0$$
$$c(x,0) = c_B \text{ for } x < 0.$$

2. At all times, a long way from the weld, there are no B atoms in A, and no A atoms in B:

$$c(x,t) \to 0 \text{ for } x \to \infty$$
$$c(x,0) \to c_B \text{ for } x \to -\infty.$$

The resulting solution of Fick's second law is similar to the thick-film solution:

$$c(x,t) = \frac{1}{2}c_B \left\{ 1 - \text{erf}\left(\frac{x}{2\sqrt{Dt}}\right) \right\}.$$

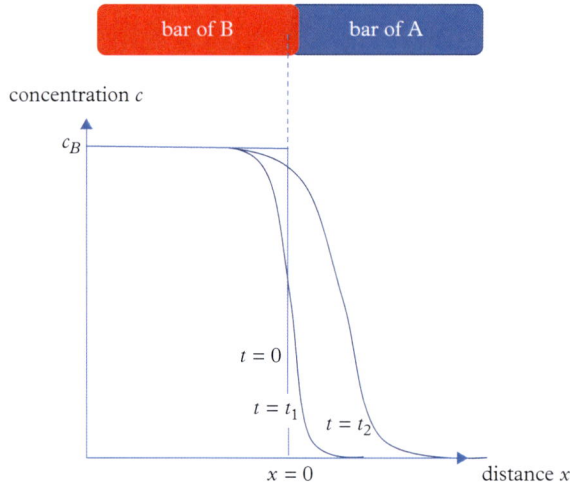

Figure 7.6 *Welded-bars solution of Fick's second law*

In all these cases, the diffusion coefficient is assumed to be constant, i.e. independent of concentration, $D \neq f(c)$, and A and B atoms are assumed to diffuse at equal and opposite rates. These assumptions are not, in general, true (as discussed later), though they may, in some cases, be a reasonable approximation.

5 Diffusion mechanisms

Vacancy diffusion

In a substitutional A–B solid solution, the A and B atoms diffuse by a *vacancy diffusion mechanism*. A vacancy is a defect in the crystal structure when an atom is missing from one of the lattice sites, as shown for a face-centred cubic (fcc) material in Figure 7.7. Vacancy diffusion takes place when an adjacent atom jumps into a vacancy, as shown in

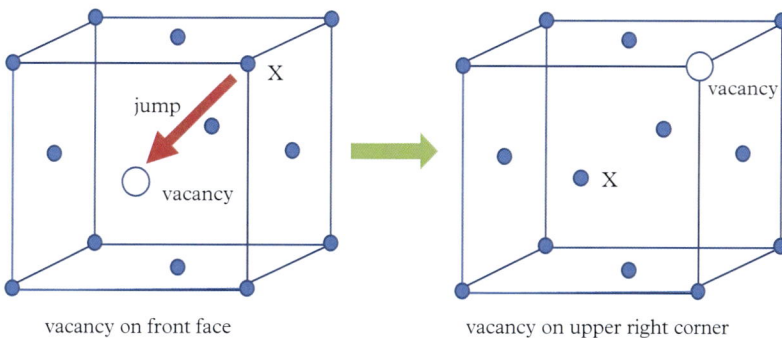

Figure 7.7 *Vacancy diffusion*

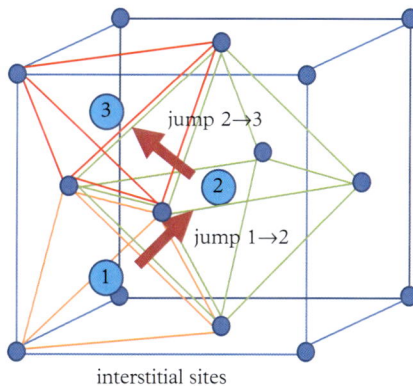

Figure 7.8 *Interstitial diffusion as a gas atom diffuses in a face-centred cubic material from a tetrahedral interstitial site 1 to an octahedral interstitial site 2 and then to another tetrahedral interstitial site 3*

Figure 7.7. As adjacent atoms continue to jump into the vacancy, it moves through the crystal, and the A and B atoms interdiffuse through the material.

Interstitial diffusion

Gas atoms are small, and when they diffuse through a solid material, they usually do so by an *interstitial diffusion* mechanism, i.e. by jumping through the interstitial sites in the crystal, as shown in Figure 7.8 for an fcc material.

6   Random walk

Diffusion is a *stochastic* process. It takes place by the *random walk* of atoms throughout the material. Einstein showed that when an atom undergoes a random walk, the average distance $\bar{x}$ it moves away from its starting position in time t is given by $\bar{x}^2 = Dt$.

In a crystalline material, atoms migrate either by random jumping into adjacent vacancies, as shown in Figure 7.7, or by random jumping between interstitial sites, as shown in Figure 7.8. Individual atom jumps can take place in all directions, both in the forward direction down the concentration gradient and in the backward direction up the concentration gradient. This means that the diffusion rate dn/dt in Fick's first law is the overall rate, i.e. the difference between forward and backward rates of atoms jumping.

Consider two adjacent planes X and Y perpendicular to the concentration gradient in an A–B simple cubic substitutional solid solution, as shown in Figure 7.9. The planes are separated by the lattice spacing a. The number of B atoms in the X plane N_{BX} is the concentration c multiplied by the volume of each lattice plane aA, where A is the cross section of the material through which diffusion can take place:

$$N_{BX} = caA.$$

And the number of B atoms in the Y plane N_{BY} is different because of the concentration gradient:

$$N_{BY} = \left(c + a\frac{dc}{dx}\right)aA.$$

The frequency with which solute B atoms jump from plane X into plane Y is η_x, and the frequency with which solute B atoms jump in the reverse direction from plane Y into plane X is η_{-x}. Both are equal to the mean jumping frequency η multiplied by the proportion of jumps in the x direction $\frac{1}{6}$:

$$\eta_x = \eta_{-x} = \frac{1}{6}\eta.$$

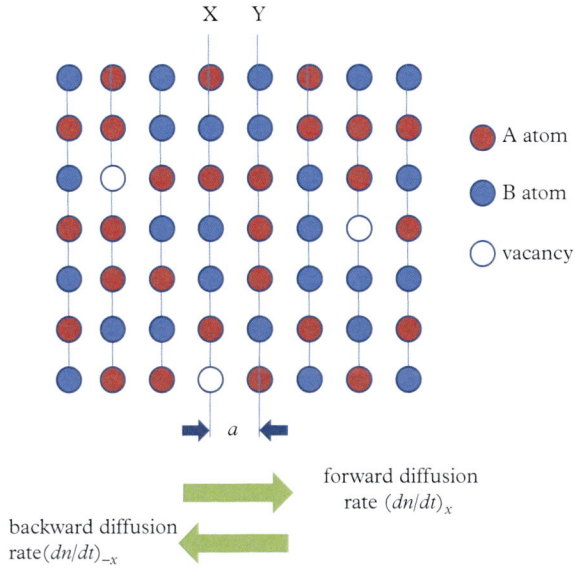

A atom

B atom

vacancy

a

forward diffusion
rate $(dn/dt)_x$

backward diffusion
rate $(dn/dt)_{-x}$

Figure 7.9 *Forward and backward diffusion rates*

The diffusion rates in the forward x and backward $-x$ directions are, then, given by

$$\left(\frac{dn}{dt}\right)_x = \eta_x N_{BX} = \frac{1}{6}\eta c a A$$

$$\left(\frac{dn}{dt}\right)_{-x} = \eta_{-x} N_{BY} = \frac{1}{6}\eta\left(c + a\frac{dc}{dx}\right)aA,$$

and the overall diffusion rate is

$$\frac{dn}{dt} = \left(\frac{dn}{dt}\right)_x - \left(\frac{dn}{dt}\right)_{-x} = \frac{1}{6}\eta c a A - \frac{1}{6}\eta\left(c + a\frac{dc}{dx}\right)aA = -\frac{1}{6}\eta a^2 A\frac{dc}{dx}.$$

Comparing with Fick's first law shows that the diffusion coefficient D is given by

$$D = \frac{1}{6}\eta a^2$$

or, more generally,

$$D = \alpha\eta a^2.$$

In other words, the diffusion coefficient D is proportional to a geometric factor α ($\alpha = \frac{1}{6}$ for a simple cubic lattice or $\alpha = \frac{1}{4}$ for an fcc lattice), the square of the jump distance a, and the mean jump frequency η.

7   The Kirkendall effect and Darken's equations

The diffusion rate in a substitutional solid solution depends on the rate at which the atoms jump into vacancies. In general, different kinds of atoms jump into vacancies at different rates, and they migrate, therefore, with different diffusion rates. We define an *intrinsic diffusion coefficient* D_i for each component, given by Fick's first law:

$$\frac{dn_i}{dt} = -D_i A \frac{dc_i}{dx},$$

where i represents the ith component.

Consider the two bars of materials A and B welded together, as shown in Figure 7.6, and then heated to promote interdiffusion of A and B. There are two intrinsic diffusion coefficients D_A and D_B:

$$c\frac{dn_A}{dt} = -D_A A \frac{dc_A}{dx}$$
$$\frac{dn_B}{dt} = -D_B A \frac{dc_B}{dx}.$$

Assume that A atoms jump into vacancies faster than B atoms, so the diffusion rate of A is faster than B, $D_A > D_B$. After some time, the number of A atoms that have diffused into B will be greater than the number of B atoms that have diffused into A. There is an overall mass flow towards B, and a corresponding overall flow of vacancies towards A. This leads to four effects, shown in Figure 7.10:

1. The boundary between the bars migrates towards A. This is called the *Kirkendall effect*. It can readily be demonstrated experimentally by inserting inert wire markers at the boundary between the two bars when they are welded together. After heating to allow interdiffusion, the wire markers are found to have moved relative to the bar ends because of the overall mass flow associated with the different A and B diffusion rates. The wire markers move closer to the A end and further from the B end. Effectively, the excess mass flow of A atoms towards B and past the inert markers pushes them back towards the A end.

2. Porosity develops within A as a result of coalescence of the excess vacancies that have moved into A to counteract the excess mass flow in the opposite direction.

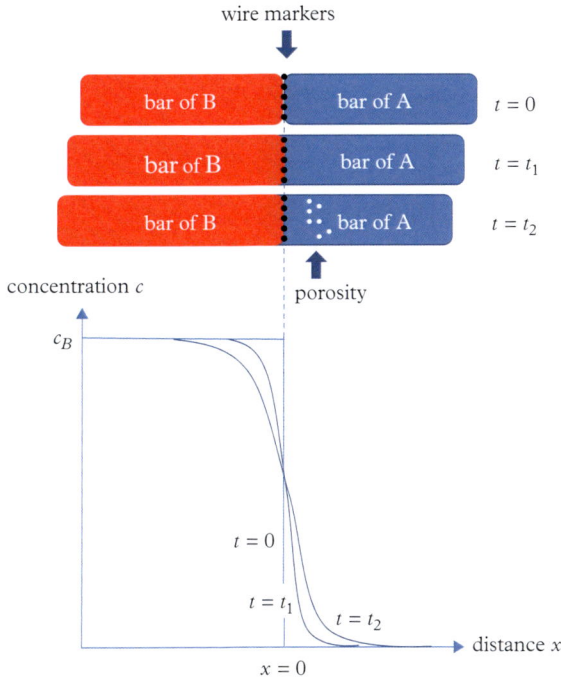

Figure 7.10 *The Kirkendall effect*

3. The overall interdiffusion process is described by *Darken's equations.* Darken's first equation gives the speed of the inert markers v:

$$v = (D_A - D_B) \frac{\partial x_A}{\partial x} = (D_B - D_A) \frac{\partial x_B}{\partial x}.$$

Darken's second law defines the *interdiffusion coefficient* $\tilde{D}$:

$$\tilde{D} = x_B D_A + x_A D_B,$$

where $x_A = n_A / (n_A + n_B)$ and $x_B = n_B / (n_A + n_B)$ are atom fractions of A and B.

4. Analytic solutions to Fick's second law, the diffusion equation, such as discussed previously, are clearly not valid in general because (a) the x origin moves at a velocity v with the inert markers at the initial boundary between A and B; (b) the diffusion rates of A and B are not equal and opposite, $dn_A/dt \neq -dn_B/dt$ and $D_A \neq D_B$; and (c) the interdiffusion coefficient is not independent of concentration $\tilde{D} = f(c)$.

8   Self-diffusion

In a two-component substitutional A–B solid solution, the A and B atoms diffuse by the vacancy mechanism, with random jumping of A and B atoms into adjacent vacancies, leading to the vacancies migrating through the crystal, and the A and B atoms interdiffusing through the material. The same mechanism also takes place in a one-component material, with random jumping of atoms into adjacent vacancies, again leading to the vacancies migrating through the crystal. This process is called *self-diffusion*. There is no overall interdiffusion effect because there is only one component, but the atoms and the vacancies are in a state of continual random migration through the crystal. The *self-diffusion coefficient D* can be measured experimentally by monitoring the diffusion of a thin film of a radioactive isotope, and it is then called the *isotope diffusion coefficient D**.

The *self-diffusion coefficient D* is found to vary with temperature T according to an Arrhenius equation:

$$D = D_o \exp - \left(\frac{Q}{RT} \right),$$

where D_o is the frequency factor, Q is the activation energy for diffusion and R is the gas constant. An Arrhenius temperature dependence can be explained by considering separately the processes of formation and migration of vacancies, and then the relation between diffusion coefficient and jump frequency.

Vacancy formation

The excess free energy dG_V associated with a number n_V of vacancies in a material at temperature T is a combination of the excess energy of forming vacancies Q_f and the excess entropy dS_V caused by the vacancies:

$$dG_V = Q_f - TdS_V = Q_f - RT \ln \left(\frac{n_V}{N} \right),$$

where Boltzmann's equation has been used for the configurational entropy w of distributing n_V vacancies and N atoms on $(n_V + N)$ lattice sites, i.e.

$$dS_V = k \ln w = k \ln \left(\frac{n_V! N!}{(n_V + N)!} \right) \approx R \ln \left(\frac{n_V}{N} \right),$$

and N is Avogadro's number. The equilibrium number of vacancies is when the free energy is minimised $dG_V = 0$, so that

$$n_V = N \exp - \left(\frac{Q_f}{RT} \right).$$

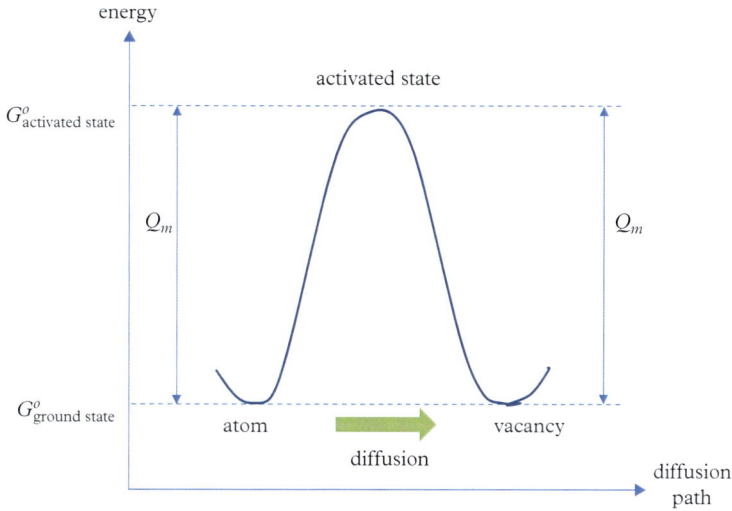

Figure 7.11 *Activation energy for vacancy migration*

Vacancy migration

The rate at which atoms jump into a vacancy η_V depends on the activation energy for vacancy migration Q_m, i.e. the activation energy for an adjacent atom to jump into a vacancy, as shown in Figure 7.11,

$$\eta_V = \eta_{VO} \exp - \left(\frac{Q_m}{RT} \right),$$

where η_{VO} is the frequency factor for vacancy migration.

Jump frequency

The overall jump frequency η is the number of vacancies n_V multiplied by the rate at which they migrate η_V,

$$\eta = n_V \eta_V = N \exp - \left(\frac{Q_f}{RT} \right) . \eta_{VO} \exp - \left(\frac{Q_m}{RT} \right) = N \eta_{VO} \exp - \left(\frac{Q_f + Q_m}{RT} \right),$$

and the self-diffusion coefficient is obtained from the jump frequency:

$$D = \alpha \eta a^2 = \alpha a^2 N \eta_{VO} \exp - \left(\frac{Q_f + Q_m}{RT} \right)$$

$$\therefore D = D_o \exp - \left(\frac{Q}{RT} \right),$$

where α is a geometric factor, a is the jump distance, and the frequency factor D_o and the activation energy for diffusion Q are given by

$$D_o = \alpha a^2 N \eta \nu_O$$
$$Q = Q_f + Q_m.$$

9   Fourier's and Ohm's laws

Fick's first and second laws apply to *mass transport*, i.e. diffusion. Analogous laws apply to the transport of heat and electricity. The equivalent laws to Fick's first law are, respectively, *Fourier's law* and *Ohm's law*.

For thermal conduction, the rate of heat flow dq/dt in a bar of material is found to be proportional to the cross sectional area A of the bar and the temperature gradient along its length dT/dx:

$$\frac{dq}{dt} = -kA\frac{dT}{dx},$$

where the constant of proportionality is called the *thermal conductivity k*. This equation is called *Fourier's law*.

For electrical conduction, *Ohm's law* is usually written for a *resistor* component in an electrical circuit as current I proportional to voltage V:

$$V = IR,$$

where the constant of proportionality is the *resistance R*. This can be re-written for a bar of material as the rate of flow of electrical charge dq/dt proportional to the cross sectional area A of the bar and the electric field gradient along its length $d\mathcal{E}/dx$,

$$\frac{dq}{dt} = -\sigma A\frac{d\mathcal{E}}{dx},$$

because the current is the rate of flow of charge $I = dq/dt$, the voltage is the electric field gradient $V = d\mathcal{E}/dx$, and the electrical conductivity σ and electrical resistivity ρ of a material are given by

$$\sigma = \frac{1}{\rho} = \frac{1}{AR}.$$

Clearly, Fick's first law of diffusion, Fourier's law of thermal conduction and Ohm's law of electrical conduction all have a similar form. They are all manifestations of movement by a random walk, of atoms, phonons or electrons in the cases, respectively, of diffusion, heat flow and electrical flow.

....

10 REFERENCES

1. Edward Shapiro. 'Adolf Fick: forgotten genius of cardiology'. *The American Journal of Cardiology* **30** (1972), 662.
2. Hugh Troidl, M. F. McKneally, David S. Mulder, A. S. Wechsler, Bucknam McPeek, and W. O. Spitzer. *Surgical Research: Basic Principles and Clinical Practice* (Springer Science and Business, Berlin, 2012), 28.
3. Adolf Fick. 'Tractatur de errore optico quodam asymetria bulbi oculi effecto' ['Treatment of some visual errors caused by eyeball asymmetry']. Dissertatio inauguralis, Marburg University, 1851.
4. Johann Gustav Droysen. 'Speech to the Frankfurt Assembly'. 1848. http://www.historyman.co.uk/unification/Droysen.html (accessed 5 February 2020).
5. 'The situation of Germany'. *New York Times*, 1 July 1866.
6. Otto von Bismarck. 'Blood and Iron Speech'. 30 September 1862. https://en.wikipedia.org/wiki/Blood_and_Iron_(speech) (accessed 5 February 2020).
7. Shapiro, n 1, 663.
8. Charles F. Code, A. E. Renolds, and W. F. Hamilton. *Handbook of Physiology: A Critical, Comprehensive Presentation of Physiological Knowledge and Concepts* (Arkose Press, 2015).
9. Leroy D. Vandam and John A. Fox. 'Adolf Fick (1829–1901) physiologist: a heritage for anesthesiology and critical care medicine'. *Anesthesiology* **88** (1998), 517.
10. Jean-Baptiste Joseph Fourier. *Théorie Analytique de la Chaleur* [*The Analytical Theory of Heat*] (Firmin Didot, Paris, 1822; re-issued Cambridge University Press, Cambridge, 2009).
11. Vandam and Fox, n 9, 516.
12. Adolf Fick. 'Ueber diffusion'. *Annalen der Physik* **170** (1855), 59–86.
13. Adolf Fick. 'On liquid diffusion'. *Philosophical Magazine* **10** (1855), 30–39.
14. Shapiro, n 1, 665.

....

11 BIBLIOGRAPHY

'Adolf Fick: Forgotten Genius of Cardiology'. Edward Shapiro. *Journal of the American College of Cardiology* **30** (1972), 662–665.
'Adolf Fick (1829–1901), physiologist: a heritage for anesthesiology and critical care medicine'. Leroy D. Vandam and John A. Fox. *Anesthesiology* **88** (1988), 514–518.
Diffusion in Solids. Helmut Mehrer (Springer-Verlag, Berlin, 2010).
Diffusion in Solids (*2nd edition*). Paul G. Shewmon (Springer International, Cham, 2016).
Physical Metallurgy Principles (*4th edition*). Reza Abbaschian, Lara Abbaschian, and Robert E. Reed-Hill (Cengage Learning, Boston, 2009).

8

The Scheil Equation

Solidification

---

1   Solidification and casting

Many materials are made initially in the liquid phase, followed by *solidification* to form either a shaped final product or an intermediate product such as an *ingot* or bar. Industrial solidification processes are called *casting*, and the resulting shaped products or intermediate ingots are called *castings*. For instance, car engines and gearboxes are made out of *cast iron* or a *cast aluminium alloy*; floor beams and wall panels are often made out of *pre-cast concrete*; kitchen sinks and toilet bowls are often made by *slip casting* alumino-silicate ceramics; and components as diverse as car bodies, railway lines, fridge and washing-machine casings, and sewing needles are made from *cast steel* ingots, which are then forged, rolled, pressed and/or drawn to shape. In any cast material, it is essential to control the composition and microstructure carefully during the solidification process to ensure that the final product has the desired material properties.

2   Partitioning

The solidification of a material is often dominated by *partitioning*, which is the redistribution of solute atoms or molecules during the solidification process. Partitioning is obviously important when we have a two-component A–B material. The solute B atoms or molecules have been added to improve the properties of the material, and we clearly want to control their distribution to make sure we have a homogeneous material with the same composition and, therefore, the same properties everywhere. The liquid phase is a good way of mixing two components initially to manufacture a homogeneous material, but partitioning makes it difficult to maintain homogeneity during solidification. Partitioning is also important in single-component materials, because even very high-purity materials contain impurities, the distribution of which must be controlled as well as possible during solidification to make sure they don't build up and affect the properties adversely.

Figure 8.1 shows an A–B phase diagram in the region of the melting point of A. When B is added to A, melting and solidification take place over a range of temperatures,

The Equations of Materials. Brian Cantor. Oxford University Press (2020). © Brian Cantor.
DOI: 10.1093/oso/9780198851875.001.0001

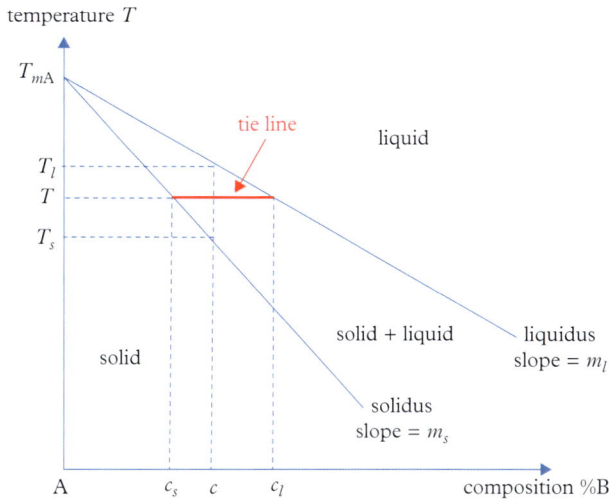

Figure 8.1 *Liquidus, solidus and partition coefficient in an A–B material*

rather than at a single melting point. A two-phase liquid-plus-solid region separates the liquid solution at high temperature and the solid solution at low temperature. At each composition c, the boundary between liquid and liquid plus solid is called the *liquidus temperature* T_l, and the boundary between solid and liquid plus solid is called the *solidus temperature* T_s, as shown in Figure 8.1. The lines showing the variation of T_l and T_s with c are called the *liquidus* and *solidus* respectively.

During solidification, an A–B material must be cooled through the liquid-plus-solid region. At each temperature T, the equilibrium composition of the solid and liquid c_s and c_l are different, and are given by the end compositions of the *tie line*, as shown in Figure 8.1. During cooling, the solute B atoms or molecules try to maintain equilibrium, with different liquid and solid compositions, i.e. they *partition* between the liquid and solid, leading to a characteristic redistribution of solute during solidifiacation.

The ratio of the solid and liquid compositions c_s and c_l is called the *partition coefficient k*:

$$k = \frac{c_s}{c_l},$$

with the value of k usually less than one because the melting point is usually depressed by adding solute. The liquidus and solidus are often approximately straight lines, as shown in Figure 8.1, with liquidus and solidus temperatures proportional to composition c:

$$T_l = T_{mA} - m_l c$$
$$T_s = T_{mA} - m_s c,$$

where T_{mA} is the melting point of pure A, m_l and $m_s > m_l$ are the liquidus and solidus slopes respectively, and m_l, m_s and k are constants, independent of composition.

3   The Scheil equation

Consider a bar of homogeneous A–B material of composition c_0, held in the liquid phase in a furnace at high temperature, and then gradually withdrawn, as shown in Figure 8.2, causing solidification to begin at one end and progress through the material until it is fully solid. The Scheil equation says that the final solid composition c varies with distance along the bar, i.e. with the fraction solidified f_s, according to

$$c = kc_0(1 - f_s)^{k-1},$$

as also shown in Figure 8.2.

The extent of partitioning and its final effects depend upon how much time is available for solute B atoms or molecules to redistribute to try to maintain equilibrium, via diffusion in the solid, and via diffusion and fluid flow in the liquid. This, in turn, depends upon the speed at which the bar is withdrawn from the furnace, which determines the speed at which the solid–liquid interface moves from one end of the bar to the other, i.e. the speed of the solidification process R.

In general, solidification takes place far too fast to maintain equilibrium liquid and solid compositions throughout the material, which would require extensive diffusion

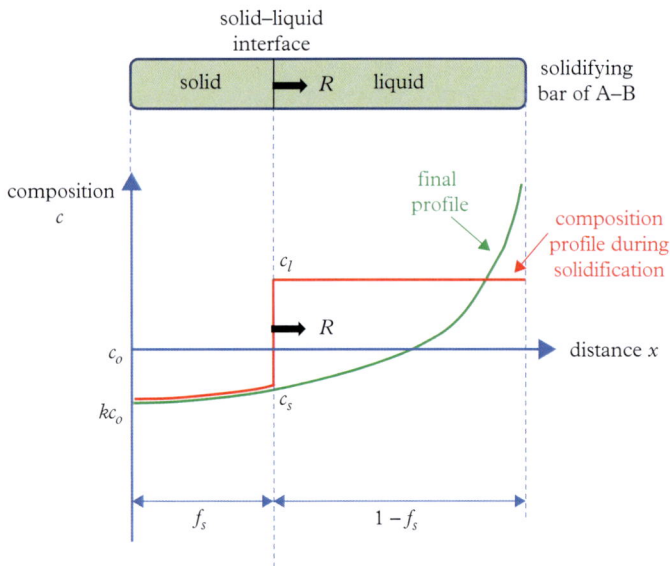

Figure 8.2 *The Scheil equation for the composition profile of a solidifying bar*

of solute atoms or molecules along the whole length of the bar. The Scheil equation corresponds to solidification fast enough to prevent any mixing in the solid, but slow enough to allow complete mixing in the liquid. This is not impossible, because mixing in the solid can only take place by slow solute diffusion, whereas mixing in the liquid can take place by much faster liquid diffusion and fluid flow.

When the temperature reaches the liquidus T_l, the initial solid to form has a composition kc_o, as shown on the Scheil composition profile in Figure 8.2. This solid composition is fixed, because there is no time for diffusive mixing in the solid. It is less than the initial homogeneous composition of the liquid c_o, so some solute atoms or molecules are rejected at the solid–liquid interface, which diffuse and flow throughout the liquid, enriching its composition above c_o. The next solid to form, therefore, has a composition a bit higher than kc_o, but still well below c_o, rejecting more solute at the solid–liquid interface and further enriching the composition of the liquid. This partitioning process continues, producing a gradually increasing final composition profile, as shown in Figure 8.2.

We can derive the Scheil equation by applying the law of conservation of mass to the B solute atoms or molecules. At a general point during the solidification process, the liquid and solid compositions at the solid–liquid interface are c_s and c_l respectively, as shown in Figure 8.2. Let the solid–liquid interface advance by a distance dx, corresponding to an increase in fraction solidified df_s. This produces an enrichment of B in the liquid dc_l. The depletion of B in the solid balances the enrichment of B in the liquid, i.e.

$$(c_l - c_s)\, df_s = (1 - f_s)\, dc_l.$$

Substituting $c_l = c_s/k$ and $dc_l = dc_s/k$,

$$\frac{c_s}{k}(1-k)\, df_s = (1-f_s)\,\frac{dc_s}{k}$$

and integrating,

$$\int_0^{f_s} \frac{df_s}{(1-f_s)} = \int_{kc_o}^{c} \frac{dc_s}{c_s\,(1-k)}$$

$$\therefore -\ln\,(1-f_s) = \frac{1}{(1-k)}\,[\ln\,c - \ln\,kc_o]$$

$$\therefore c = kc_o(1-f_s)^{k-1},$$

which is the Scheil equation. The resulting composition profile in the initially homogeneous bar, as shown in Figure 8.2, is depleted in solute at the beginning of solidification and enriched in solute towards the end of solidification. Effectively, the solute atoms or molecules have been pushed ahead of the solidifying interface until they have nowhere to go at the end of the bar.

4 Modified Scheil conditions

The redistribution of solute described by the Scheil equation and shown in Figure 8.2 can, to some extent, be ameliorated by increasing the solidification speed to prevent full mixing in the liquid. With only partial mixing in the liquid, solute atoms or molecules rejected at the solid–liquid interface cannot diffuse or flow throughout the liquid and are restricted to a region close to the interface, as shown in Figure 8.3. This still leads to a zone of solute depletion at the beginning of solidification, and a zone of solute enrichment near the end of solidification, but the majority of the bar has a final composition equal to the initial homogeneous composition c_o, as shown in Figure 8.3.

The initial solid to form again has a composition kc_o, which is less than the initial liquid composition c_o, and solute atoms or molecules are rejected at the interface. There is again, therefore, an initial zone of solute depletion. Unlike Scheil conditions, however, solute atoms or molecules cannot diffuse throughout the liquid, and they build up and increase the liquid composition just ahead of the interface. A steady state is reached, with a stable region of rejected solute ahead of the interface, where the interface liquid and solid compositions are given by $c_l = c_o/kc_l$, and $c_s = c_o$, as shown in Figure 8.3. Rejected solute continues to be pushed ahead of the solidifying interface until it solidifies and forms a solute-enriched zone at the end of the bar.

Steady-state solidification is reached when solute diffusion away from the interface is balanced by the interface advance, i.e. when the size of the solute-enriched region ahead of the interface Δ_i is given approximately by

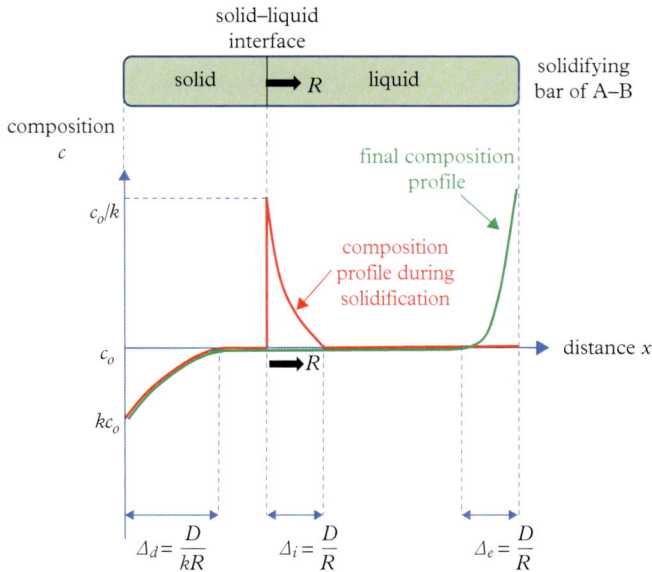

Figure 8.3 *Composition profile of a solidifying bar with partial mixing in the liquid*

$$\Delta_i \sim \sqrt{Dt} = \sqrt{\frac{D\Delta_i}{R}}$$

$$\therefore \Delta_i \sim \frac{D}{R},$$

where D is the solute diffusion coefficient in the liquid and t is the time for solute or the interface to move a distance Δ_i. The size of the final solute-enriched zone is also, therefore,

$$\Delta_e \sim \frac{D}{R}.$$

Solute is conserved, so the size of the initial solute-depleted zone Δ_d is given approximately by

$$\Delta_d (c_o - kc_o) \sim \Delta_i \left(\frac{c_o}{k} - c_o \right)$$

$$\therefore \Delta_d \sim \frac{D}{kR}.$$

Partitioning during solidification is one of the principal problems with casting materials directly into a desired shape. Solutes added to improve the material properties are inhomogeneous, leading to variable properties from place to place in the as-cast material, and impurities are concentrated into particular places, adversely affecting the properties in these regions. In industrial practice, casting conditions are chosen to reduce these effects, ideally by using materials with $k \sim 1$, and cooling rapidly to prevent excessive mixing in the liquid. Fully sound, high-quality material is often only achieved by a post-solidification combination of deformation processing by forging, rolling or drawing, and heat treatment.

5   Erich Scheil

Erich Scheil was born on 8 December 1897. There is very little biographical information available about him. He worked at the Technische Hochschüle Stuttgart (Stuttgart Technology University of Applied Sciences). He was appointed as a Scientific Member of the Max Planck Institut für Metallforschung (the Max Planck Institute for Metals Research, now the Max Planck Institute for Intelligent Systems) in Stuttgart in 1937. He published his famous paper deriving the (now-called) Scheil equation for partitioning of solute during solidification in *Zeitschrift für Metallkunde* in 1942.[1] The Scheil equation is sometimes called the *Gulliver-Scheil equation*, because of a previous, more convoluted derivation by the English metallurgist G. H. Gulliver, who worked at the University of Edinburgh.[2] Erich Scheil died, aged 64, on 2 April 1962.

6   The driving force for solidification

Figure 8.4 shows the liquid and solid free energies G_l and G_s respectively versus temperature T for a pure material:

$$G_l = H_l - TS_l$$
$$G_s = H_s - TS_s,$$

where H_l, H_s, S_l and S_s are the liquid and solid enthalpies and entropies respectively. In general, $H_l > H_s$ and $S_l > S_s$, because atoms and molecules have higher energy and are more disordered in the liquid state. The liquid and solid free energies are equal at the *melting point* T_m:

$$G_l = G_s$$
$$\therefore H_l - T_m S_l = H_s - T_m S_s$$
$$\therefore L = T_m \Delta S; \text{ and } \Delta S = \frac{L}{T_m},$$

where $L = H_l - H_s$ is the *latent heat of solidification* and $\Delta S = S_l - S_s$ is the *entropy change on solidification*.

Solidification can only take place at a temperature T below T_m, i.e. at an *undercooling* $\Delta T = T_m - T$, with a *driving force for solidification* ΔG given by

$$\Delta G = G_l - G_s = (H_l - TS_l) - (H_s - TS_s)$$
$$= L - T\Delta S = L - T\frac{L}{T_m} = \frac{L\Delta T}{T_m}.$$

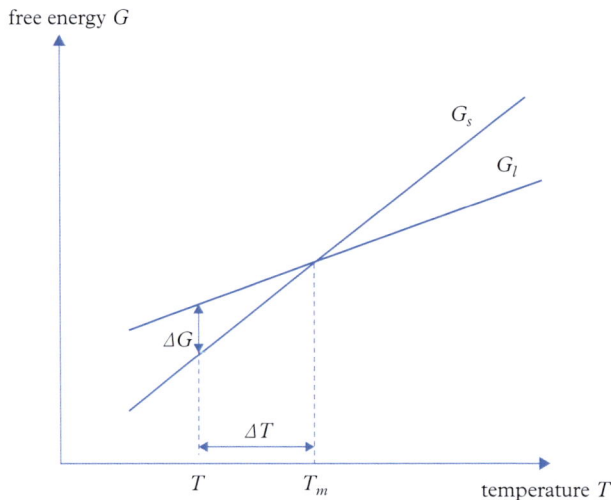

Figure 8.4 *Driving force ΔG for solidification at an undercooling ΔT below the melting point*

In other words, the driving force for solidification increases linearly with increasing undercooling below the melting point.

Solidification usually takes place in two distinct stages, called *nucleation* and *growth*. These two stages are discussed separately before they are combined to look at the overall process.

7   Nucleation

The nucleation stage of solidification is the initial formation of the first small particle or *nucleus* of solid within the liquid.

Consider cooling a mass of liquid to a temperature T below the melting point T_m. Let a small solid particle of radius r form within the bulk of the liquid, as shown in Figure 8.5. The change in free energy ΔG_r is given by the sum of two terms: a decrease in free energy caused by replacing a volume of liquid by solid, and an increase in free energy caused by creating an area of solid–liquid interface:

$$\Delta G_r = -\frac{4}{3}\pi r^3 \Delta G_V + 4\pi r^2 \gamma,$$

where $\Delta G_V = \Delta G/v_m$ is the driving force for solidification per unit volume, v_m is the molar volume, and γ is the solid–liquid surface free energy. ΔG_r varies with r, as shown in Figure 8.6. For small values of r, the surface area of the particle is large compared to its volume, and ΔG_r is positive. For large values of r, the surface area of the particle is small compared with its volume, and ΔG_r is negative. As a small particle forms, it must initially increase its free energy because of the need to create the solid–liquid interface. As it grows, it reaches a maximum increase in free energy ΔG^*, called the *work of nucleation*, at a radius r^*, called the *critical nucleus size*, and then its energy decreases.

The work of nucleation is, effectively, an energy barrier to the nucleation process, a barrier to the formation of an initial solid particle. An initial particle must somehow attain an excess free energy ΔG^* and reach the critical nucleus size r^* before it can begin to grow with a continuous reduction in free energy. Particles with $r < r^*$ are called *sub-critical nuclei*, particles with $r = r^*$ are called *critical nuclei*, and particles with $r > r^*$ are

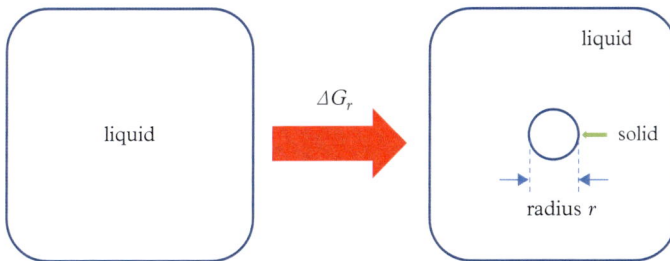

Figure 8.5 *Nucleation of a particle of solid in a liquid*

free energy change ΔG_r

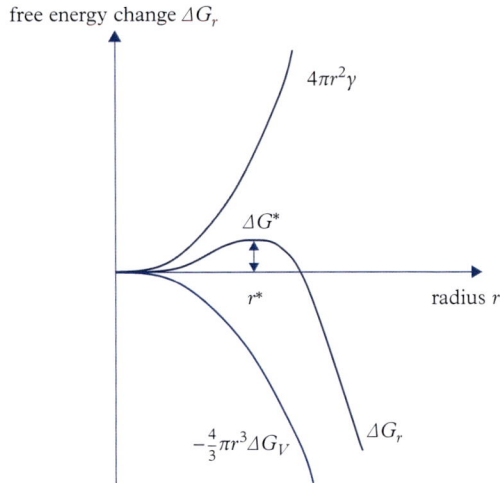

Figure 8.6 *Work of nucleation ΔG^* and critical radius for nucleation r^**

called *super-critical nuclei*, or just nuclei. The energy barrier and the critical radius are obtained by differentiating ΔG_r with respect to r and putting $d\Delta G_r/dr = 0$:

$$\Delta G^* = \frac{16\pi}{3}\frac{\gamma^3}{\Delta G_V^2}$$

$$r^* = \frac{2\gamma}{\Delta G_V}.$$

Both ΔG^* and r^* increase sharply with increasing surface energy γ, and decrease sharply with increasing driving force ΔG_V, i.e. with increasing undercooling ΔT below the melting point.

How can a nucleus form? Atoms and molecules in the liquid are moving and vibrating at high temperatures, and, from time to time, groups of atoms fluctuate into solid-like configurations or *clusters*, effectively forming a population of sub-critical nuclei. For very small particles $r < r^*$, as shown in Figure 8.6, a sub-critical nucleus is more likely to decay back to the liquid than to continue to grow in size. However, when a sufficiently large fluctuation allows a particle to reach the critical radius r^*, the critical nucleus can grow with a continuous reduction in free energy.

The *rate of nucleation i* is usually defined as the rate at which critical nuclei grow into super-critical nuclei, i.e. the number of critical nuclei N^* multiplied by the frequency ν with which an additional adjacent atom vibrates, and can, therefore, add on to a critical nucleus at the solid–liquid interface:

$$i = N^*\nu = N_o\exp\left(-\frac{\Delta G^*}{kT}\right)\nu_o\exp\left(-\frac{Q}{kT}\right)$$

$$= i_o\exp\left(-\frac{\Delta G^* + Q}{kT}\right),$$

where $N^* = N_o \exp(-\Delta G^*/kT)$ is given by a Boltzmann expression for the probability of a cluster with excess energy ΔG^*, N_o is Avogadro's number, v_o and Q are the frequency factor and activation energy for diffusion, and $i_o = N_o v_o$ is the frequency factor for nucleation.

This process is called *homogeneous nucleation*, i.e. nucleation of the sold phase within the bulk of the liquid without any other influence. Because ΔG^* is proportional to ΔG_V^{-2} and, therefore, to ΔT^{-2}, the rate of nucleation i is very low, effectively zero, for a wide range of temperatures below the melting point, and then increases dramatically when the undercooling reaches a critical value of approximately $\Delta T \sim 0.2 T_m$.

In fact, nucleation takes place much more readily, and such large undercoolings are almost never seen. This is because nucleation is usually aided by impurities floating in the liquid or by the container wall, a process called *heterogeneous nucleation*. Initial solid particles or clusters form in contact with impurity particles or the container wall, reducing the amount of new solid–liquid interface that needs to be created, and allowing the particles to reach the critical radius $r^\star$ with a reduced work of nucleation:

$$\dot{i}_{\text{hetero}} = i_o \exp\left(-\frac{\Delta G^*_{\text{hetero}} + Q}{kT}\right),$$

$$\Delta G^*_{\text{hetero}} < \Delta G^*_{\text{homo}}; \ \dot{i}_{\text{hetero}} > \dot{i}_{\text{homo}},$$

as shown in Figure 8.7. This means that, in practice, nucleation usually takes place close to the melting point at low undercoolings on the container wall and on floating impurity particles in the body of the liquid.

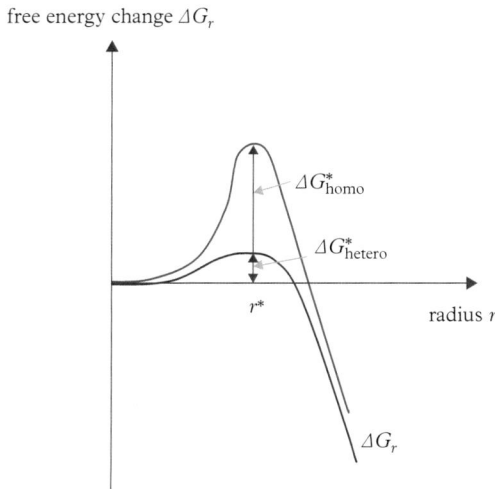

Figure 8.7 *Comparison of the work of homogeneous and heterogeneous nucleation*

8   Growth

After a solid particle has nucleated, it grows by the addition of atoms or molecules from the liquid at the interface. There are two different kinds of atomic or molecular attachment mechanism, corresponding to two different kinds of solid–liquid interface structure, and two different kinds of thermal condition controlling and influencing the attachment process. We examine these one at a time.

Non-faceted solid–liquid interface

For some materials, atoms or molecules can add on very easily to the growing solid. This is the case when the material has a simple atomic or molecular structure, with non-directional bonding, e.g. most metals and van der Waals solids. The speed of the interface R is then proportional to the driving force ΔG and, therefore, to the *interface undercooling* ΔT_i:

$$R = \mu \Delta T_i,$$

where the constant of proportionality μ is called the *solid–liquid interface mobility*. The mobility is high, because addition of atoms or molecules is easy. The interface moves rapidly and growth is fast even at small interface undercoolings. There are no significant orientation effects, and the interface is *non-faceted*, i.e. atomically diffuse and non-crystallographic.

Faceted solid–liquid interface

For other materials, atoms or molecules cannot add on so easily to the growing solid. This is the case when the material has a complex molecular structure with strong directional bonding, e.g. most ceramics and polymers. The speed of the interface is again proportional to the driving force and the interface undercooling, but the mobility is low, because addition of atoms or molecules is not easy. The interface moves slowly and growth is slow even at high interface undercoolings. There are strong orientation effects, and the interface is *faceted*, i.e. atomically sharp and crystallographic.

Positive thermal gradient in the liquid

Latent heat of solidification is liberated at the solid–liquid interface as atoms or molecules are attached and growth takes place. The liberated latent heat must be removed as growth takes place to maintain the interface undercooling and the speed of solidification. Otherwise, the interface temperature rises and continued solidification is stifled. In many cases, latent heat is removed through the existing solid and there is a positive temperature gradient in the liquid ahead of the interface, as shown in Figure 8.8. If a perturbation grows ahead of the interface, it grows into hotter liquid, the local solidification speed

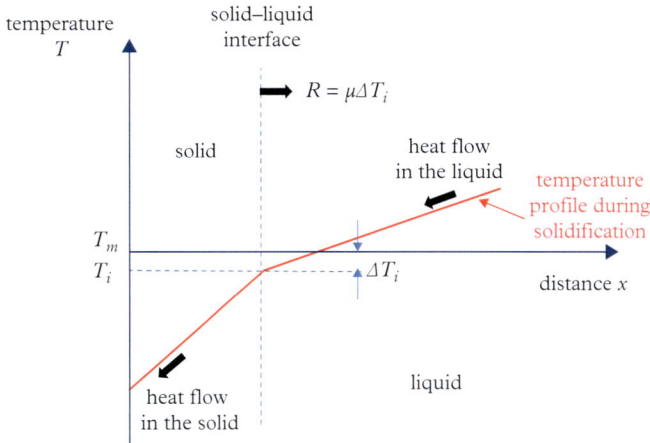

Figure 8.8 *Solidification with a positive temperature gradient in the liquid, leading to planar or faceted growth*

falls, and the perturbation is caught up by the rest of the interface. In other words, the solid–liquid interface is stable, and is constrained to remain approximately parallel to the melting point isotherm and normal to the heat flow direction. For a non-faceted material, this leads to *planar growth*, and for a faceted material, it leads to *faceted growth*.

Negative temperature gradient in the liquid

In many cases, the solid grows into an undercooled liquid, and the latent heat is also removed through the liquid, as shown in Figure 8.9. If a perturbation grows ahead of the interface, it grows into colder liquid, the local solidification speed rises, and the perturbation accelerates away ahead of the rest of the interface. This happens everywhere, all along the interface, which is *destabilised*, leading to an irregular solid–liquid interface and unstable, non-planar growth. The same effect also happens through *constitutional undercooling*, when the increased undercooling and destabilisation is caused by a build-up of rejected solute atoms or molecules ahead of the interface. For a non-faceted material, destabilisation is non-crystallographic and follows the isotherms, with a characteristic *dendritic* or tree-like morphology. For faceted materials, destabilisation is crystallographic, with a faceted *needle-like* morphology.

Planar, faceted, dendritic and needle-like growth

Planar, faceted, dendritic and needle-like solid–liquid interface growth forms are shown schematically in Figure 8.10.

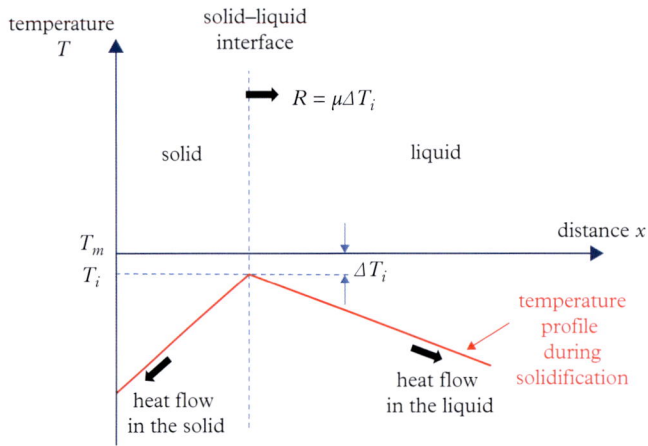

Figure 8.9 *Solidification with a negative temperature gradient in the liquid, leading to dendritic or needle-like growth*

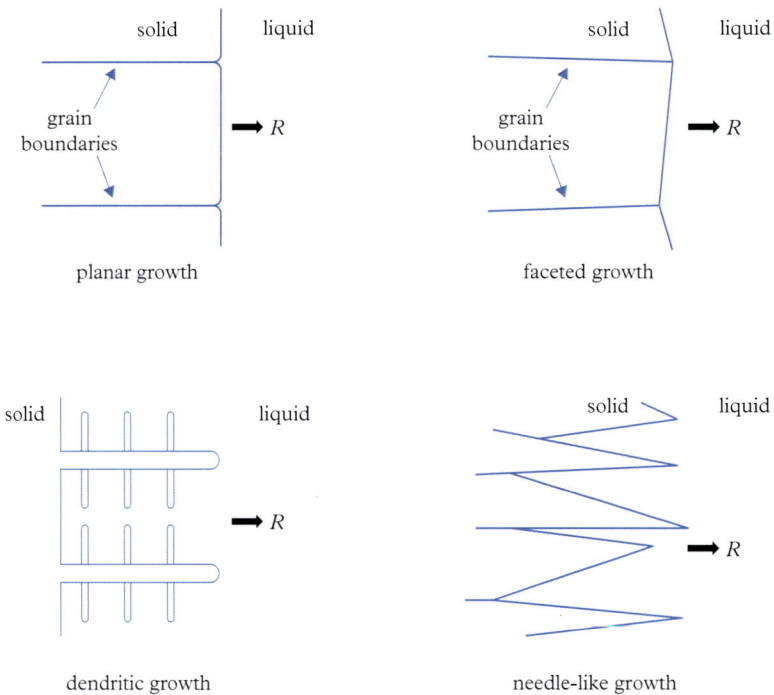

Figure 8.10 *Planar, faceted, dendritic and needle-like solid–liquid interface growth modes*

9 Eutectic solidification

When a material with a eutectic composition c_e is cooled from the liquid, two solids are formed at the same time. The solidification reaction is

$$\mathrm{L} \rightarrow \alpha + \beta,$$

and the two solids α and β have different structures and compositions, either side of the liquid eutectic composition, i.e. $c_\alpha < c_e < c_\beta$ as given by the *eutectic tie line*, as shown in Figure 8.11. The solids α and β can solidify in a side-by-side fashion, as shown in Figure 8.12. Solute B atoms or molecules migrate laterally away from the α phase and towards the β phase in the liquid just ahead of the solid–liquid interface, as shown in Figure 8.12. At the same time, solvent A atoms or molecules migrate laterally away from the β phase and towards the α phase in the liquid just ahead of the solid–liquid interface, as also shown in Figure 8.12. This leads to a *coupled growth* solidification process, leading to fine-scale distributions of the α and β phases, usually in a *lamellar (plate-like) eutectic* structure or a *rod-like eutectic* structure.

For materials with an *off-eutectic* composition c, i.e. where $c \neq c_e$, initial solidification of *primary α* or *primary β* is followed by subsequent solidification of the fine-scale $\alpha + \beta$ eutectic:

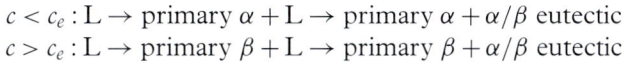

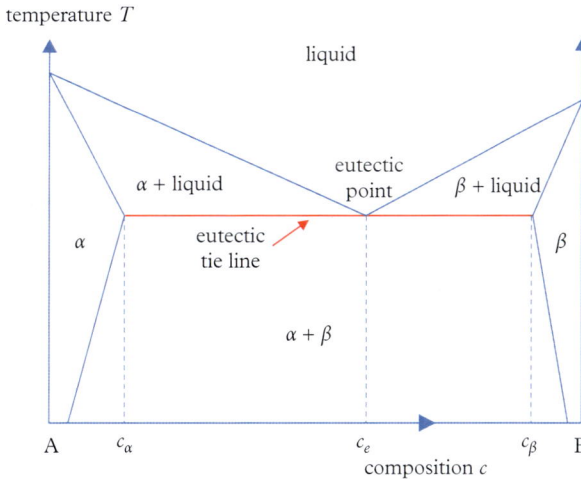

$$c < c_e : \mathrm{L} \rightarrow \text{primary } \alpha + \mathrm{L} \rightarrow \text{primary } \alpha + \alpha/\beta \text{ eutectic}$$
$$c > c_e : \mathrm{L} \rightarrow \text{primary } \beta + \mathrm{L} \rightarrow \text{primary } \beta + \alpha/\beta \text{ eutectic}$$

Figure 8.11 *Two-phase eutectic solidification*

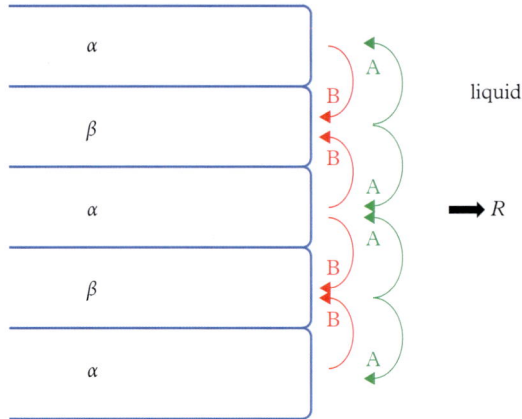

Figure 8.12 *Coupled growth during eutectic solidification*

10   Casting

Common forms of the solidification process include *directional solidification, ingot casting, continuous casting* and *shape casting*. We examine these one at a time.

Directional solidification

In *directional solidification*, a bar of liquid material is pulled slowly from a furnace so that a solid–liquid interface runs from one end of the bar to the other, as shown in Figures 8.2 and 8.3, gradually solidifying the whole bar. The temperature gradient in the liquid is positive, as shown in Figure 8.8, so non-faceted and faceted materials solidify with planar or faceted morphologies respectively, as shown in Figure 8.10. This type of solidification is used to grow single crystals, with a single crystal seed at one end of the bar and with very slow solidification rates to prevent the incorporation of defects, in particular to prevent the formation of grain boundaries. This is important in the manufacture of single-crystal semiconductors for integrated circuits and turbine blades for jet aeroengines.

Ingot casting

Many materials are initially cast as ingots before secondary processing by forging, rolling and drawing to make the final shape. The liquid is poured into a *mould* and the material solidifies slowly as heat is extracted slowly from the mould sides and base. The typical microstructure in a cast ingot is shown schematically in Figure 8.13. It consists of

(1) a *chill zone*: an initial solidification layer of very small grains, nucleated along the relatively cold mould walls;

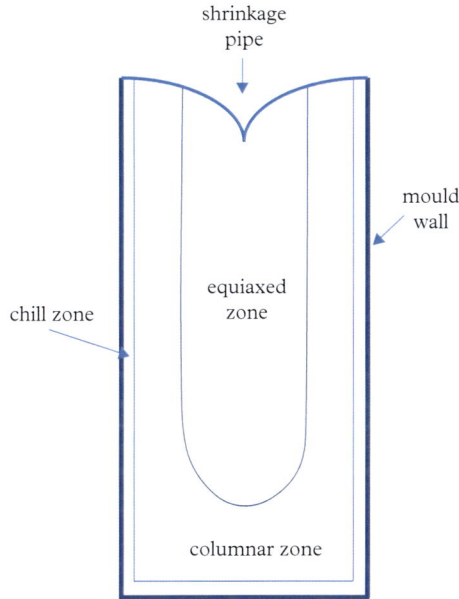

Figure 8.13 *Ingot microstructure*

(2) a *columnar zone*: growth of columnar grains into the middle of the mould, with a positive temperature gradient in the liquid, as shown in Figure 8.8, but with partitioning leading to constitutional supercooling, so the columnar grains usually solidify with a dendritic growth structure;

(3) an *equiaxed zone*: large equiaxed grains in the centre of the ingot, nucleated from impurities in the liquid, which again solidify with a dendritic growth structure;

(4) *macro-segregation*: large-scale differences in composition between the edges and the centre of the ingot caused by partitioning during solidification of the ingot;

(5) *micro-segregation*: fine-scale differences in composition between the dendritic spines and inter-dendritic regions caused by local partitioning during solidification;

(6) *pipe*: large-scale shrinkage at the top and centre of the ingot caused during the final stages of solidification as the last of the liquid is sucked into the inter-dendritic regions, because of the higher density of the solid compared to the liquid; and

(7) *micro-porosity*: caused by inter-dendritic pockets of liquid that become isolated, creating pores when they solidify and reduce their volume, again because of the higher density of the solid compared to the liquid.

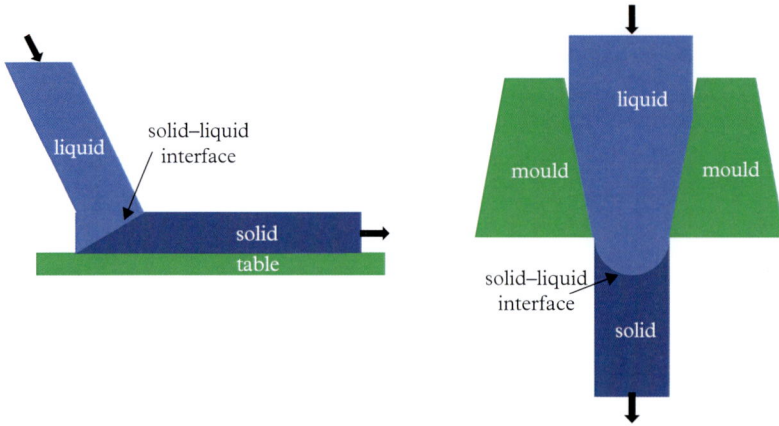

Figure 8.14 *Continuous casting*

The resulting complex variations in microstructure, composition and density in an ingot are usually removed during subsequent processing, in part by mechanical working (forging, rolling and drawing to create the final shape) and in part by thermal *heat treatments* to soften the material between different stages of working. Collectively, these secondary processes *homogenise* the material, creating a high-quality product that is sound and has good properties everywhere.

Continuous casting

In *continuous casting*, long ingots are manufactured by pouring liquid into an open mould or onto a moving table so the solidifying material can be withdrawn continuously, as shown in Figure 8.14. As with ingot casting, this is usually a precursor to subsequent working and homogenisation to create a final high-quality product.

Shape casting

In *shape casting*, complex shaped components are manufactured by pouring the liquid material into a complex shaped mould. There are many different ways of making the mould, out of different mould materials such as sand, ceramics and metals. *Homogenisation* is not possible via subsequent working, and heat treatment alone does little to remove variations in composition, microstructure and density. Mould material, shape and manufacture, and liquid filling and cooling conditions are all controlled as carefully as possible to try to minimise such variations and minimise associated defects and variations in properties. In general, this is difficult to achieve completely.

....

11 REFERENCES

1. E. Scheil. 'Bemurkungen zur schichtkristalbildung' ['Notes on the formation of crystals by layers']. *Zeitschrift für Metallkunde* **34** (1942), 70–72.
2. G. H. Gulliver. 'The quantitative effect of rapid cooling upon the constitution of binary alloys'. *Journal of the Institute of Metals* **9** (1913), 120–157.

....

12 BIBLIOGRAPHY

Castings. John Campbell (Butterworth-Heinemann, Oxford, 2003).
Principles of Metal Casting. Mahi Sahoo and Sam Sahu (McGraw-Hill, New York, 2014).
Principles of Solidification. Bruce Chalmers (Krieger, Malabar, 1977).
Principles of Solidification. Martin E. Glicksman (Springer, Berlin, 2011).
Solidification Processing. Merton C. Flemings (McGraw-Hill, New York, 1974).

9

The Avrami Equation

Phase Transformations

1   The Avrami equation

When a material is heated or cooled, we often find that its atomic or molecular structure changes. The material is said to undergo a *phase transformation*. Such phase transformations are frequently used to control the *microstructure* and, therefore, the properties of the material. For instance, when a mild carbon steel alloy with a composition of, say, Fe-0.1% C is cooled from 1000°C to room temperature, the face-centred cubic (fcc) *austenite* phase transforms to a mixture of body-centred cubic (bcc) *ferrite* and a carbide with a complex cubic crystal structure called *cementite* Fe_3C. The time, temperature, and heating and cooling rates during *heat treatment* determine the speed and extent of the transformation, which in turn determine the microstructure and properties of the steel.

Consider a general phase transformation, with one solid phase α transforming at a constant temperature, i.e. *isothermally*, into another solid phase β:

$$\alpha \rightarrow \beta.$$

The rate at which the transformation takes place is often described by the *Avrami equation*, which is also sometimes called the *Johnson-Mehl-Avrami equation*, the *JMA equation*, the *Johnson-Mehl-Avrami-Kolmogorov equation* or the *JMAK equation* (it was discovered independently, and at roughly the same time, by Avrami, by Mehl and Johnson, and by Kolmogorov). The volume fraction of the new phase, i.e. the *fraction transformed f*, is given by the Avrami equation as

$$f = 1 - \exp\left(-kt^n\right)$$
$$= 1 - \exp\left[-(t/\tau)^n\right],$$

where k is called the *rate constant*, $\tau = k^{-\frac{1}{n}}$ is called the *time constant* and n is called the *Avrami exponent*. The Avrami exponent is often a small integer or simple fraction in the range one to four.

Figure 9.1 shows a typical variation of fraction transformed with time according to the Avrami equation. It is a *sigmoidal* or *S-shaped* curve, i.e. the transformation accelerates

The Equations of Materials. Brian Cantor. Oxford University Press (2020). © Brian Cantor.
DOI: 10.1093/oso/9780198851875.001.0001

initially, at low times when $t < \tau$, but then decelerates later on, at long times when $t > \tau$. The transformation accelerates initially because more and more particles of the new phase form and grow in the old phase, but it decelerates later because the old phase becomes exhausted, leaving nowhere for the new phase to form or grow. The time constant is the time $t = \tau$ at which the fraction transformed is $f = 1 - \exp\,[-(1^n)] = 1 - (1/e) \approx 0.37$.

Taking logs twice, the Avrami equation can also be written as

$$\ln\,[-\ln\,(1-f)] = \ln\,k + n \ln\,t.$$

In other words, plotting $\ln\,[-\ln\,(1-f)]$ versus $\ln\,t$ on an *Avrami plot* gives the Avrami exponent as the gradient and the rate constant as the intercept, as shown in Figure 9.2.

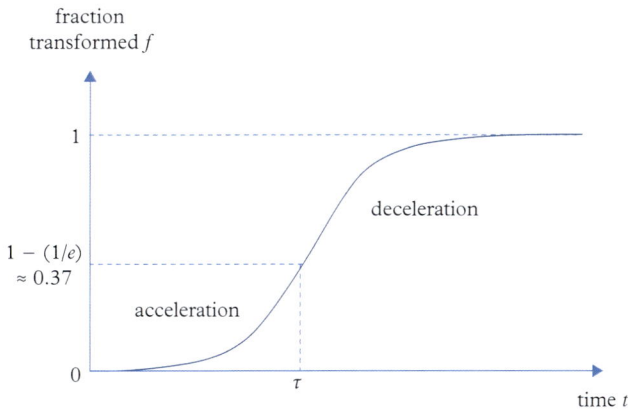

Figure 9.1 *Fraction transformed versus time according to the Avrami equation*

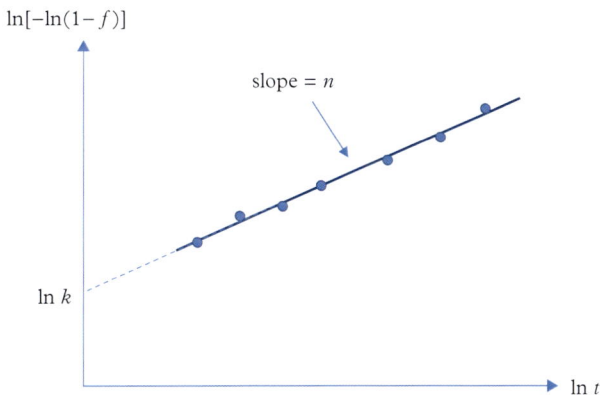

Figure 9.2 *Avrami plot to obtain rate constant k and Avrami exponent n*

2   Melvin Avrami, Robert Mehl and Andrei Kolmogorov

The Avrami equation was derived independently in three pieces of work: by the little-known and somewhat mysterious American physicist Melvin Avrami, published in three papers entitled 'Kinetics of Phase Change I, II and III' in the *Journal of Chemical Physics* in 1939, 1940 and 1941 respectively[1–3]; by the influential American metallurgist Robert Franklin Mehl, working with his graduate student W. Johnson at Carnegie Tech, published with the title 'Reaction Kinetics in Processes of Nucleation and Growth' in 1939 in *Transactions of the American Institute of Mining, Metallurgical and Petroleum Engineers*[4]; and by the famous Russian mathematician and physicist Andrei Kolmogorov, published with the title 'A Statistical Theory for the Recrystallization of Metals' in 1937 in the Russian journal *Izvestiya Akademii Nauk SSSR Seriya Materialicheskaya (Reports of the USSR Academy of Sciences: Materials Series)*.[5] The Avrami equation is, therefore, also sometimes called the *Johnson-Mehl-Avrami equation*, the *JMA equation*, the *Johnson-Mehl-Avrami-Kolmogorov equation* or the *JMAK equation*, but the shorter and simpler name, the Avrami equation, is still widely used, despite it not giving due recognition to Johnson, Mehl or Kolmogorov. The Avrami equation has been applied to a variety of other phenomena outside the field of phase transformations in materials, including animal population growth, the development of languages and vocabulary, and the culturing of biological cells.

Melvin Avrami

Melvin Avrami

Melvin Avrami was born in Palestine in 1913. His parents were American citizens who were visiting Palestine at the time. His ancestors were Jewish, originally from Spain, but tracing a line through England, Scandinavia, the Baltic countries, Palestine and the United States. At first, Melvin was named Moshe Yoel, after his maternal grandfather, Yoel Moshe Solomon, a writer and a printer, one of the leaders of a group of European religious pioneers who founded the city of Petah-Tiqva (Gate of Hope), also known as Em HaMoshavot (Mother of all Settlements), the first modern Jewish agricultural settlement in what was then Ottoman Southern Syria, and which has grown to be one of the most populous urban centres and the second most industrial conurbation in Israel, about 10 km east of Tel Aviv. On returning to the United States, Moshe Yoel was renamed Melvin, and his parents also changed the family name to Avrami to try to avoid anti-Jewish sentiments.

Little is known of Avrami's early life. He registered to study theoretical physics in 1930 at the University of Chicago and was awarded his BS, MS and PhD degrees in 1933, 1935 and 1938 respectively. He worked through the last year or two of his doctoral studies at US Steel's Gary Works in nearby Gary, Indiana, on the shores of Lake Michigan, for many years the largest steel plant in the world, and still the largest integrated steel mill in North America. He began by working eight-hour rotating shifts as an apprentice metallurgist, mainly monitoring the temperature of the steel as it progressed through the rolling mills, but he worked his way up to become a researcher in the company's Metallurgical Laboratory attached to the Works. He developed an efficient method for optimising heat treatments using a temperature gradient furnace, designed using a novel Fourier transform technique, and his last research work at US Steel was on the evolution of hydrogen in cooling steel and the formation of shatter cracks, with the objective of preventing subsequent premature fracture of heavy steel components such as train wheels.

Avrami's research achievements at the US Steel plant led to a strong recommendation from his employers when he applied for a teaching job at Columbia University in Manhattan. There was no opening in the Physics Department, but he was welcomed with open arms by Eric Jette, Head of the Columbia School of Mines, who was looking for someone to teach fundamental metal physics to budding applied metallurgists. Avrami was appointed initially in 1938 as an Instructor in Metallurgy, was promoted to be an Associate in Metallurgy in 1945, and was further promoted to be Assistant Professor and then Associate Professor in 1946 and 1947 respectively. The mid-1940s were an important time in another way for Melvin. As well as being promoted to his professorship, he met and married his wife Sophia in 1946, and their two sons, Jonathan and Paul, were born in the following two years.

In his early years at Columbia, from 1938 to 1942, Jette asked Melvin to investigate the martensitic hardening of steels and, with his physics background, this led Melvin to concentrate on studying the theoretical aspects of phase transformations, including developing the Avrami equation. As he put it himself, 'with the generalising tendency that was very strong in me, I was led to develop the papers Kinetics of Phase Changes I,

II and III in J Chem Phys and the related Geometry and Dynamics of Populations in J Philos Science'.[6]

During the Second World War, Columbia was at the centre of the US effort to develop an atomic bomb, led by the physicists Enrico Fermi and Leo Szilard. Jette's laboratory was taken over by the Chemical War Service and Jette himself joined the Manhattan Project, later moving to Los Alamos to work on the metallurgy of plutonium, one of the two major radioactive fuels for atomic bombs. Avrami was, as he put it, 'lease-lended'[7] to the Physics Department when many of the physics faculty also went to Los Alamos. There, Melvin enjoyed working alongside Willis Lamb, who later won the Nobel Prize for his studies of the fine structure of the hydrogen spectrum. Otherwise, however, none of this was to Avrami's taste and it led to a growing discomfort, unhappiness and disillusionment, culminating in his resignation from Columbia in 1948 and his apparent disappearance from academia and scientific research. In fact, he went through a deep personal crisis and dropped out altogether, changing his name and living a near-hermit's existence for four years before returning to professional life at Princeton.

Melvin explained what happened in a letter to a Japanese scientific friend, Kyozi (Ken) Sekimoto at Kyushu University in Fukuoka, written many years later and dated August 1988, after he had retired to Santa Barbara:

> [T]he dropping of the nuclear bomb on the cities of Japan, with which I totally disagreed, feeling as I did that it is better to die singly than kill by the million, has had a powerful effect on my life; it was one of the two principal reasons for my resigning my associate professorship at Columbia and leaving academic life for four years. ... I was beginning to consider resigning my tenured position at Columbia and leaving academic life for a time. I felt a revulsion against civilization and the application of science to the killing of many people, and a pull to a more spiritual existence. ... We lived for four years, 1948–1952, with our savings and some meager earnings as a caretaker, on the island of Orcas in Puget Sound.[8]

Orcas is the largest of the San Juan Islands, a rural backwater of approximately 57 square miles of beautiful countryside, with a sparse population of about 5,000 residents, situated roughly halfway between Seattle and Vancouver in Puget Sound, the sea passage on the edge of the Pacific Ocean, just off the far north-western coast of Washington state. It is about as far from Manhattan as it is possible to get in America, both in distance and in lifestyle.

Avrami changed his name to Mael Avrami Melvin. He was determined to make a step change away from his previous life, and to make further efforts to avoid anti-Jewish and also anti-Italian prejudice. In his letter he explains,

> My two younger brothers [changed] their name [i.e. surname] to something more American ... 'Avery', but I chose to perform a reflection operation on my name [i.e.

replace Melvin Avrami with Avrami Melvin] … to spare our two sons being saddled with a name foreign to America … also [with] a strong impulse to break with my previous life … I reversed the first and last names which I had carried from childhood to age 34, and added a new first name closely related to the original first name given to me at birth.[9]

He chose his new first name, Mael, in part because it telescoped his original two first names Moshe Yoel, in part because it is pronounced the same as his previous first name Mel and in part because the two syllables 'Ma-el' in Hebrew mean 'What is God?', which he said 'could hardly be bettered as a statement of my new lifelong search for what is reality'.[10] One might wonder whether the name Mael Avrami Melvin is, in fact, noticeably less Jewish than Melvin Avrami, but he was, nevertheless, clearly happier once the change was made.

In 1952, having, at least to some extent, worked his way out of his scientific and existential crisis, he returned to mainstream academia. He applied successfully, under his new name, for a Guggenheim Fellowship grant to work on the application of generalised symmetry to electrodynamics and quantum physics, based on some ideas he had developed while on Orcas. He took up his Guggenheim Fellowship initially at Princeton University in New Jersey, but was quickly recommended by Paul Ewald, Hans Bethe and Eugene Wigner to a full academic position as Professor of Physics at Florida State University in the state capital, Tallahassee, in the Florida Panhandle. With those referees he could hardly fail.

He stayed at Florida State for 14 years, including visiting professorships with Per-Olov Löwdin's quantum chemistry group at Uppsala University in Sweden in 1956, and with the atomic energy group at the Institute of Physics in Bariloche, Argentina, in 1959. He was Chair of the Florida Academy of Sciences in 1954 and was elected a Fellow of the American Physical Society in 1964. In 1966 he moved to a new job as Professor of Physics at Temple University in Philadelphia, where he stayed until his retirement in the mid-1970s, when he and Sophia moved to Santa Barbara on the Pacific West Coast in California, and where he remained active as a researcher and continued to publish scientific papers. While at Temple, he again held visiting professorships, at Princeton, the Tata Institute in India, Kyushu and Kyoto Universities in Japan, and the Jet Propulsion Lab at Caltech in Pasadena, California.

During the later, rejuvenated part of his career, Avrami avoided metallurgy and materials, and instead worked on many aspects of theoretical physics, including elementary particles and fields, the fundamentals of symmetry, quantum cosmology, neutrinos, plasmas, pulsars, and electric and magnetic stars. He suffered a stroke in 2009 and died peacefully at his home in Santa Barbara, attended by his wife, two sons and five grandchildren, on 1 October 2014, having reached the grand old age of 101. As well as physics, Avrami was passionate about the history of science, about seeing the world spiritually as well as scientifically, and about healthy eating, which may have contributed to his longevity.

Robert Mehl

Robert Mehl

Robert Franklin Mehl was born in Lancaster, Pennsylvania, on 30 March 1898. His paternal grandfather had emigrated to the United States from Munich following the revolution of 1848. His father was manager of a department store. His mother also came from a German immigrant background. His parents encouraged Robert's education at Franklin and Marshall College, and allowed him to have a small laboratory in the basement of their home. He began doing research in his final year and graduated near the top of his class. As well as science, Mehl enjoyed athletics, art, languages and literature. In 1920, he was successful in applying for a research assistantship at Princeton, and received his PhD in 1924 for a thesis on the topic 'The Electronic Properties of Aluminium–Magnesium Alloys'.[11] While doing his doctoral research, he also taught chemistry at Juanita College. He married Helen Charles in 1923, and later had three children: Robert Jr, Marjorie and Gretchen. In 1925 he was appointed to a research fellowship at Harvard, working on the compressibility of metal alloys with Theodore William Richards, the first American to win a Nobel Prize in Chemistry for his work on determining the atomic weights of the elements.

   In 1927 Robert was appointed as the first Head of the new Division of Physical Metallurgy at the US Naval Research Laboratory (NRL) just outside Washington, DC.

Before moving to the NRL, Robert had become interested in *Widmanstätten structures* in meteorites and in commercial steels. These are beautiful needle- and plate-like structures formed by solid-state precipitation, when the precipitate and matrix phases exhibit a strong *orientation relationship*, i.e. when there is a pronounced matching of particular planes and directions in the precipitate and matrix crystal structures. At the NRL, he employed Charles S. Barrett from Chicago, and they became famous for their studies of precipitation phenomena using the new technique of *X-ray diffraction* to measure and explain orientation relationships and their role in the important industrial topic of precipitation hardening in steels and in aluminium alloys. In 1931 Robert left the NRL to become Assistant Director of the Research Labs of the American Rolling Mill Company in Middletown, Ohio. It was not a good time to be in industry, which was suffering from the effects of the Great Depression. Robert was not happy and left a year later to become Professor of Metallurgy and Director of the Metals Research Lab at Carnegie Tech, later part of Carnegie Mellon University, in Pittsburgh. Although he believed strongly in the importance of industrial applications in metallurgy, an academic position proved to be much more congenial, mainly because of the terrible effect on industry of the stock market crash in 1929 and the resulting industrial depression of the 1930s. He stayed at Carnegie Tech for 28 years, until his retirement in 1960.

At Carnegie Tech, Mehl cemented his reputation as being one of the father figures in the development of metallurgical science in the United States. He made important contributions to our understanding of many fundamental phenomena such as diffusion, recrystallisation, precipitation, plastic deformation and oxidation, and in particular developed theories of nucleation and growth, and the kinetics of solid-state phase transformations. He received many honours, including the Matthewson and Gold Medals of the American Institute of Mining, Metallurgical and Petroleum Engineers, the Howe and Gold Medals of the American Society of Metals, the Le Châtelier Medal of the Société Française de Metallurgie, and election to the US National Academy of Sciences in 1958. According to his biographers, Cyril Stanley Smith and William W. Mullins, 'At one time about a quarter of the heads of metallurgy and materials science departments in the United States and Canada were his former students or faculty colleagues'.[12] Robert was a great lecturer, widely sought for talks all round the world: 'His delivery was smooth, theatrical and inspiring'.[13] He was strong minded, hard working and authoritarian, insisting on high standards, disliking too much originality from his students, and impatient if they deviated from his own clear views. He believed in metallurgy as a single discipline and he did not like the intrusion of metal physics or the development of a unified materials science. He opposed the concentration in universities on fundamental research, and believed fiercely that 'in this union [of science and application] lies the metallurgical mystique'.[14] He waged unsuccessful campaigns against the concept of dislocations in explaining plastic flow, and against the concept of vacancies in explaining atomic diffusion, both of which he regarded as 'fanciful inventions of physicists intruding into his domain of metallurgy'.[15]

After retiring from Carnegie Tech in 1960, Mehl became a consultant to US Steel and lived in Zurich, liaising between US and European metallurgists and industrialists. He returned to the United States in 1966, as a Visiting Professor at the University of

Delaware, and then went back to Pittsburgh. In later life he suffered from diabetes, and was confined to bed and a wheelchair when he had to have both his legs amputated. He died aged 78 in Pittsburgh on 29 January 1976.

Andrei Kolmogorov

Andrei Kolmogorov

Andrei Nikolaevich Kolmogorov was born on 25 April 1903 in the city of Tambov, at the confluence of the Tsna and Studenten Rivers, about 450 km southeast of Moscow, roughly halfway between Moscow and Volgograd. His mother, Mariya Yakovlevna Kolmogorova, was staying with a friend on her way back from the Crimea to her home in Yaroslavl, on the Volga about 250 km northeast of Moscow. Sadly, she died during Andrei's birth, and he was taken back to Yaroslavl and then on to his maternal grandfather's house in the nearby village of Tunoshna, where he was brought up by his aunt, Vera Yakovlevna Kolmogorova, who adopted him.

The early part of the 20th century was a tumultuous period of time in Russia's history, with a decadent royal family, ineffective attempts to install a parliamentary democracy, and the gradual growth and ultimate success of revolutionary forces. Andrei's father, Nikolai Matveevich Kataev, was the son of a clergyman and had been exiled from St Petersburg to Yaroslavl (because of his revolutionary activities) to become a *Zemstvo* (local council) statistician and agronomist. In fact, the whole of Andrei's family were well-known revolutionaries, notably the three sisters: Mariya (Andrei's natural mother), Vera (his aunt and adoptive mother) and Nadezhda, who were from an aristocratic family. Earlier, both Mariya and Vera had been arrested and held at a detention centre in St Petersburg; the sisters operated a secret printing press in Tunoshna, promulgating seditious material; and Vera was an active member of the Yaroslavl revolutionary underground. After the Bolsheviks' Red October revolution in 1917, which led to the installation of Lenin's communist government, Andrei's father became head of the training section of the *Narkomzem* (Peoples' Commissariat of Agriculture). The following two years saw a vicious civil war as the counter-revolutionary White Army under Anton Denikin tried to wrest power back from Lenin and his Red Army, led by Leon Trotsky. Andrei's father died in 1919 on the Southern Front, during the initially successful White Army advance on Moscow, which was ultimately repulsed and defeated by Trotsky.

Despite the turbulent political times, Andrei spent his early childhood relatively peacefully in Tunoshna. He wrote later, 'I experienced the joy of mathematical discovery ... when I noticed at the age of five or six the pattern $1 = 1^2$, $1 + 3 = 2^2$, $1 + 3 + 5 = 3^2$, $1 + 3 + 5 + 7 = 4^2$ and so on'[16] (he was clearly somewhat precocious). In 1910, Vera and Andrei moved to Moscow, where he studied at the E. A. Repman private gymnasium (secondary school), which had originally been founded by a group of democratic intellectuals and was taken over after the revolution as the 23rd secondary school of the communist state. Andrei wrote,

> [F]inding a serious and useful purpose for myself I owe to family tradition and ... the wonderful E. A. Repman private gymnasium where I studied. My enthusiasm for science stemmed from the teachers of this school. ... In 1918–1920 [however] life in Moscow was not easy. Only the most persistent studied seriously in school. ... I had to leave together with other students to work on the construction of the Kazan-Ekaterinburg railway.[17]

In 1920 he enrolled in the Physics and Maths Department of Moscow State University, which was open to anyone without examination, and the maths section of the prestigious Mendeleev Institute of Chemical Engineering, which required him to pass a tough examination. He was fascinated by metallurgy and history, but concentrated on mathematics, attending courses by well-known mathematicians such as Nikolai Luzin, Pavel Aleksandrov and Pavel Uryson. He was fiercely talented, correcting some of his lecturers' mathematical errors and beginning to publish original papers as a young undergraduate.

In 1922 Andrei gained international recognition for discovering a Fourier series that diverges everywhere, a completely unexpected result. He stayed at Moscow State University, graduating in 1925, getting his doctorate in 1929 and becoming a professor in 1931. Beginning in 1935, he was at different times Head of the Departments of Probability, Statistics, Mechanics, and Mathematical Logic, and he was elected to the USSR Academy of Sciences in 1939. He began a lifelong friendship with Aleksandrov, saying of him, 'He was really the most amazing person with regard to the richness and breadth of his views not only here but in the whole world. His knowledge of music and painting, and his sincere regard for people were extraordinary'.[18] In 1936

Kolmogorov and Aleksandrov were involved in the political persecution of Luzin, as part of Stalin's Great Purge, a repression of intellectuals and dissenters, leading in total to perhaps a million deaths. Luzin was denounced as a counter-revolutionary, accused initially of plagiarism, blocking the promotion of loyalists (including Kolmogorov), and then of disloyalty for publishing in France not Russia. Kolmogorov and Aleksandrov were counter-accused of being homosexuals. During the Second World War, Andrei contributed to the Russian war effort by developing a statistical scheme for deploying barrage balloons to protect Moscow from German bombers. In 1942 he married Anna Dmitrievna Egorova, and they remained married till his death 45 years later.

He researched very widely, with profuse interests, investigating many different mathematical problems in fields such as trigonometric series, Markov chains, probability theory, stochastic processes, generalising differentiation and integration, the laws of large numbers, measure theory, turbulence, classical mechanics, Brownian motion, ergodics, topology, Mendelian genetics, solar activity and algorithmic complexity. He is famous, amongst other things, for the Chapman-Kolmogorov equations for the joint probability of linked variables, Kolmogorov complexity theory for textual analysis, the Kolmogorov-Arnold-Moser theorem for quasiperiodic motion, the Kolmogorov-Sinai entropy of dynamical systems, Kolmogorov widths for approximating function classes, the Kolmogorov dimension for fractals, Kolmogorov microscales of turbulence, topological Kolmogorov spaces and, of course, the Johnson-Mehl-Avrami-Kolmogorov equation, which he derived for the crystallisation of materials by nucleation and growth.

In the 1960s and '70s, Andrei's interests broadened even further. He worked actively to improve schooling and pedagogy, founding the Kolmogorov Boarding School, revising teaching methods, writing textbooks and chairing the joint Committee of the Academy of Sciences and Academy of Pedagogical Sciences, tasked with determining the content of Russian secondary maths education. In 1960 he founded the Statistical Methods Lab at Moscow State University. He began to study network theory to compute the size of the brain from the number of neural connections. In 1961 he used a combinatorial approach to define the amount of information transmitted by speech, concentrating on how to describe the flexibility and rhythm of speech, and applying his ideas to the study of poetry. In 1971 he joined a major oceanographic expedition aboard the research vessel *Dmitri Mendeleev*. He was awarded the Stalin Prize in 1941, the Balzan Prize in 1962, the Lenin Prize in 1965, the Wolf Prize in 1980 and the Lobachevsky Prize in 1986. He received honorary doctorates, fellowships and prizes in France, Romania, Poland, Great Britain, Germany, the United States, Sweden, India, Hungary and Finland.

For most of Kolmogorov's life, he was healthy and robust, fit and active, tremendously hard working, a great companion and larger than life. He loved climbing, skiing, swimming and running, as well as more sedate activities such as reading and classical music. The last decade of his life was beset by serious illness, though at first he ignored this and remained mercurial and irrepressible. He had problems with his eyesight and his usual 40-km ski tours along the Vorya and Skalba Rivers in the Oblast region were reduced to a mere 20 km. According to his colleague and friend Vladimir Arnold, 'But even then during our last ski tour, the almost completely blind Kolmogorov leapt over the bank with his skis onto the ice of the Klyaz'ma river'.[19] Andrei and Anna moved into the Uzkoe, a 17th-century estate southwest of Moscow, which was originally owned by Prince Gagarin, was taken over as part of Gagarin's wife's dowry by the famous general and statesman Count Pyotr Aleksandrovich Tolstoy, who had opposed Napoléon, and was then, after the revolution, appropriated and used by the Russian Academy of Sciences as a sanatorium. He showed Arnold how he escaped from the strict supervision of his wife and doctors to climb over the fence and go swimming in the pond. Later, he developed Parkinson's disease, which gradually became worse. He was only able to speak a few words an hour, and could only get around by being carried in someone else's arms. His wife, nurse and students had to watch over him 24 hours a day, and this continued for several years. In the summer of 1987 he entered a special clinic to try to improve his eyesight, but his illness progressed rapidly and his lungs were badly affected, so he was transferred to the pulmonary section. His student Albert Nikolayevich Shiryaev describes his last-ever conversation with Kolmogorov:

> 'Where am I now'? he asked.
> 'In the pulmonary section'.
> 'Why'?
> 'They have found something wrong with your lungs'.
> 'What will come of this'?
> 'You'll have to stay so you can return home with healthy lungs'.
> 'Well then, that's all right'.[20]

In the next few days, his temperature began to oscillate wildly, his blood pressure dropped sharply and breathing became difficult. His heart stopped during the afternoon of 20 October 1987 and he died aged 84. His remains are buried in the Novodevichy cemetery in Moscow, alongside many other famous Russian authors, musicians, playwrights, poets, statesmen and scientists. His official obituary reads: 'Andrei Nikolaevich Kolmogorov's whole life was an unparalleled feat in the name of science. He was a model of nobility, unselfishness and moral purity. ... [He] has entered the galaxy of great Russian and world scholars'.[21]

It is remarkable to compare and contrast the lives of these three scientists: Melvin Avrami, Robert Mehl and Andrei Kolmogorov. Avrami went to the United States from a peripatetic Jewish family background, and his professional life flowered during the Second World War, causing him to suffer a crisis of conscience from his closeness to, and revulsion from, both the development and use of the first atomic bombs and the omnipresence of anti-Semitism. Mehl, like Avrami, also grew up in the United States from an immigrant background, in his case German protestant, and at an earlier period, suffering and overcoming the deprivations of the stock market crash of 1929 and the Great Depression of the 1930s, to become one of the father figures of the then-new science of metallurgy. And Kolmogorov grew up during the Russian revolution and survived the great Stalin purges of intellectuals to become a famed mathematician, contributing to a bewildering array of fundamental and applied mathematical fields. All three, despite their different backgrounds, different skills and different temperaments, arrived, amongst other achievements, at the simplicity of the Avrami equation, perhaps better described as the Johnson-Mehl-Avrami-Kolmogorov equation, which is now used on a daily basis by materials scientists worldwide to describe and explain the speed of solid-state transformations, which control the structure and properties of so many of our material artefacts.

3   The driving force for transformation

Consider a general phase transformation with a high-temperature solid phase α transforming isothermally at a temperature T into a low-temperature solid phase β:

$$\alpha \to \beta.$$

Figure 9.3 shows the free energies of the two phases G_α and G_β versus temperature T:

$$G_\alpha = H_\alpha - TS_\alpha$$
$$G_\beta = H_\beta - TS_\beta,$$

where $H_\alpha, H_\beta, S_\alpha$ and S_β are the α and β enthalpies and entropies respectively. In general, $H_\alpha > H_\beta$ and $S_\alpha > S_\beta$ because atoms and molecules have higher energy and are more disordered in the high-temperature α state. The α and β free energies are equal at the *equilibrium temperature* T_e:

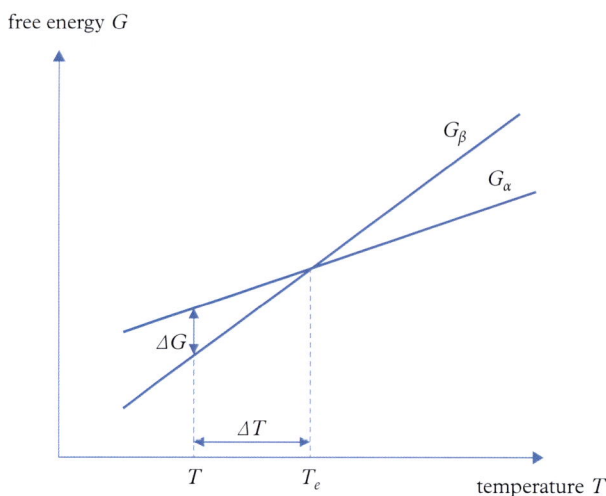

Figure 9.3 *Driving force ΔG for an $\alpha \to \beta$ phase transformation at an undercooling ΔT below the equilibrium temperature*

$$G_\alpha = G_\beta$$

$$\therefore H_\alpha - T_e S_\alpha = H_\beta - T_e S_\beta$$

$$\therefore L = T_e \Delta S; \text{ and } \Delta S = \frac{L}{T_e},$$

where $L = H_\alpha - H_\beta$ is the *latent heat of the transformation* and $\Delta S = S_\alpha - S_\beta$ is the *entropy change of the transformation*.

The phase transformation can only take place at a temperature T below T_e, i.e. at an *undercooling* $\Delta T = T_e - T$, with a *driving force for transformation* ΔG given by

$$\Delta G = G_\alpha - G_\beta = (H_\alpha - TS_\alpha) - (H_\beta - TS_\beta)$$

$$= L - T\Delta S = L - T\left(\frac{L}{T_e}\right) = \frac{L\Delta T}{T_e} = \Delta S \Delta T.$$

In other words, the driving force for the transformation increases linearly with increasing undercooling below the equilibrium temperature, and the constant of proportionality is the entropy change of the transformation or its latent heat divided by the equilibrium temperature.

Phase transformations usually take place in two distinct stages, called *nucleation* and *growth*. We examine these two stages separately before combining them to look at the overall process.

4   Nucleation

The nucleation stage of a phase transformation is the initial formation of the first small particle or *nucleus* of the new phase. Consider cooling a mass of α from its stable region above T_e to a temperature T in its unstable region below T_e. Let a small β particle of radius r form within the bulk of α, as shown in Figure 9.4. The change in free energy ΔG_r is given by the sum of three terms: (1) a decrease in free energy caused by replacing a volume of α by β, (2) an increase in free energy caused by creating an area of $\alpha - \beta$

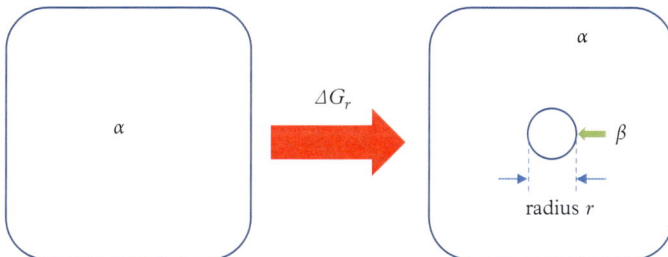

Figure 9.4 *Nucleation of a particle of β in a matrix of α*

free energy change ΔG_r

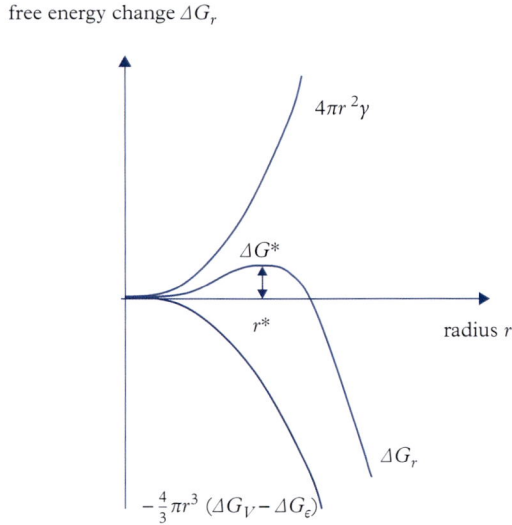

Figure 9.5 *Work of nucleation ΔG^* and critical radius for nucleation r^**

interface and (3) a further increase in free energy caused by the expansion or shrinkage strain associated with the difference in densities of the two phases:

$$\Delta G_r = -\frac{4}{3}\pi r^3 \Delta G_V + 4\pi r^2 \gamma + \frac{4}{3}\pi r^3 \Delta G_\epsilon$$

$$= \frac{4}{3}\pi r^3 (\Delta G_V - \Delta G_\epsilon) + 4\pi r^2 \gamma,$$

where $\Delta G_V = \Delta G/v_m$ is the driving force per unit volume, v_m is the molar volume, γ is the $\alpha - \beta$ surface energy and ΔG_ϵ is the strain energy per unit volume. ΔG_r varies with r, as shown in Figure 9.5. For small values of r, the surface area of the particle is large compared with its volume, and ΔG_r is positive. For large values of r, the surface area of the particle is small compared with its volume, and ΔG_r is negative. As a small particle forms, it must initially increase its energy because of the need to create the $\alpha - \beta$ interface. The effect of the strain energy is to reduce the driving force available to create the new surface. As the particle grows, it reaches a maximum increase in energy ΔG^*, called the *work of nucleation*, at a radius r^*, called the *critical nucleus size*, and then its energy decreases.

The work of nucleation is, effectively, an energy barrier to the nucleation process, a barrier to the formation of an initial β particle. An initial particle must somehow attain an excess free energy ΔG^* and reach the critical nucleus size r^* before it can begin to grow with a continuous reduction in free energy. Particles with $r < r^*$ are called *sub-critical nuclei*, particles with $r = r^*$ are called *critical nuclei*, and particles with $r > r^*$ are called *super-critical nuclei*, or just nuclei. The energy barrier and the critical radius are obtained by differentiating ΔG_r with respect to r and putting $d\Delta G_r/dr = 0$:

$$\Delta G^* = \frac{16\pi}{3}\frac{\gamma^3}{(\Delta G_V - \Delta G_\epsilon)^2}$$

$$r^* = \frac{2\gamma}{(\Delta G_V - \Delta G_\epsilon)}.$$

The energy barrier and the critical radius both increase sharply with increasing surface energy γ, and decrease sharply with increasing driving force ΔG_V, i.e. with increasing undercooling ΔT below the equilibrium temperature.

How can a nucleus form? Atoms and molecules in the α phase are moving and vibrating at high temperatures, and, from time to time, groups of atoms fluctuate into β-like configurations or *clusters*, effectively forming a population of sub-critical nuclei. For very small particles $r < r^*$, as shown in Figure 9.5, a sub-critical nucleus is more likely to decay back to α than to continue to grow in size. However, when a sufficiently large fluctuation allows a particle to reach the critical size, with a radius r^*, it can begin to grow with a continuous reduction in free energy.

The *nucleation rate i* is usually defined as the rate at which critical nuclei grow into super-critical nuclei, i.e. the number of critical nuclei N^* multiplied by the frequency ν with which an additional adjacent atom adds on to a critical nucleus by a diffusive jump at the α–β interface:

$$i = N^*\nu = N_o\exp\left(-\frac{\Delta G^*}{kT}\right)\nu_o\exp\left(-\frac{Q}{kT}\right)$$

$$= i_o\exp\left(-\frac{\Delta G^* + Q}{kT}\right),$$

where $N^* = N_o\exp(-\Delta G^*/kT)$ is given by a Boltzmann expression for the probability of a cluster with excess energy ΔG^*, N_o is Avogadro's number, ν_o and Q are the frequency factor and activation energy for diffusion respectively, and $i_o = N_o\nu_o$ is the frequency factor for nucleation.

This process is called *homogeneous nucleation*, i.e. nucleation of the new phase β within the bulk of the old phase α without any other influence. In fact, nucleation is often aided by defects in the α phase such as dislocations, grain boundaries or impurity particles. This is called *heterogeneous nucleation*. Initial β particles form in contact with the defects, reducing the amount of new $\alpha - \beta$ interface that needs to be formed and removing some or all of the strain energy of the defect, allowing the particles to reach the critical radius r^* with a reduced work of nucleation:

$$i = i_o\exp\left(-\frac{\Delta G^*_{\text{hetero}} + Q}{kT}\right),$$

$$\Delta G^*_{\text{hetero}} < \Delta G^*_{\text{homo}}; \; i_{\text{hetero}} > i_{\text{homo}}.$$

5 Growth

For a transformation $\alpha \rightarrow \beta$ that involves a change of crystal structure but no change in composition, the growth process is *interface controlled* because it depends only on the attachment of atoms or molecules from the α phase onto the β phase at the interface. The *growth rate g* of a β particle is given by the rate at which the atoms or molecules can attach at the interface, and this is approximately proportional to the diffusion rate:

$$g \propto D = D_o \, \exp - \left(\frac{Q}{kT} \right),$$

where D is the diffusivity and D_o and Q are the frequency factor and activation energy for diffusion respectively. The growth rate is independent of time as the particle grows, so the particle size increases linearly with time:

$$r \propto t.$$

Typical concentration profiles and growth rate are shown in Figure 9.6. The growth rate, like the diffusivity, varies strongly with the temperature at which the transformation is taking place.

For a transformation $\alpha \rightarrow \beta$ that involves a change of composition as well as a change in crystal structure, the growth process is *diffusion controlled* because it depends on the rate at which the atoms or molecules diffuse from the α phase to the interface. In a two-component A–B material, if α is A rich and β is B rich, the growth rate of a β particle slows down with time because the B atoms or molecules have to migrate to the interface from further and further away. The particle size increases *parabolically*, i.e. with the square root of time:

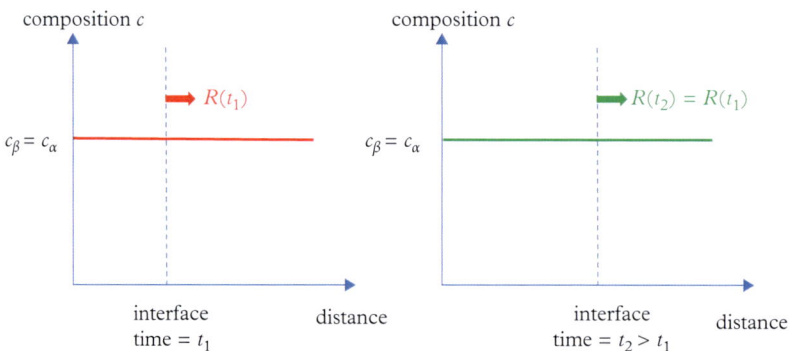

Figure 9.6 *Composition profiles for interface-controlled growth*

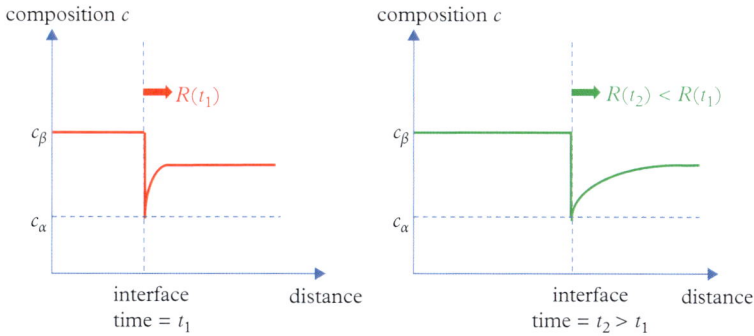

Figure 9.7 *Composition profiles for diffusion-controlled growth*

$$r \propto t^{\frac{1}{2}}.$$

Typical concentration profiles and growth rate are shown in Figure 9.7. The growth rate, like the diffusivity, again varies strongly with the temperature at which the transformation is taking place.

There is often a strong orientation relationship between the α and β phases, depending on the detailed nature of the α and β crystal structures. The structure and mobility of the $\alpha - \beta$ interface are, therefore, often highly anisotropic, leading to plate-like or needle-like growth forms.

6   The Avrami exponent

The overall transformation rate and, therefore, the Avrami equation can be obtained by combining the nucleation and growth stages of a phase transformation. Consider a transformation in which (1) nuclei form at random throughout the material, with a nucleation rate per unit volume i, and (2) nuclei grow *isotropically*, i.e. equally in all directions, with a constant growth rate g. First, let us ignore the problem of exhausting the old phase, i.e. let us ignore the problem that transformation cannot take place in already-transformed material. To be more accurate, we will at first ignore two facts: (1) that nuclei cannot form in places where previous nuclei have already formed and are growing; and (2) that, after some time, adjacent growing nuclei will impinge, preventing their continued growth. This is obviously unrealistic. Double nucleation and complete non-impingement are clearly impossible, but we will first allow them and then correct for the mistake later.

If we let nucleation continue throughout the whole volume of the material V, ignoring the presence of any previously formed and growing nuclei, then the number of nuclei dN^* formed during the time period between t^* and $t^* + dt^*$ is

$$dN^* = iV dt^*.$$

If we let growth of all these nuclei continue, ignoring any impingement of adjacent-growing nuclei, then at a later time t they all have the same radius r and the same volume v:

$$v = \frac{4}{3}\pi r^3 = \frac{4}{3}\pi g^3 (t - t^*)^3.$$

The volume transformed at a later time t, which nucleated between t^* and $t^* + dt^*$, is obtained by multiplying the number of nuclei dN^*, which formed in the nucleation period dt^*, by their volume v:

$$dV^* = v dN^* = \frac{4}{3}\pi r^3 iV dt^* = \frac{4}{3}\pi g^3 (t - t^*)^3 iV dt^*,$$

and the total volume transformed at t that nucleated at any time is obtained by integrating dV^* over all nucleation periods dt^* between $t^* = 0$ and $t^* = t$:

$$V^* = \int_0^t dV^* = \int_0^t \frac{4}{3}\pi g^3 (t - t^*)^3 iV dt^*$$

$$= \left[-\frac{1}{3}\pi g i V (t - t^*)^4 \right]_0^t = \frac{1}{3}\pi g^3 iV t^4.$$

This is called the *extended volume* V^*, i.e. the volume transformed ignoring the exhaustion of the old phase, and ignoring the fact that transformation cannot take place in already-transformed material, i.e. allowing nucleation to take place anywhere independent of previous existing nuclei, as well as allowing continued growth in all directions without any impingement. The corresponding *extended fraction transformed* f^* is given by

$$f^* = \frac{V^*}{V} = \frac{1}{3}\pi g^3 it^4.$$

This calculation has clearly overestimated the volume and fraction transformed by ignoring the gradual exhaustion of the old phase and the reduction in material available for transformation, i.e. the reduction in number of nuclei formed and the extent of their growth. We can correct for this overestimate quite easily. In any time increment dt, the real change in fraction transformed df is smaller than the change in extended fraction transformed df^* by the fraction $1 - f$, since this is the fraction untransformed at that time, i.e. this is the only fraction of the material that is available for transformation when we prevent nucleation in already-transformed material and when we prevent growth when adjacent nuclei impinge:

$$df^* = \frac{df}{(1 - f)}$$

$$\therefore f^* = \int_0^t \frac{df}{(1 - f)} = -\ln (1 - f)$$

$$\therefore f = 1 - \exp (-f^*) = 1 - \exp \left(-\frac{1}{3}\pi g^3 it^4 \right) = 1 - \exp (-kt^n),$$

which is the Avrami equation, with the rate constant $k = \frac{1}{3}\pi g^3 i$ and the Avrami exponent $n = 4$.

For the case described here, nucleation is random and continuous at a constant rate, characteristic of either *homogeneous nucleation* or *heterogeneous nucleation* on an inexhaustible number of identical heterogeneities; and growth is isotropic and also continuous at a constant rate, characteristic of *interface-controlled growth*, with a change of structure at the interface but no change in composition. The Avrami exponent n is a pure number that is independent of temperature and describes the *dimensionality* of the transformation. The fraction transformed increases in part because the number of nuclei increases linearly with time $N \propto t^1$, and in part because growth of the nuclei is linear in time in each of three dimensions $v \propto t^3$. The extent of the transformation depends on the amount of nucleation multiplied by the amount of growth, and the Avrami exponent gives the overall time dependency, which is the sum of the time exponents of nucleation and growth $n = 1 + 3 = 4$. On the other hand, the rate constant $k = \frac{1}{3}\pi g i$ is a temperature-dependent material parameter, a function of a geometric factor $\frac{1}{3}\pi$, and the individual rate constants for nucleation and growth $k_N = i$ and $k_G = g$.

Similar behaviour is found to apply to a wide variety of different transformations, with different mechanisms and morphologies of nucleation and growth. In all cases, the Avrami exponent is independent of temperature and describes the overall dimensionality or time exponent of the combination of nucleation and growth. For instance, consider a transformation similar to the one just described, but with a plate-like growth form, i.e. growth is not fully isotropic and is limited to just two dimensions. The Avrami exponent again gives the overall time dependency, which is the sum of the time exponents of nucleation and growth, but in this case $n = 1 + 2 = 3$. For a transformation again similar to the one considered earlier, but with a needle-like growth form, i.e. growth is not isotropic and is limited to just one dimension, the Avrami exponent is $n = 1 + 1 = 2$. Alternatively, consider transformations involving a change in composition. Growth is then *diffusion controlled*, depending on the transport of solute towards or away from the interface, and the growth rate decreases proportional to the square root of time in each growing dimension. The Avrami exponent is then $n = 1 + \frac{3}{2} = 2.5$ for spherical growth, $n = 1 + \frac{2}{2} = 2$ for plate-like growth and $n = 1 + \frac{1}{2} = 1.5$ for needle-like growth. As a final example, consider different nucleation behaviour, with heterogeneous nucleation on a small, instantaneously saturated number of heterogeneities, so the number of nuclei N is effectively fixed at the beginning of the transformation and remains fixed throughout. The time exponent for nucleation is now zero, so the Avrami exponent is $n = 3, 2$ and 1 for interface-controlled spherical, plate-like and needle-like growth respectively, and is $n = 1.5$, 1 and 0.5 for diffusion-controlled spherical, plate-like and needle-like growth respectively. In all cases, the rate constant is temperature dependent, and combines a geometric factor and the individual rate constants for nucleation and growth k_N and k_G.

It is common to track the progress of a phase transformation in a material by measuring the emission or absorption of latent heat in a calorimeter, or the evolution of the electrical resistivity of the material. This gives the fraction transformed f versus time t. The Avrami exponent and rate constant can then be obtained from an Avrami plot of $\ln[-\ln(1-f)]$ versus $\ln t$, indicating the kinetics and, therefore, the underlying mechanisms of nucleation and growth. Direct observation of the transformation morphology by optical and/or electron microscopy is usually used to back up the kinetic results.

7　Precipitation reactions

Figure 9.8 shows a *eutectic* phase diagram in a two-component A–B phase diagram. There is *limited solubility* in the two *terminal solid solutions*, i.e. there is limited solubility of B in the A-rich α phase and limited solubility of A in the B-rich β phase. A material of composition c is cooled from a high temperature T_1 in the single-phase α region above the α *solvus* to a lower temperature T_2 in the two-phase $\alpha + \beta$ region below the α solvus to form a material consisting of β particles in an α matrix.

On cooling to T_2, the α matrix has its initial composition c, which is above the equilibrium solvus composition c_e, i.e. it is *supersaturated* in B atoms or molecules. The excess B atoms or molecules diffuse to form β particles, gradually removing the supersaturation, and finally leaving the α matrix with its equilibrium solvus composition c_e. This process of forming β particles from a supersaturated α matrix is called a *precipitation reaction* and the particles are called β *precipitates*.

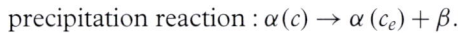

$$\text{precipitation reaction} : \alpha(c) \rightarrow \alpha(c_e) + \beta.$$

Figure 9.9 shows the variation of Gibbs free energy with composition for the α and β phases at the lower temperature T_2. On cooling to T_2, the material consists initially of supersaturated α, with a Gibbs free energy G_α. When precipitation of β is complete, the material consists of a mixture of α and β, with a Gibbs free energy $G_{\alpha+\beta}$. The driving force for the precipitation reaction is $G_\alpha - G_{\alpha+\beta} = \Delta G$.

Precipitation usually takes place by heterogeneous nucleation of β particles on vacancies, dislocations, grain boundaries, impurity particles or pre-existing clusters of B atoms or molecules in the α matrix, followed by diffusion-controlled growth as B

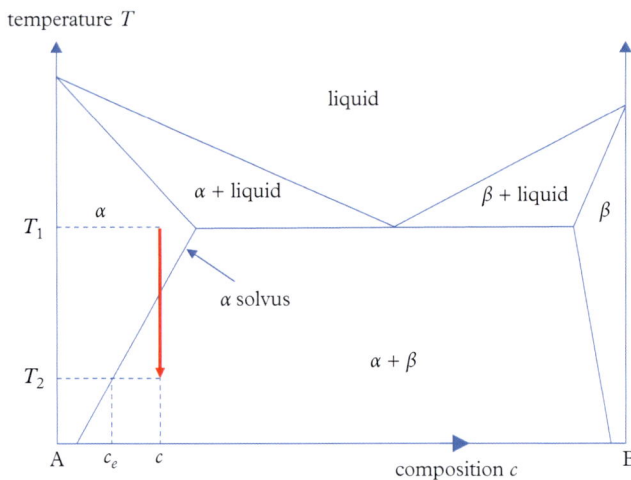

Figure 9.8 *Precipitation of β from α*

Gibbs free
energy G

α

β

temperature
$= T_2$

G_α^o

$G_\alpha(c)$

G_α

ΔG

common
tangent

$G_\alpha(c_e)$

$G_{\alpha+\beta}$

α

$\alpha + \beta$

β

A c_e c

composition c

c_β B

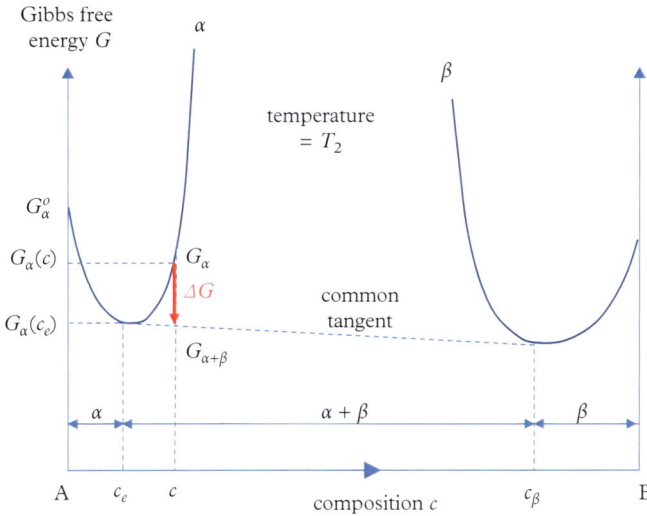

Figure 9.9 *Gibbs free energy–composition curves for α and β*

atoms or molecules diffuse gradually from the supersaturated α matrix to the α–β interface. The time exponent for diffusion-controlled growth is a half for each growth dimension. When heterogeneous nucleation sites are plentiful, the time exponent for nucleation is one, and the overall Avrami exponent is 1.5, 2 or 2.5 for needle, plate or spherical precipitate morphologies respectively. When heterogeneous nucleation sites are limited and are saturated rapidly, the time exponent for nucleation is zero, and the overall Avrami exponent is 0.5, 1 or 1.5 for needle, plate and spherical precipitate morphologies respectively.

8   Eutectoid reactions

Figure 9.10 shows a *eutectoid* phase diagram in a two-component A–B phase diagram. A material of *eutectoid composition* c_e is cooled from a high temperature T_1 in the single-phase γ region above the *eutectoid temperature* T_e to a lower temperature T_2 in the two-phase $\alpha + \beta$ region just below T_e to form a two-phase material consisting of $\alpha + \beta$. The two new phases form at the same time, and the *eutectoid reaction* or *eutectoid decomposition* is

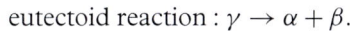

$$\text{eutectoid reaction} : \gamma \to \alpha + \beta.$$

The two phases α and β have different structures as well as different compositions, either side of the eutectoid composition of the γ phase, i.e. $c_\alpha < c_e < c_\beta$ as given by the *eutectoid tie line*, as shown in Figure 9.10. Figure 9.11 shows the variation of Gibbs free energy

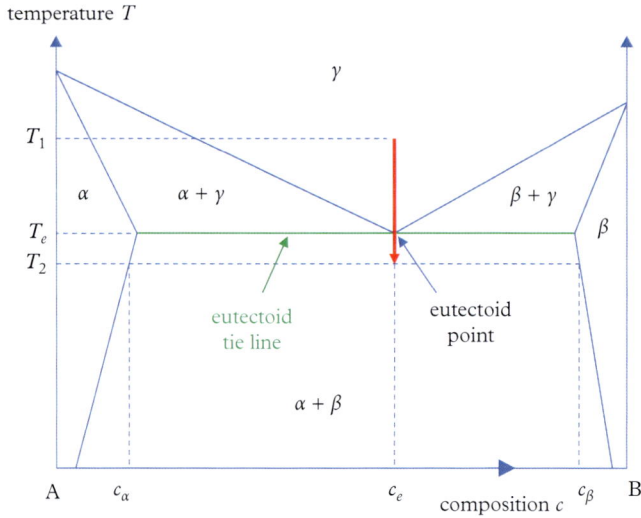

Figure 9.10 *Eutectoid reaction in a two-component A–B material*

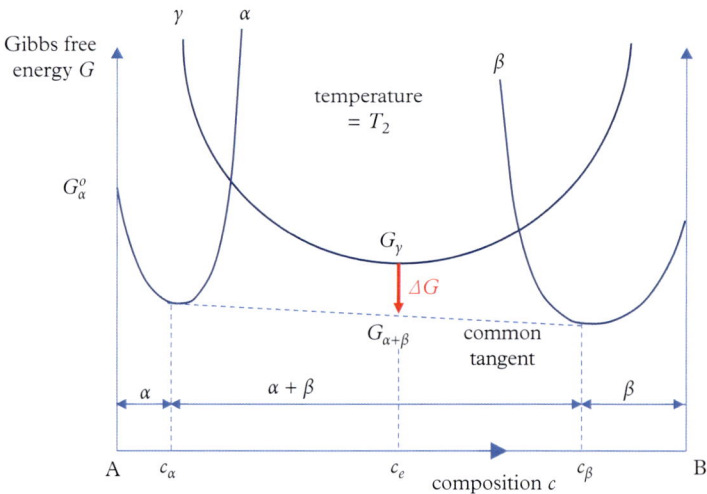

Figure 9.11 *Gibbs free energy–composition curves for α, β and γ*

with composition for the α, β and γ phases at the lower temperature T_2. On cooling to T_2, the material consists initially of γ, with a Gibbs free energy G_γ. When the eutectoid reaction is complete, the material consists of a fine-scale mixture of α and β, with a Gibbs free energy $G_{\alpha+\beta}$. The driving force for the eutectoid reaction is $G_\gamma - G_{\alpha+\beta} = \Delta G$.

The two phases α and β form in a side-by-side fashion, as shown in Figure 9.12. Solute B atoms or molecules migrate laterally away from the α phase and towards the β phase,

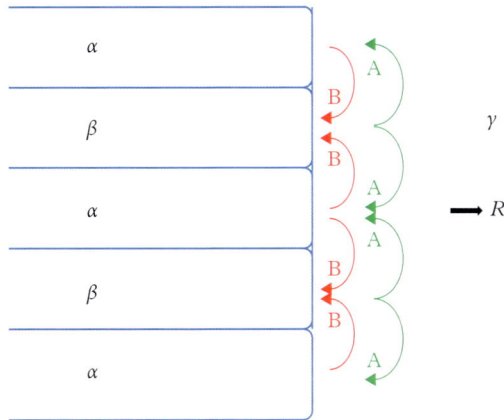

Figure 9.12 *Coupled growth during a eutectoid reaction*

just ahead of the interface, as shown in Figure 9.12. At the same time, solvent A atoms or molecules migrate laterally away from the β phase and towards the α phase, just ahead of the interface, as also shown in Figure 9.12. This leads to a *coupled growth* process, with a fine-scale distribution of the α and β phases, usually in a *lamellar (plate-like) eutectoid* or *rod-like eutectoid* structure.

Eutectoid reactions usually take place, like precipitation, by heterogeneous nucleation on dislocations, grain boundaries or impurity particles, so the time exponent for nucleation is again one or zero, depending on whether the heterogeneous nucleation sites are plentiful or not. Eutectoid reactions are not diffusion controlled because the coupled α and β phases have the same overall composition as the parent γ phase, so growth does not involve long-range diffusion. Eutectoid growth is usually spherical, given the complexity of orientation relationships between the three phases. The time exponent for growth is, therefore, one in each growth dimension, i.e. 3, and the overall Avrami exponent is 4 or 3, depending on whether nucleation is continuous or not.

9   Martensite reactions

Precipitation and eutectoid reactions are examples of phase transformations in which the starting phase is *chemically unstable*, i.e. after cooling (or otherwise) its chemical Gibbs free energy has become higher than an alternative phase or combination of phases, as shown in Figures 9.8 and 9.11. In other (somewhat less common) phase transformations, the starting phase is *mechanically unstable*, i.e. after cooling (or otherwise) its lattice and structure collapse into an alternative phase.

In a *martensite reaction*, the starting phase becomes mechanically unstable and shears to form a new phase, often in the form of thin laths or plates, and often containing internal twins or dislocations, both of these caused by mechanical constraints associated with the difficulty of shearing a lattice. Martensite transformations are sometimes called *athermal* because they are very fast, but do not always go to completion. The amount transformed is, therefore, independent of time and determined only by temperature.

The most famous example of a martensite reaction is *martensitic steel*. Steels are iron–carbon Fe–C alloys, which transform on slow cooling from a single high-temperature fcc *γ-austenite* phase into a two-phase mixture of low-temperature bcc *α-ferrite* and an iron carbide phase Fe_3C, called *cementite*, in part by precipitation and in part by a eutectoid reaction. On fast cooling or *quenching*, however, the atomic diffusion required for precipitation or a eutectoid reaction is suppressed, and the high-temperature austenite shears to form a single supersaturated body-centred tetragonal (bct) phase, called *martensite*. The supersaturation with excess carbon, the associated lattice deformation from bcc to bct, the thin martensite laths and plates, and the internal twins and dislocations all combine to produce an exceptionally hard material. For final use, martensitic steels are usually softened slightly by heating or *tempering*, and they are then called *quenched and tempered steels*.

10 Time–temperature–transformation curves

Figure 9.13 shows a typical set of *isothermal time–temperature–transformation curves*, or *TTT curves*, for a phase transformation that takes place on cooling below the equilibrium temperature T_e. The time t required to reach a given degree $x\%$ of transformation after quenching below T_e and holding isothermally is plotted for different temperatures T. The start of the transformation is taken as, say, 1% transformed, and the end as, say,

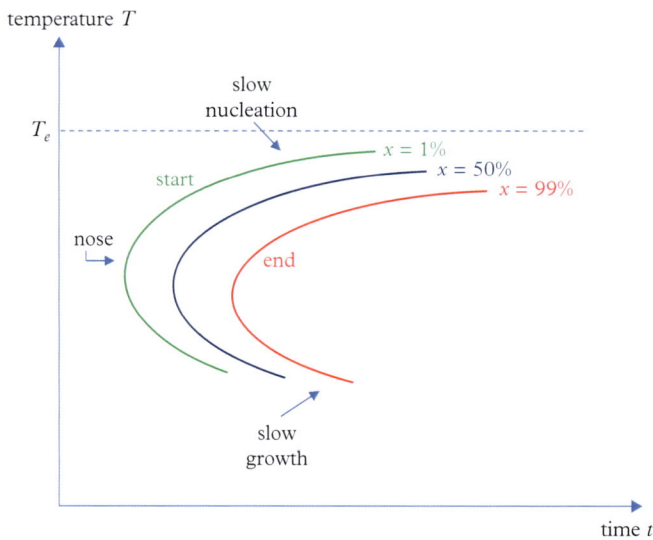

Figure 9.13 *Isothermal time–temperature–transformation (TTT) curves for a phase transformation*

99% transformed, because it is hard to measure the exact start and end points. Figure 9.13 shows curves for $x = 1\%$, 50% and 99% transformed.

Nucleation and growth transformations such as precipitation or a eutectoid reaction have TTT curves shaped like a capital C, which are often, therefore, called *C curves*, as shown in Figure 9.13. This is because nucleation is slow at high temperatures just below the equilibrium point, and growth is slow at low temperatures a long way below the equilibrium point. The fastest transformation and, therefore, the shortest transformation time is at intermediate temperatures, the so-called *nose* of the TTT curve, when both nucleation and growth are reasonably fast, as shown in Figure 9.13.

....

11 REFERENCES

1. M. Avrami. 'Kinetics of phase change I: general theory'. *Journal of Chemical Physics* **7** (1939), 1103–1112.
2. M. Avrami. 'Kinetics of phase change II: transformation-time relations for random distribution of nuclei'. *Journal of Chemical Physics* **8** (1940), 212–224.
3. M. Avrami. 'Kinetics of phase change III: granulation, phase change and microstructure'. *Journal of Chemical Physics* **9** (1941), 177–184.
4. W. A. Johnson and R. F. Mehl. 'Reaction kinetics in processes of nucleation and growth'. *Transactions of the American Institute of Mining, Metallurgical and Petroleum Engineers* **135** (1939), 416–458.
5. N. Kolmogorov. 'A statistical theory for the recrystallization of metals'. *Izvestiya Akademii Nauk SSSR Seriya Materialicheskaya* [*Reports of the USSR Academy of Sciences: Materials Series*] **3** (1937), 355–359.
6. Mael Avrami Melvin. Letter to Kyozi Sekimoto, July 1988. Columbia University archive; and Kyoto University archive. *Physical Property Research* **52** (1989), 109.
7. Ibid.
8. Ibid. 106–108.
9. Ibid. 107.
10. Ibid.
11. Robert F. Mehl. 'The electronic properties of aluminium-magnesium alloys'. PhD diss., Princeton, 1924.
12. Cyril Stanley Smith and William W. Mullins. 'Robert Franklin Mehl', vol. 78, 129–145. In: *Biographical Memoirs* (National Academy of Sciences, Washington, DC, 2000), 136.
13. Ibid.
14. Ibid. 141.
15. Ibid. 135.
16. A. N. Shiryaev. 'Andrei Nikolaevich Kolmogorov: a biographical sketch of his life and creative paths', 1–89. In: *Kolmogorov in Perspective* (American Mathematical Society, Providence, 2000), 4.
17. Ibid. 5.
18. Ibid. 21.
19. V. I. Arnol'd. 'On A. N. Kolmogorov', 89–109. In: *Kolmogorov in Perspective* (American Mathematical Society, Providence, 2000), 105.
20. Shiryaev, n. 16, 84.
21. See ibid. 85.

· ·

12 BIBLIOGRAPHY

Kolmogorov in Perspective. (American Mathematical Society, Providence, 2000).

Phase Equilibria, Phase Diagrams and Phase Transformations: Their Thermodynamic Basis. Mats Hillert (Cambridge University Press, Cambridge, 2007).

Phase Transformations in Metals and Alloys (3rd edition). David A. Porter, Kenneth E. Easterling, and Mohamed Y. Sherif (Taylor and Francis, Boca Raton, 2009).

Robert Franklin Mehl. C. S. Smith and W. W. Mullins. *US National Academy of Sciences Biographical Memoirs* 78 (2000), 130–135.

10

Hooke's Law

Elasticity

1   Hooke's law

If you pull a piece of elastic, it gets longer. If you pull it harder, it gets longer still. If you stop pulling, it returns to its original length. These effects, not surprisingly, are called *elasticity*. More surprisingly, perhaps, it turns out that all solid materials exhibit elasticity. Pieces of elastic are stretchy, so it is easy to see their extension, but with most other materials the extension is small and not so easy to see.

Figure 10.1 shows a solid bar pinned to a fixture at the top, with a force F applied by hanging a weight on the bottom. The force F is called a *tensile force*, because it acts in such a way as to stretch or extend the bar. *Hooke's law* says that the extension of the bar x is proportional to the applied force F:

$$F \propto x = kx.$$

The constant of proportionality k is called the bar's *stiffness*. It is the force F required to produce a unit extension $x = 1$. The inverse of the bar's stiffness is called its *compliance* $c = 1/k$. It is the extension x produced by a unit force $F = 1$:

$$\text{stiffness: } k = \frac{F}{x}$$

$$\text{compliance: } c = \frac{1}{k} = \frac{x}{F}.$$

A piece of elastic has low stiffness and high compliance, i.e. the force required to extend it is relatively small and the extension produced is relatively large. In other words, elastic is rather stretchy. Conversely, a piece of metal or ceramic has high stiffness and low compliance, i.e. the force required to extend it is relatively large, and the extension produced is relatively small. In other words, metals and ceramics are not very stretchy.

The Equations of Materials. Brian Cantor. Oxford University Press (2020). © Brian Cantor.
DOI: 10.1093/oso/9780198851875.001.0001

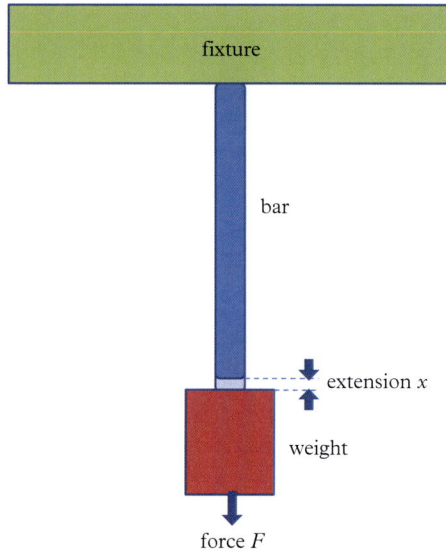

Figure 10.1 *Tensile extension of a bar*

In general, a material is said to exhibit *elastic behaviour* when

1. its response to an applied force is *linear*, i.e. it obeys Hooke's law; and
2. its response to an applied force is *reversible*, i.e. when the force is removed the material returns to its original size and shape.

It is easier to stretch a thin bar rather than a thick one, and a long bar rather than a short one. For a given material, in other words, the stiffness of the bar increases in proportion to its cross sectional area A and decreases in proportion to its length l:

$$F = EA\frac{x}{l},$$

where $E = kl/A$ is a constant characteristic of the material called the *tensile modulus* or *Young's modulus*.

We define the *tensile stress* in a material σ as the applied tensile force per unit cross sectional area, $\sigma = F/A$, and the *tensile strain* in a material ε as the resulting fractional extension, i.e. the extension per unit initial length $\varepsilon = x/l$. Hooke's law can then be written

$$\frac{F}{A} = E\frac{x}{l}$$

or

$$\sigma = E\varepsilon.$$

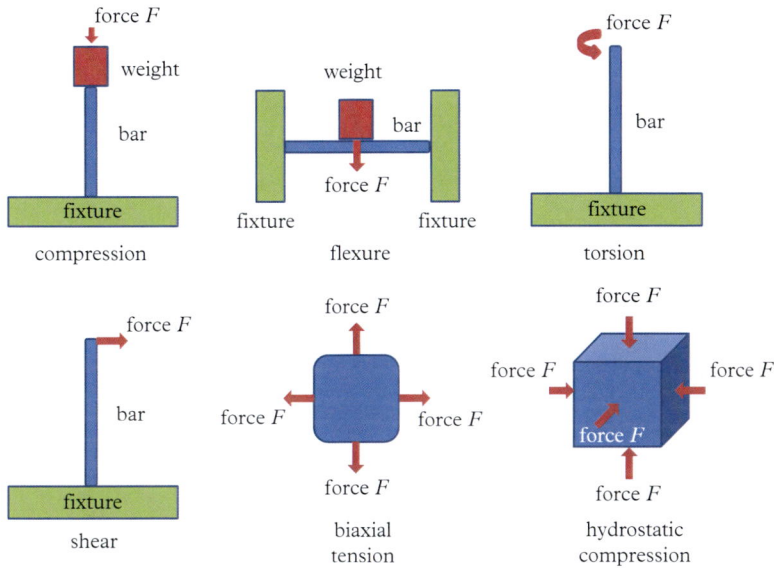

Figure 10.2 *Different loading modes*

This form of Hooke's law says that, when a material is subjected to an applied tensile force, the resulting tensile strain is proportional to the applied tensile stress, with the Young's modulus as the constant of proportionality.

Pieces of elastic are stretchy, so it is quite easy to see their extension when they are stretched. With most materials, however, we have to use special measurement techniques, such as a magnifying telescope with a cross wire to monitor the position of the end of the bar, or an *extensometer* or *strain gauge* attached to the surface, producing an electrical signal when the bar is stretched.

Hooke's law is ubiquitous. It applies to all materials, as long as the applied force is not so large as to cause permanent damage. And it applies to all *loading modes*, i.e. different ways in which forces can be applied to a solid. Figure 10.2 shows some examples. In the first a bar is pinned at its bottom end, and a *compressive force F* is applied by sitting a weight on its top; in the second a bar is pinned at both ends, with a *bending* or *flexural force F* applied by sitting a weight on its middle; in the third a bar is pinned at one end and a twisting *torsional force* or *torque F* is applied to its other end; in the fourth a bar is also pinned at one end and a *shear force F* is applied to its other end, at right angles to the axis of the bar; in the fifth, a *biaxial tensile force F* is applied equally in two dimensions to a square plate; and in the sixth, a *hydrostatic pressure* is applied equally in all dimensions to a solid cube.

Hooke's law is obeyed in all cases, with the *compressive, flexural, torsional, shear, biaxial* or *bulk deflection x* in each case proportional to the applied force *F*:

$$F \propto x = kx$$

$$\frac{F}{A} = E\frac{x}{l}$$

$$\sigma = E\varepsilon,$$

where σ, ε, k, $1/k$ and E are now the *compressive, flexural* (or *bending*), *torsional, shear, biaxial* and *hydrostatic* (or *bulk*) *stress, strain, stiffness, compliance* and *modulus* respectively. In general, different moduli for different geometries are not the same even for the same material. For most materials, however, the uniaxial tensile and compressive moduli are equal and are called the *Young's modulus*.

2   Robert Hooke

On 28 April 1686, the Fellows of the Royal Society gathered in their usual meeting place in Gresham College in the City of London for one of the most important scientific events ever: the first public presentation of Book I of Isaac Newton's *Philosophiae Naturalis Principia Mathematica (Mathematical Principles of Natural Philosophy)*.[1] Newton's *Principia,* according to a letter from him to the Society earlier that month, was to explain 'all phaenomena of celestial motions by only the supposition of a gravitation towards the center of the sun decreasing as the squares of the distances therefrom'.[2]

Newton had shown that the law of inertia (bodies continue to move at a constant velocity unless impacted on by a force), combined with his famous inverse square law of gravitational attraction, explained Kepler's experimental laws of planetary motion, i.e. that planets move in ellipses, that they sweep out constant areas with time and that the square of their orbiting period is proportional to the cube of their elliptical axis. The Fellows of the Royal Society were excited and impressed, fully recognising the significance of Newton's work, which finally solved what was seen as the most important scientific problem of the age, namely the motion of the sun and planets.

However, there was a dissenting voice. Robert Hooke, one of the founders of the Society and its first Curator of Scientific Works, complained bitterly that he had had the idea first, that Newton had just worked out the maths and that he (Newton) had failed to recognise his (Hooke's) contribution. Edmund Halley, newly appointed Clerk to the Society, wrote to Newton on 22 May, describing the book's reception as an 'incomparable treatise',[3] and mentioning, somewhat reluctantly, Hooke's complaints:

> There is one thing more that I ought to informe you of, viz, that Mr Hook has some pretensions on the invention of ye rule of the decrease of Gravity, being reciprocally as the squares of the distances from the Center. He sais you had the notion from him. ... Mr Hook seems to expect you should make some mention of him, in the preface. ...[4]

At first, Newton replied to Halley in a conciliatory way, but, after hearing direct accounts of the meeting, Newton wrote again, much more angrily:

Now is this not very fine? Mathematicians that find out, settle and do all the business must content themselves with being nothing but dry calculators & drudges & another that does nothing but pretend & grasp at all things must carry away all the invention. ... And why should I record a man for an Invention who founds his claim upon an error therein & on that score give me trouble? ... Should a man who thinks himself knowing, & loves to shew it in correcting & instructing others, come to you ... to boast that he taught you all he spake & oblige you to acknowledge it & cry out injury and injustice if you do not, I believe you would think him a man of strange unsociable temper.[5]

Moreover, Newton then removed Hooke's name from all his publications wherever he possibly could, for instance striking out his original acknowledgement to Hooke at the beginning of his yet-to-be-submitted Book III of the *Principia*.

The argument between Newton and Hooke is one of the most notorious scientific disputes of all time. History has judged decisively (and correctly) in Newton's favour. Hooke was undoubtedly the first to propose a *centripetal* rather than *centrifugal* gravitational force, in a lecture to the Royal Society in 1666 and in a letter to Newton in 1679 (the source of Newton's original acknowledgement of Hooke). Hooke also wrote to Newton later, in 1679, proposing that gravitational attraction decreased with distance according to an inverse square law. Newton had already discovered the inverse square law, but he did not say so in his response. They both independently had the idea of an inverse square law, though both thought they were the first. The clinching issue, however, was that Hooke never took the time, and did not have the mathematical ability, to demonstrate that these ideas led to Kepler's experimental laws. Hooke's views were a *hypothesis* whereas Newton's were fully substantiated mathematically and experimentally.

This was not the first time that Newton and Hooke had argued scientifically. In 1669, aged just 27, Newton took over from his tutor, Isaac Barrow, as the second Lucasian Professor of Maths at Cambridge University. Two years later, he submitted his first two communications to the Royal Society. One was about a small 6-inch (15-cm) reflecting telescope that the Royal Society's assessment committee found to produce larger images than a refracting telescope four times the size. The second was a letter describing experiments on light and colours, demonstrating and explaining how white light was composed of several different colours. These were Newton's first forays into public debate over scientific discoveries. Hooke regarded himself as an expert on telescopes and the nature of light, and didn't take kindly to the younger man's success. He claimed to have discovered Newton's results previously. As with the dispute over gravitation, there were elements of truth in this. Hooke had indeed studied reflecting telescopes, but discarded them in favour of refracting ones, and had also been the first, in his famous book *Micrographia*,[6] to show how a thin separation between two plates can refract light into its component colours. Again, however, as with the dispute over gravitation, Hooke did not pursue either topic to a conclusion, in part because he was too busy with many other projects and in part because of not having the mathematical ability to analyse his results. Newton and Hooke also maintained a long-running dispute over the trajectories (caused by gravitation) of objects dropped from a height.

Hooke also argued with many other scientists at this time. He argued with Christiaan Huygens about clocks and watches; he argued with Johannes Helvelius over telescopes and astronomical measurements; he argued with Constantijn Huygens and Antoni van Leeuwenhoek over biological microscopy; he argued with Gottfried Wilhelm Leibnitz over arithmetical calculating machines. He *almost* made many great scientific discoveries, which were subsequently made by others: Boyle's law of gas pressure, Huygen's theory of pendulums, Harrison's longitudinal timekeeper, Lavoisier's theory of combustion, Towneley's measuring micrometer. He worked on scientific subjects as diverse as semaphore messaging, printing presses, dyes, oil lamps, the design of carriages, respiration, barometers, microscopic organisms, musical instruments, comets, arches, hydraulics, animal poisons, plant growth, windmills, acoustics and many others. History has judged, in this case perhaps unfairly, in favour of Newton's jibe about Hooke as being 'a man of strange unsociable temper'.[7] In truth, Hooke worked simultaneously on hundreds of different scientific problems, usually not finishing one before rushing onto the next. He often kept half-worked-out inventions and ideas secret and, when other scientists presented papers on the topic, declared suddenly that he had already done the work, and done it better. His argumentativeness was a side effect of wanting to be recognised for what he had achieved. In fact he was paranoid from many years spent with no money and in hiding from the revolutionary Roundheads as a 'hated' Catholic Royalist. Personally, he was in fact very sociable, loved and supported over many years by his inner circle of friends and colleagues, spending, after the restoration of the monarchy, much time with them in his favourite resting place, the coffeehouses of London.

Robert Hooke was born in July 1635 in Freshwater on the Isle of Wight; he died in March 1703, aged 67, at his lodgings in Gresham College in the City of London. Between these dates, he lived through one of the most tumultuous periods in English history including: three civil wars, in 1642 to 1646, 1648 to 1699 and 1649 to 1651; the execution of King Charles I in 1649; the exile of his son King Charles II in 1651; the Commonwealth and Protectorate Roundhead governments under Oliver Cromwell in 1653 to 1659; the restoration of the monarchy under Charles II in 1660; the Great Plague of London in 1665; the Great Fire of London in 1666; and the glorious revolution in 1688, deposing King James II and installing William of Orange as King William III.

Hooke's father was a clergyman, and not well off. Robert was a sickly child, expected to follow his father into the church. However, the English civil wars were raging, and this had a decisive impact on his life. In November 1647, King Charles I escaped his captors at Hampton Court and fled to take refuge in Carisbrook Castle on the Isle of Wight, which was a major Royalist centre. In October 1648 the King was re-arrested by the Governor of the Isle of Wight, and just a few days later Robert's father died unexpectedly. Robert's family was high Anglican and Royalist, so Robert had to flee. Aged just 12 years old, with a tiny inheritance of £40, he left for London to be apprenticed to Sir Peter Lely, the Dutch-born royal portrait painter. Royalist connections continued to protect him. He became a student at Westminster School, where he learnt Latin, Greek, geometry, mechanics, and how to play the organ, under the protection of the Headmaster, Dr Richard Busby, another strong Royalist. He moved to Oxford University, under the tutelage of Dr John Wilkins, Warden of Wadham College and a Royalist cleric, becoming a student and chorister at Christ Church College. Money was a problem, but he had a genius for building scientific instruments, and he gained employment as a laboratory assistant to Sir Thomas Willis and then Robert Boyle.

At Oxford, Hooke became a member of an extraordinary group of young intellectuals, the cream of the country's up-and-coming talent, many with High Church, aristocratic and Royalist backgrounds, all excited by the opportunities for scientific and medical understanding, and most, unlike Hooke, well off. They included Robert Boyle, a wealthy Anglo-Irish aristocrat; Thomas Willis, who invented the field of neuroscience; John Locke, materialist philosopher; John Dryden, who became the first Poet Laureate; William Petty, inventor of statistics and census techniques; and Sir Christopher Wren, friend of the King, later appointed Royal Architect.

Following a lecture by Wren in November 1660, just after the restoration of Charles II, Hooke was involved with Boyle, Wilkins, Wren and others in setting up the Royal Society, which received its Royal Charter in June 1662. In November 1663, Hooke was appointed to be Curator of Experiments at the Royal Society, with the job of presenting each of the weekly meetings with exciting experiments. He needed a job: unlike the other founding members, he was too poor to treat the Royal Society as a dilettante pastime. The following year he was made Curator for life with an increased salary (£30 per annum), and was then appointed as Gresham's Cutlerian Lecturer on the History of Trades and Gresham's Professor of Geometry, in each case with the duty of giving educational lectures to the public. He was given lodgings at Gresham's College, where the Royal Society also held its meetings, and where he stayed until his death in 1703. In 1665 Hooke published his famous work on microscopy, *Micrographia*.[8]

The Great Plague of London in 1665 and the Great Fire in 1666 were cataclysmic events. The Great Fire came within one block of burning down Gresham's College, including all Hooke's (and the Royal Society's) scientific equipment, library and papers. It destroyed 12,000 houses and many public buildings, and left 65,000 people homeless. It led to the Royal Society temporarily abandoning Gresham's College. However, Hooke was in the right place at the right time, and he was appointed City Surveyor for the rebuilding by the Mayor of London. The King appointed Christopher Wren as Royal Architect to rebuild all the churches, and his close friend, Hooke, became second-in-command, as First Officer in Wren's architectural firm. Wren and Hooke worked together on many rebuilding projects, including St Pauls, the Royal College of Physicians, Bedlam (Bethlehem) Hospital for the Insane, Montagu House (later the first home of the British Museum) and perhaps as many as 30 or 40 churches. Through this work, Hooke gradually built up a large number of private clients for his own design and surveying practice. Hooke at last became quite well off, but, as with much of his scientific work, he got little long-term credit. Wren's son worked hard after their death to promote his father's name to the exclusion of Hooke. It is widely believed and often stated that Wren rebuilt London after the Great Fire, ignoring Hooke's contribution.

By the 1670s, Hooke had finally shaken off money concerns and was living comfortably at Gresham's College. His energy, enthusiasm and drive were prodigious. He had taken on an almost unbelievable range of intellectual activities and he worked at them with a furious, almost manic, intensity. He was doing, in parallel, at least five different jobs, all nominally full time and all incredibly demanding: Lab Assistant and general amanuensis to Robert Boyle; Curator of Experiments at the Royal Society; Gresham's Lecturer and Professor; City Surveyor; and First Officer in Wren's architectural firm. His diary describes a frenetic round of meetings, lectures, discussions and communications. He was, however, not intrinsically a strong or a well man, and he was increasingly beset with ailments. He was an insomniac and a hypochondriac, and he treated his own physical complaints with a wide range of completely unsuitable quack remedies and drugs, all meticulously recorded in his diaries, as if his body was a suitable field for medical experimentation. He paid a toll for this. The vibrant, over-achieving intellectual dynamo of the 1660s and 1670s had become, by the 1680s and 1690s, a weak, emaciated, haggard, paranoid, cantankerous old man. He developed a phobia against spending any money. Finally, he died in 1703, and his money and goods were stolen and fought over, ingloriously, by members of his family and scientists and members of the Royal Society.

Robert Hooke is, of course, best known for Hooke's law of elasticity. Hooke's interest in springs stemmed from his long-standing attempts to solve the problem of accurate timekeeping to determine longitude at sea. He developed the first balanced spring watch, similar to (but less accurate than) Harrison's successful solution to this problem in the next century. In 1675, Hooke published *A Description of Helioscopes and some other instruments*,[9] describing many of his inventions over the previous few years and their underlying principles. Some of the work was not fully revealed and was described only in an anagrammatic code. This was a (highly infuriating) publishing method Hooke had developed and liked to use to conceal details so that others could not steal his ideas. There were 10 chapters: one (Chapter 3) was entitled 'The True Theory of Elasticity or Springiness, and a Particular Explication Thereof in Several Subjects in Which It Is to be Found: And the Way of Computing the Velocity of Bodies Moved by Them. ceiiinossssttuu'[10]; and another (Chapter 9) was entitled 'A New Sort of Philosophical Scales, of Great Use in Experimental Philosophy. cdeiinnoopssssttuu'.[11] The anagrams, translated in Latin, as Hooke revealed in his Cutler Lecture two years later, stood for *ut tensio sic, vis*, meaning 'as the extension, so the force', and *ut pondus sic tensio*, meaning 'as the weight, so the extension'.

In July 1678 Hooke wrote up his theory of springs, entitled *Lectures de Potentia Restitutiva: Or of Springs*[12]; in August, he delivered his lecture on the topic to the Royal Society; in September, he got Wren's approval; in October, the book went to the printer; and in November, he distributed marbled or gilt copies to his special friends and colleagues, including Wren and Boyle. In the book, Hooke claims to have discovered the theory of springs 18 years earlier, when developing his spring-regulated watch. As well as describing the proportionality of applied force and resulting extension, Hooke devised a theory of matter to explain the behaviour. As described in Inwood's biography of Hooke, he claimed that materials consist of

> vibrating particles of the same or a harmonious magnitude and velocity … defending their space and their shape from the invasion of other particles. … If an elastic body were compressed to (say) half its former size, the distance between each vibrating particle would be halved and the number of collisions doubled. … The increase in the number of

collisions gives the compressed body a tendency to expand … an outward spring which is directly proportionate to the degree of its compression. … The dilation or stretching of a springy body would have precisely the opposite effect, creating longer vibrations, fewer collisions, and a tendency to retreat to its previous shape. So the longer the spring is extended, the greater the weight needed to prevent it from returning to its original shape, and the more a gas is compressed the greater the pressure it exerts on the vessel surrounding it.[13]

What a tour de force, giving the microscopic basis for Hooke's law and presaging Boltzmann's kinetic theory of gases as well as Einstein's and Debye's theories of atomic vibrations in solids.

3   Interatomic forces

Figure 10.3 shows the potential energy φ of two atoms separated by a distance a. The potential energy is at a minimum when the atoms are at their equilibrium interatomic separation a_o, with interatomic forces between the two atoms in balance. The potential energy increases when the atoms are displaced from their equilibrium separation by applying either a tensile force to stretch them apart, $a > a_o$, or a compressive force to push them together, $a < a_o$. Figure 10.3 also shows the applied force F obtained by differentiating the potential energy:

$$F = \frac{d\varphi}{da}.$$

The applied force is zero at the equilibrium separation a_o, positive (tensile) when the atoms are stretched apart $a > a_o$, and negative (compressive) when the atoms are pushed together $a < a_o$.

   The effect of changing the separation of the two atoms depends on their chemical interaction, i.e. on the interactions of their constituent electrons. Electronic interactions vary a lot from case to case, depending on the nature of the atoms. In general, however, there are two kinds of interaction: attractive forces as the outer electrons of the atoms interact to form covalent, ionic, metallic or van der Waals bonds; and repulsive forces when their inner core electrons overlap and repel each other because they cannot occupy the same inner core atomic space. As shown in Figure 10.3, attractive forces are dominant at large separations, repulsive forces are dominant at small separations, and the equilibrium separation and minimum potential energy correspond to attractive and repulsive forces being in balance.

   The nature of the attractive and repulsive forces are different, and the shape of the potential energy curve is not, therefore, symmetrical about the equilibrium separation, as shown in Figure 10.3. Nevertheless, the shape of the potential well can be regarded as symmetrical as long as displacements from the equilibrium separation are not too large, i.e. as long as $u = a - a_o$ is small. This is called the *harmonic approximation*. It comes from a *Taylor expansion* of the shape of the potential energy curve. In other words, when u is small, a is close to a_o, and we can write the potential energy at a separation a as a function of the potential energy and its derivatives at the equilibrium separation a_o:

potential
energy φ

repulsive forces as
electron clouds
overlap at low
separation

attractive forces
as outer electrons
interact at
high separation

equilibrium
potential φ_o

interatomic
separation a

equilibrium
interatomic
separation a_o

interatomic
force F

attractive forces
as outer electrons
interact at
high separation

interatomic
separation a

equilibrium
interatomic
separation a_o

repulsive forces as
electron clouds overlap
at low separation

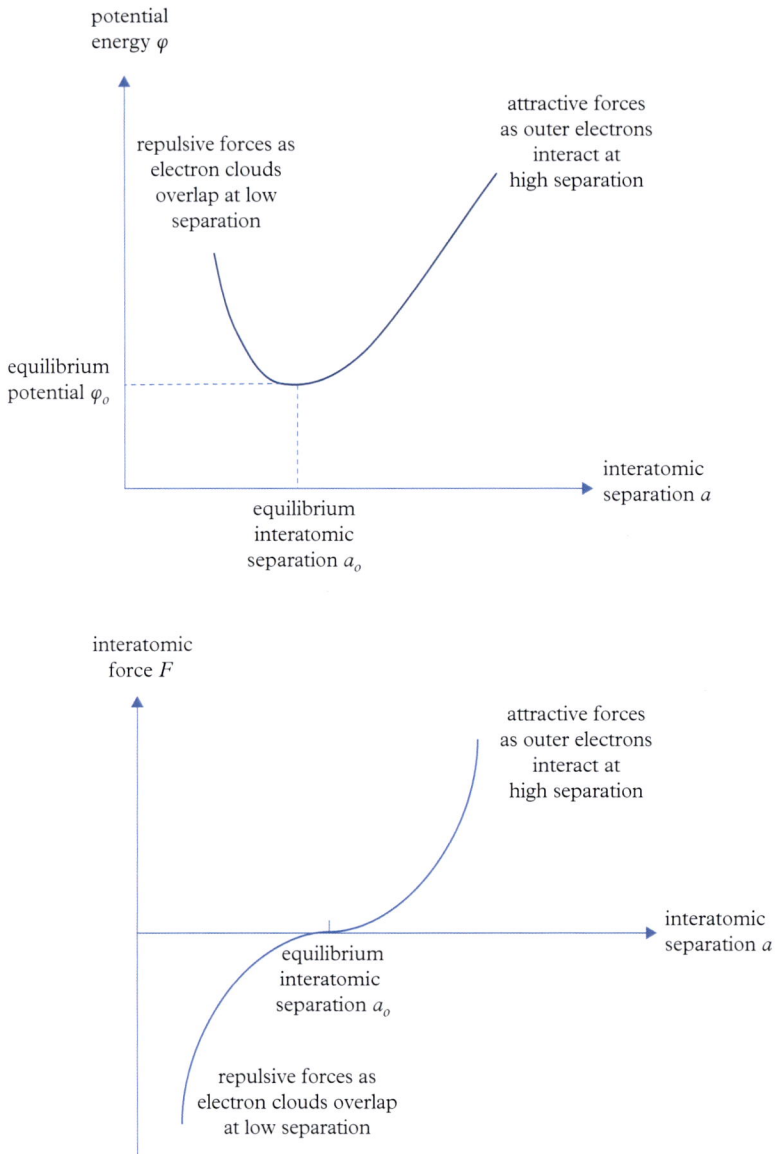

Figure 10.3 *Potential energy and interatomic force versus interatomic separation*

$$\varphi(a) = \varphi(a_o) + u\frac{d\varphi}{da}(a_o) + \frac{u^2}{2!}\frac{d^2\varphi}{da^2}(a_o) + \frac{u^3}{3!}\frac{d^3\varphi}{da^3}(a_o) + \dots$$

$$\approx \varphi(a_o) + \frac{1}{2}u^2\frac{d^2\varphi}{da^2}(a_o),$$

with the gradient $d\varphi/da = 0$ at a_o, and ignoring terms in u^3 and above.

Differentiating φ with respect to u gives Hooke's law,

$$F = \frac{d\varphi(a)}{du} = u\frac{d^2\varphi}{da^2}(a_o) = ku,$$

where the stiffness k is equal to the curvature of the potential energy $d^2\varphi/da^2$ at the equilibrium separation a_o:

$$k = \frac{d^2\varphi}{da^2}(a_o).$$

This explains, in broad terms, why elasticity is linear and reversible, why Hooke's law applies universally for small displacements, and why the stiffness and modulus are the same in tension and compression.

4   Stress and strain tensors

In a real material, we don't, of course, have just two interacting atoms as shown in Figure 10.3. Instead we have a large number of atoms and molecules, all interacting with each other in a complex geometric crystalline or amorphous structure.

When we apply external forces, therefore, to a real piece of material, the atoms and molecules in the material are all displaced from their equilibrium positions in complex ways, depending on

(1) the magnitudes of the external applied forces;

(2) the three-dimensional geometry of how they are applied to the material (as, for instance, in the different examples in Figure 10.2); and

(3) the three-dimensional network of internal forces of interaction among all the atoms and molecules in the material, which resist the effect of the external applied forces.

We use the concept of a *stress tensor* to describe the internal forces set up within a material by the application of a general set of external forces. And we use the concept of a *strain tensor* to describe the corresponding displacements of the material and its constituent atoms.

Consider a small unit cube within the bulk of a material subjected to a set of external applied forces. Excluding unbalanced forces, which translate or rotate the material as a whole, in general nine different internal stresses σ_{ij} are set up, as shown in Figure 10.4. Each stress σ_{ij} is a force per unit area, i.e. an internal force applied to one of the cube faces. The first subscript $i (= x, y$ or $z)$ gives the cube face to which the force is applied, normal to either the x-, y- or z-axis; and the second subscript $j (= x, y$ or $z)$ gives the direction of the force, parallel to either the x-, y- or z-axis. In other words, the stress σ_{ij} is the force applied to the face perpendicular to axis i in the direction parallel to axis j. The nine stresses are all balanced, i.e. there are equal and opposite stresses on the relevant front and back faces of the cube because there is no overall translation or rotational motion of the material.

The collection of nine stresses is called the *stress tensor*, which can be written in short form as $\{\sigma_{ij}\}$ or in full form as

$$\text{stress tensor} = \{\sigma_{ij}\} = \begin{array}{ccc} \sigma_{xx} & \sigma_{xy} & \sigma_{xz} \\ \sigma_{yx} & \sigma_{yy} & \sigma_{yz} \\ \sigma_{zx} & \sigma_{zy} & \sigma_{zz} \end{array} .$$

Only six of the nine components of the stress tensor are independent. There are three independent *uniaxial* stresses, σ_{xx}, σ_{yy} and σ_{zz}, which, for obvious reasons, are called the *diagonal* components of the stress tensor, and are tensile when positive and compressive when negative. And there are three independent *shear* stresses, $\sigma_{xy} = \sigma_{yx}, \sigma_{xz} = \sigma_{zx}$ and $\sigma_{yz} = \sigma_{zy}$, which are the *off-diagonal* components of the stress tensor. In each case, $\sigma_{ij} = \sigma_{ji}$ to prevent overall rotation of the cube, and we say that the stress tensor is *symmetric*. Figure 10.4 shows how uniaxial stresses extend or compress the cube and how shear stresses distort it.

Different internal states of stress are described by different stress tensors, with typical examples as follows:

uniaxial tension or compression:
$$\begin{array}{ccc} \sigma_{xx} & 0 & 0 \\ 0 & 0 & 0 \\ 0 & 0 & 0 \end{array}$$

biaxial tension or compression:
$$\begin{array}{ccc} \sigma_{xx} & 0 & 0 \\ 0 & \sigma_{xx} & 0 \\ 0 & 0 & 0 \end{array}$$

hydrostatic tension or compression:
$$\begin{array}{ccc} \sigma_{xx} & 0 & 0 \\ 0 & \sigma_{xx} & 0 \\ 0 & 0 & \sigma_{xx} \end{array}$$

pure shear:
$$\begin{array}{ccc} 0 & \sigma_{xy} & 0 \\ \sigma_{yx} & 0 & 0 \\ 0 & 0 & 0 \end{array} .$$

internal tensile stresses

internal shear stresses

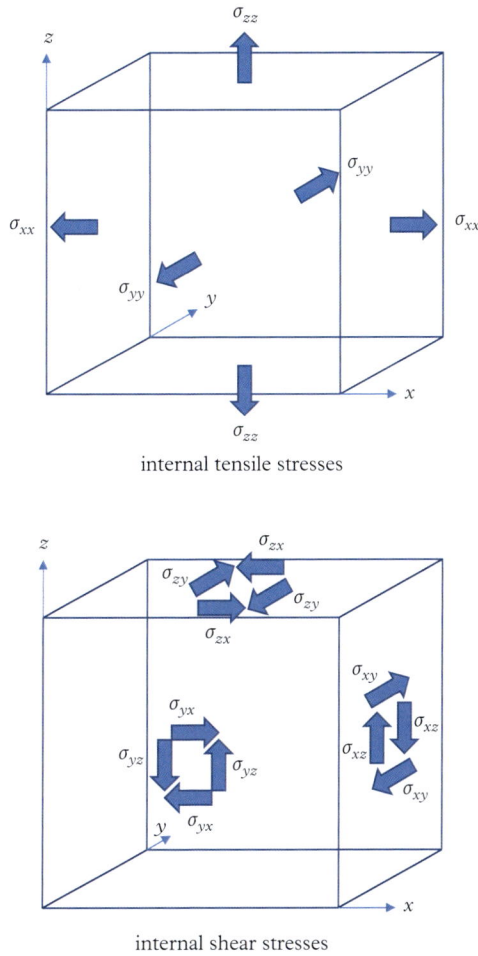

Figure 10.4 *Definition of internal tensile and shear stresses σ_{ij}*

Consider again our small unit cube within the bulk of a material, subjected to a set of external applied forces. Excluding displacements that translate or rotate the material as a whole, in general nine different internal strains ε_{ij} give the resulting change in shape, as shown in Figure 10.5. Each strain ε_{ij} is an extension per unit length, i.e. an internal displacement applied to one of the cube faces along the unit length of one of the cube axes. The first subscript $i (= x, y$ or $z)$ gives the cube face to which the displacement is applied, normal to either the x-, y- or z-axis; and the second subscript $j (= x, y$ or $z)$ gives the direction of the displacement, parallel to the x-, y- or z-axis. In other words, the strain ε_{ij} is the displacement of the face perpendicular to axis i in the direction parallel to axis j.

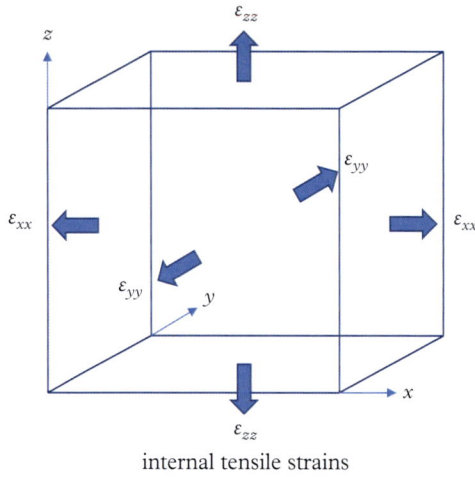

internal tensile strains

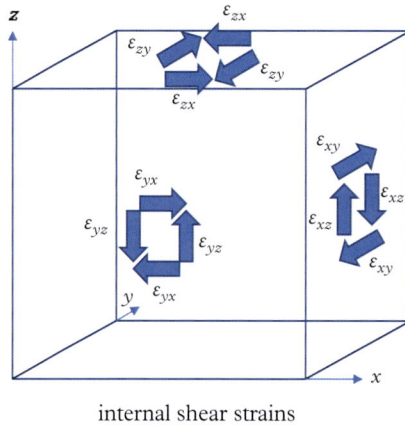

internal shear strains

Figure 10.5 *Definition of internal tensile and shear strains ε_{ij}*

The collection of nine strains is called the *strain tensor*, which can be written in short form as $\{\varepsilon_{ij}\}$, or in full form as

$$\text{strain tensor} = \{\varepsilon_{ij}\} = \begin{matrix} \varepsilon_{xx} & \varepsilon_{xy} & \varepsilon_{xz} \\ \varepsilon_{yx} & \varepsilon_{yy} & \varepsilon_{yz} \\ \varepsilon_{zx} & \varepsilon_{zy} & \varepsilon_{zz} \end{matrix} \ .$$

As with the stress tensor, only six of the nine components of the strain tensor are independent. There are three independent *uniaxial* strains, ε_{xx}, ε_{yy} and ε_{zz}, which are the *diagonal* components of the strain tensor, and are tensile when positive and compressive when negative. And there are three independent *shear* strains, $\varepsilon_{xy} = \varepsilon_{yx}$, $\varepsilon_{xz} = \varepsilon_{zx}$ and

$\varepsilon_{yz} = \varepsilon_{zy}$, which are the *off-diagonal* components of the strain tensor. In each case, $\varepsilon_{ij} = \varepsilon_{ji}$ to prevent overall rotation of the cube, and we say that the strain tensor is *symmetric*. Figure 10.5 shows how uniaxial strains are extensions or compressions, and shear strains are distortions.

5   Generalised Hooke's law

In the most general case, a combination of external applied forces leads to a distribution of internal forces described by a stress tensor with nine components, six of which are independent, and a corresponding distribution of local displacements described by a strain tensor also with nine components, six of which are independent. Each component of stress depends on each strain, and for a *linear elastic material*, the dependences are all linear, i.e. obey Hooke's law:

$$\sigma_{ij} = \sum_{k,l=x,y,z} c_{ijkl}\varepsilon_{kl},$$

where c_{ijkl} are individual *elastic constants*. For σ_{xx} for example,

$$\begin{aligned}
\sigma_{xx} = &\, c_{xxxx}\varepsilon_{xx} + c_{xxxy}\varepsilon_{xy} + c_{xxxz}\varepsilon_{xz} \\
&+ c_{xxyx}\varepsilon_{yx} + c_{xxyy}\varepsilon_{yy} + c_{xxyz}\varepsilon_{yz} \\
&+ c_{xxzx}\varepsilon_{zx} + c_{xxzy}\varepsilon_{zy} + c_{xxzz}\varepsilon_{zz},
\end{aligned}$$

with similar equations for all the other nine stress components σ_{ij}. There is a total of $9 \times 9 = 81$ elastic constants c_{ijkl}, with 21 independent in the most general case. For a material with a cubic crystal structure, this reduces further to only three independent elastic constants:

$$\begin{aligned}
c_{11} &= c_{iiii} = c_{xxxx} = c_{yyyy} = c_{zzzz} \\
c_{12} &= c_{iijj} = c_{xxyy} = c_{yyzz} = c_{zzxx} \\
c_{44} &= c_{ijij} = c_{xyxy} = c_{yzyz} = c_{zxzx},
\end{aligned}$$

with all the other elastic constants being zero. Polycrystalline solids are usually elastically isotropic (i.e. the same in all directions) because the anisotropies of the individual crystals cancel out. This reduces the number of independent elastic constants further, to just two, c_{11} and c_{12}, with $c_{44} = \frac{1}{2}(c_{11} - c_{12})$.

6   Elastic moduli

Uniaxial tension

Under conditions of uniaxial tension, the stress and strain tensors are

$$\begin{pmatrix} \sigma & 0 & 0 \\ 0 & 0 & 0 \\ 0 & 0 & 0 \end{pmatrix}$$

$$\begin{pmatrix} \varepsilon & 0 & 0 \\ 0 & -\nu\varepsilon & 0 \\ 0 & 0 & -\nu\varepsilon \end{pmatrix},$$

where $v \sim 0.5$ is the *Poisson's ratio*. In other words, a uniaxial tensile stress σ in the x direction leads to a tensile strain ε in the x direction, where $\varepsilon = \sigma/E$, and compressive strains $-v\varepsilon$ in both the y and z directions. The material behaviour is determined by its Young's modulus E and the Poisson's ratio v.

Hydrostatic pressure

Under conditions of hydrostatic pressure, the stress and strain tensors are

$$
\begin{matrix}
p & 0 & 0 \\
0 & p & 0 \\
0 & 0 & p
\end{matrix}
$$

$$
\begin{matrix}
\varepsilon & 0 & 0 \\
0 & \varepsilon & 0 \\
0 & 0 & \varepsilon
\end{matrix} .
$$

In other words, a uniform pressure p in the x, y and z directions leads to a uniform strain ε in the x, y and z directions, where $\varepsilon = p/K$ and K is the *bulk modulus*. The material behaviour is determined by its bulk modulus K.

Pure shear

Under conditions of pure shear, the stress and strain tensors are

$$
\begin{matrix}
0 & \sigma & 0 \\
\sigma & 0 & 0 \\
0 & 0 & 0
\end{matrix}
$$

$$
\begin{matrix}
0 & \gamma & 0 \\
\gamma & 0 & 0 \\
0 & 0 & 0
\end{matrix} .
$$

In other words, a pure shear stress σ in the x and y directions leads to a pure shear strain γ in the x and y directions, where $\gamma = \sigma/\mu$ and μ is the *shear modulus*. The material behaviour is determined by its shear modulus μ.

Because there are only two independent elastic constants in an elastically isotropic material, there are relationships between the different moduli:

$$
K = \frac{E}{3(1-2v)}
$$

$$
\mu = \frac{E}{2(1+v)}.
$$

Table 10.1 shows some typical values of Young's modulus for a variety of materials.

Table 10.1 *Typical Values of Young's Modulus*

Material	Young's modulus E (psi)
Alumina	50
Steel	200
Glass	10
Concrete	2
Wood	1.5
Polyethylene	0.02
Rubber	0.005

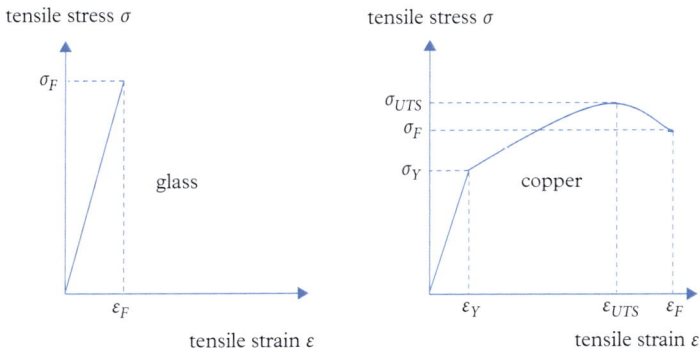

Figure 10.6 *Tensile stress–strain curves for brittle glass and ductile copper*

7   Elastic limits

Hooke's law is obeyed for almost all materials for relatively small applied forces or stresses and correspondingly small displacements or strains. At higher levels of applied force, a material exceeds its *elastic limit*, i.e. it suffers permanent damage from the effect of the applied force, and its response becomes *non-linear* and *irreversible*. The nature of the permanent damage varies quite a lot for different materials and for different geometries of the applied forces.

The fundamental behaviour of a material under mechanical loading is often shown by plotting a *stress–strain curve*, i.e. a plot of the applied stress versus the corresponding induced strain, often under simple tensile loading conditions (i.e. similar to that shown in Figure 10.1). Figure 10.6 shows some examples of tensile stress–strain curves for different materials.

Table 10.2 *Typical Values of Yield and Ultimate Tensile Strengths*

Material	Yield strength σ_Y (MPa)	Ultimate tensile strength σ_{UTS} (MPa)
Mild steel	250	850
High-strength steel	2,200	2,500
Aluminium	15–20	50
Aluminium alloy	414	448
Copper	117	210
High-density polyethylene	30	37
Nylon	450	750
Concrete	—	2–5
Rubber	—	16
Glass	—	30
Glass fibre	—	3,500–5,000
Carbon nanotube	—	11,000–63,000
Graphene	—	130,000

Ceramic materials such as quartz or glass show elastic behaviour with increasing tensile load, i.e. the induced strain is proportional to the applied stress, until they reach their *fracture strength* σ_F and *fracture strain* ε_F, with $\sigma_F = E\varepsilon_F$. This is the point at which they fracture, with minute pre-existing cracks beginning to propagate catastrophically, running across the material and causing it to break into pieces. This is called *brittle fracture* or *brittle failure*.

Metallic materials such as aluminium or copper also show elastic behaviour with increasing tensile load until they reach their *yield strength* σ_Y and *yield strain* ε_Y, with $\sigma_Y = E\varepsilon_Y$. This is the stress and strain at which they begin to respond *plastically*, i.e. the strain begins to rise non-linearly with increasing applied stress, and the materials *deform* with a permanent set after unloading. The stress reaches a maximum, called the *ultimate tensile strength* or *UTS* at σ_{UTS} and ε_{UTS}. Finally the materials reach their *fracture strength* σ_F and *fracture strain* ε_F. This is the point at which they fracture with ductile tearing of the material. Metals often show very extensive deformation, with total plastic strain often reaching several tens of percent. The permanent deformation is produced by groups of atoms sliding over each other. The extensive deformation of metals explains why they can be manufactured into complex shapes such as car bodies or aeroplane wings.

Polymers such as polyethylene and nylon also show extensive deformation, but in this case it is because their constituent long-chain molecules disentangle, stretch and re-entangle with a permanent set.

Table 10.2 shows typical values of yield and ultimate tensile strength (for brittle materials, the ultimate tensile strength is equal to the fracture strength).

· ·

8 REFERENCES

1. Isaac Newton. *Philosophiae Naturalis Principia Mathematica* [Mathematical Principles of Natural Philosophy] (Royal Society, London, 1687; translated and re-printed, University of California Press, Oakland, 2016).
2. Thomas Thomson. *History of the Royal Society: From Its Institution to the End of the Eighteenth Century* (Cambridge University Press, Cambridge, 2011), xvii.
3. Lisa Jardine. *The Curious Life of Robert Hooke: The Man Who Measured London* (Harper, London, 2003), 3.
4. Ibid. 4.
5. Ibid. 6.
6. Robert Hooke. *Micrographia: Or Some Physiological Descriptions of Minute Bodies Made by Magnifying Glasses, with Observations and Inquiries Thereupon* (J. Martyn and J. Allestry, London, 1655; re-printed, General Books, Memphis, 2010 and Cosimo Classics, New York, 2007).
7. Jardine, n. 3, 6.
8. Hooke, n. 6.
9. Robert Hooke. *A Description of Helioscopes and some other instruments* (John Martyn, London, 1676). (Gale, Farmington Hills, 2010; re-printed in *The Posthumous Works of Robert Hooke*).
10. Ibid.Chapter 3, 31.
11. Ibid.Chapter 9, 32.
12. Robert Hooke. *Lectures de Potentia Restitutiva: Or of Springs: Exploring the Power of Springing Bodies* (John Martyn, London, 1678).
13. Stephen Inwood. *The Man Who Knew Too Much: The Strange and Inventive Life of Robert Hooke 1635–1703* (Macmillan, London, 2002), 274–275.

· ·

9 BIBLIOGRAPHY

Micrographia Restaurata. Robert Hooke (Gale ECCO, Detroit, 2010).
Robert Hooke and the English Renaissance. A. Chapman (Gracewing, Leominster, 2005).
The Curious Life of Robert Hooke: The Man Who Measured London. Lisa Jardine (Harper, London, 2003).
The Man Who Knew Too Much: The Strange and Inventive Life of Robert Hooke 1635–1703. Stephen Inwood (Macmillan, London, 2002).
The Mechanical Properties of Matter. A. H. Cottrell (Wiley, New York, 1964).

11

The Burgers Vector

Plasticity

1 Dislocations

When a material is stretched it gets bigger, and when the stretching force is removed it goes back to its original size. This is called *elasticity*. There are two features of elasticity: first, the extension of the material is proportional to the force used to stretch it, i.e. it obeys *Hooke's law*; and second, the extension is reversible. In some materials, reversibility continues up until the material breaks. These are called *brittle materials*. In other materials, reversibility is only seen up to a certain maximum applied force and corresponding maximum extension, called the *elastic limit* or *yield point*. For greater applied forces and greater extensions, the material does not return fully to its original size when the force is removed, and there is a permanent residual extension, i.e. a permanent set. This is called *plasticity* or *ductility*, and these materials are called *ductile materials*.

How does a material develop a permanent set? There are several mechanisms, but the most important one is the movement of *dislocations*, which are *line defects* in the crystal structure. Dislocations are characterised by their *Burgers vector*. We examine dislocations and their Burgers vectors first, before we consider how their motion causes plasticity.

A dislocation in a material is defined as follows. Consider a piece of solid material. Make a partial cut in the material up to a line (which will become the dislocation line). Displace the two cut faces relative to each other by a vector $\boldsymbol{b}$. Repair the material across the cut faces. The displacement vector $\boldsymbol{b}$ is the *Burgers vector*, and a unit vector $\boldsymbol{l}$ along the dislocation line, i.e. along the end of the cut, is the *dislocation line vector*. If the displacement is perpendicular to the dislocation line, $\boldsymbol{b} \perp \boldsymbol{l}$, it is called an *edge dislocation*; if the displacement is parallel to the dislocation line, $\boldsymbol{b} \parallel \boldsymbol{l}$, it is called a *screw dislocation*; if the displacement is neither perpendicular nor parallel to the dislocation line, it is called a *mixed dislocation*:

$$\text{edge: } \boldsymbol{b} \cdot \boldsymbol{l} = 0$$
$$\text{screw: } \boldsymbol{b} \cdot \boldsymbol{l} = b$$
$$\text{mixed: } \boldsymbol{b} \cdot \boldsymbol{l} = b \sin \theta,$$

The Equations of Materials. Brian Cantor. Oxford University Press (2020). © Brian Cantor.
DOI: 10.1093/oso/9780198851875.001.0001

where b is the magnitude of the Burgers vector b, and θ is the angle between the Burgers vector b and line vector l.

Figure 11.1 shows an example of an *edge dislocation* in a simple cubic crystal. The edge dislocation can be described as inserting a half plane of atoms above the horizontal plane *x-x* (or removing a half plane of atoms below *x-x*). The *edge dislocation line* is the row of atoms at the end of the half plane.

Figure 11.2 shows an example of a *screw dislocation*, also in a simple cubic crystal. The screw dislocation can be described as a helical twisting of the rows of atoms. The *screw dislocation line* is the central axis of the helical twist.

In crystalline materials, the Burgers vector can be identified by drawing a *Burgers circuit*, a clockwise closed circuit of lattice points drawn in the original crystal, as shown in Figure 11.3. After the dislocation is created, the closure of the circuit is broken, and the vector required to complete closure is the Burgers vector b, as also shown in Figure 11.3. Dislocations are *topological* defects in the crystal structure, and the Burgers vector is

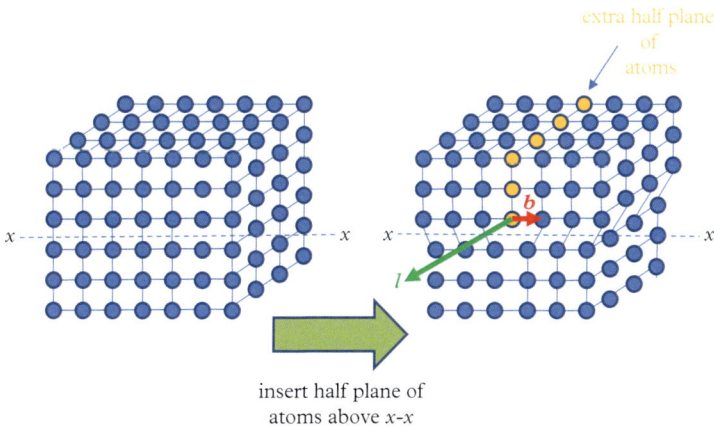

extra half plane of atoms

insert half plane of atoms above *x-x*

Figure 11.1 *Edge dislocation*

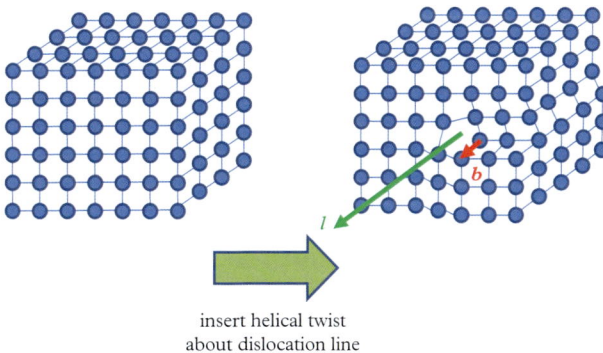

insert helical twist about dislocation line

Figure 11.2 *Screw dislocation*

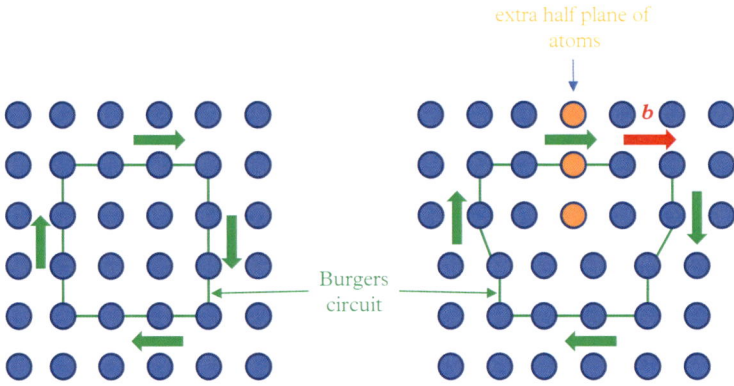

Figure 11.3 *Burgers circuit in perfect and dislocated crystal*

Table 11.1 *Common Dislocations and Slip Planes in Simple Crystal Structures*

Crystal	Burgers vector	Slip plane
Simple cubic	$a<100>$	$\{100\}$
Face-centred cubic	$a/2<110>$	$\{111\}$
Hexagonal close packed	$a/3 < 10\bar{1}0 >$	$\{0001\}$
Body-centred cubic	$a/2<111>$	$\{110\}$

conserved along the line of the dislocation, as can be seen in Figures 11.1 and 11.2. Effectively, there is a Burgers equation:

$$\boldsymbol{b} = \text{constant along } \boldsymbol{l}.$$

Burgers vectors are usually described in the form

$$\boldsymbol{b} = \frac{a}{n} < hkl >,$$

where a is the lattice parameter, $<hkl>$ is the crystallographic direction of the Burgers vector, and $1/n$ is the fractional distance of its magnitude b across the unit cell along $<hkl>$.

In principle, dislocations can have any Burgers vector $\boldsymbol{b}$, i.e. can have displacements of any magnitude b, in any direction $<hkl>$. In practice, however, materials exhibit only a few kinds of dislocations, with Burgers vectors usually corresponding to simple lattice translation vectors. The most common dislocations have Burgers vectors equal to the interatomic spacing. Typical Burgers vectors for some common dislocations in simple crystal structures are shown in Table 11.1.

2   Jan Burgers

Jan Burgers

On 25 September 1933 the famous physicist Paul Ehrenfest visited his 15-year-old son Wassik, who had suffered since birth from Down's syndrome, and was living and being cared for in the Waterink Institute for Afflicted Children in Amsterdam. He met his son in the waiting room of the Institute. He was carrying a pistol. He shot Wassik in the head, then immediately turned the gun and shot and killed himself. Wassik survived his father by a few hours. It was a shocking episode. Ehrenfest had been seriously depressed and suicidal for years. The previous year he had written to his first doctoral student, Jan Burgers, to say that he was going to commit suicide during the following year, despite the 'financial and moral damage'[1] it would cause his family. He bid Burgers farewell and begged him, 'Please forgive me for my weakness as a human being'.[2] He had been sending similar letters to his ex-doctoral students over the previous decade and more. This time, clearly, he meant to go through with it. The physicist and science writer Paul Halpern tells a macabre story about his thesis advisor at Stony Brook University, Max Dresden, coming into the classroom and announcing dramatically: 'I am very troubled. Ludwig Boltzmann committed suicide. His student Paul Ehrenfest committed suicide. Ehrenfest's student George Uhlenbeck was my advisor. If Uhlenbeck commits suicide, I'm next'.[3]

Johannes Martinus ('Jan') Burgers was born in Arnhem in the eastern part of the Netherlands on 13 January 1895, the eldest child of Johannes Burgers and Johanna Romijn. His father was a postman, not well educated but a lover of science literature and scientific instruments. He gave amateur scientific lectures and demonstrations to his friends, often aided by his son Jan, and built up impressive collections of microscopes and minerals, now in the Boerhaave Museum in Leiden. He took Jan and his younger brother Willy on rambles throughout the beautiful countryside surrounding Arnhem, instilling in them a love of natural flora and fauna and a desire to study them. Jan wrote,

> My father planted in me reverence for the wonders of nature. ... I have inherited his desire to see things as well as his broadness of interest. ... I have always been an absorber of knowledge, and one who likes to reproduce thoughts in a re-arranged form.[4]

At the Hogere Burgerschool (high school), he quickly outgrew and surpassed his father's tutelage, attending the Arnhem Physical Society and studying Latin and Greek so he could apply to go to the University of Leiden, which he finally joined to study physics in October 1914, just after the outbreak of the First World War.

Leiden was a powerhouse of physics. The long-standing Professor of Theoretical Physics was Hendrik Lorentz, who famously derived the transformation equations underpinning Einstein's theory of relativity and the invariance of the speed of light, and who won the second Nobel Prize for Physics for explaining the Zeeman effect (named after his student Peter Zeeman), whereby atomic spectral lines are split by a magnetic field. Lorentz had been appointed at Leiden in 1877 when he was just 24 years old, and had resigned 35 years later in 1912, but continued as an extraordinary professor giving occasional lectures. Lorentz's replacement was the charismatic and gregarious Paul Ehrenfest, who took up the young Burgers and invited him with a group of other outstanding students to weekly seminars at his house. Ehrenfest had an enormous impact on Burgers, who wrote,

> It is Ehrenfest who has had the greatest influence upon my [scientific] development, and who introduced me into the spirit of real scientific inquiry in physics. ... Ehrenfest taught us how to read scientific papers, to look for the assumptions made by the authors, and to hunt them out when they were not given explicitly. His powerful analytical mind opened our eyes to many subtleties in physical theory.[5]

According to Burgers biographers, Nieuwstadt and Steketee, 'Ehrenfest's room almost became his [Burgers'] home and his teacher began to replace his father'.[6] As Burgers himself said, 'No longer was it my father's maxims which took first place in my thinking. Ehrenfest's influence became the stronger one'.[7] In the end, however, Ehrenfest's influence became overpowering and intrusive, both scientifically and socially. Ehrenfest did not approve of Jan's girlfriend and fiancée, Nettie Roosenschoon. The two men began to grow apart, and they became estranged after Jan and Nettie's marriage in 1919. Burgers later wrote of Ehrenfest,

His analytical mind stirred up everything. . . . Ehrenfest would question every aspect. . . . In later years I understood that I possessed something that Ehrenfest missed: a conviction of the meaning of life, a faith in the sense of all things around us, which I owe to my father. This has helped me over difficulties that Ehrenfest has not been able to conquer.[8]

Paul Ehrenfest was born in Vienna on 18 January 1880. His parents, Sigmund Ehrenfest and Johanna Jellinek, were grocers and were Jewish, though not particularly religious. He studied theoretical physics with Ludwig Boltzmann at the University of Vienna and Felix Klein at the University of Göttingen, where he met and married his Russian-born co-student, Tatyana Afanyeseva. They had four children, their youngest, with Down's syndrome, Vassily or 'Wassik'. They moved to St Petersburg in 1907, which they loved, but as an Austrian Jew he had no chance of a permanent job, so they moved in 1912 to Leiden, where he took over the professorship held previously by Lorentz. He was a close friend of Einstein and Bohr, and in his laboratory at Leiden had a series of outstanding students such as Burgers, George Uhlenbeck, Samuel Goudsmit and Niko Tinbergen, and hosted foreign visitors such as Enrico Fermi, Robert Oppenheimer, Werner Heisenberg and Paul Dirac. He made major contributions to the theory of phase transitions and quantum mechanics. He was an enormous extrovert and an outstanding lecturer who loved to perform dramatic demonstrations to explain the physics underlying different phenomena. He lived his life very close to his students, probably oppressively so. His depression was brought on by a variety of factors: he was sympathetic to the Russian Revolution and its underlying communist and socialist principles, but he became dispirited by the harshness of the Russian government and its increasing link with the Nazis in Germany; he began an affair with the art historian Nelly Posthumus Meyjes, putting great strain on his marriage; and he worked himself into the ground, in his physics, his politics and his personal life. According to Dirk van Delft, Director of the Boerhaave Museum, he subjected himself to 'merciless self-criticism'.[9] The thought of another scientific paper or talk, wrote Ehrenfest, 'plunges me into self doubt. I'd rather be dead. The pressure I'm living under is inconceivable'.[10] And to cap it all he was, of course, continually oppressed with the effort and financial burden of caring for his severely disabled youngest son, Wassik.

In January 1916 Burgers took a job as an assistant to Kamerlingh Onnes, who had discovered superconductivity in 1911. This was experimental work, and Jan realised he preferred theory so he left the following year. Early in 1918 he completed his doctoral thesis and left Leiden to take a new job as Custodian of the Teylers Stichting (or Foundation) in Haarlem, recommended by Lorentz, who was Curator of the Foundation. This was a prestigious appointment. Teylers was founded well over a hundred years earlier by the 18th-century cloth merchant and banker Pieter Teyler van der Hulst to encourage worship, science and art. Burgers duties were to give scientific lectures and oversee the collection of scientific instruments. He enjoyed the regular contact with Lorentz, but found the job dull and missed the intellectual environment in Leiden. In September 1918 he left Teylers and Haarlem to become Professor of Aerodynamics, Hydrodynamics and Their Applications in the Department of Mechanical Engineering,

Shipbuilding and Electrical Engineering at the Technische Hogerschool Delft (later the Technische Universiteit Delft or just TU Delft). This was a remarkable appointment. Burgers only passed his doctoral examination two months later, and his thesis was on the rather obscure but important mathematical issue of adiabatic invariants in the quantum mechanics recently developed by the Danish scientist Niels Bohr. Suddenly he was appointed as professor at a technical university, in an applied field, in a subject he had never previously studied. He got the job by having three great referees: Lorentz, Onnes and Ehrenfest. He left quantum mechanics because he felt he had 'insufficient phantasy for making fruitful advances in Bohr's theory'.[11] And he judged, rightly, that the new applied fields of solid mechanics, aerodynamics, hydrodynamics and fluid flow would benefit from application of fundamental and mathematical theory. The time was right in the Netherlands for the growth of these subjects. Aeroplanes were, literally and metaphorically, taking off commercially after their successful deployment during the First World War; the Dutch government founded the Rijks Studiedienst voor de Luchtvaart (Institute for Research in Aerial Navigation, or RSL) in 1918; the Fokker Aviatik company, originally founded by the Dutch aeroengineer Anthony Fokker in Berlin, moved its main production plant to Amsterdam in 1919; and TU Delft decided to invest in Burgers' new professorship.

Burgers was to stay at Delft for 37 years, until leaving for America in 1955. Although he was, in many ways, very successful, he was, in the end, never fully supported by the University, and he never felt fully comfortable with his range of duties. He was clear from the beginning that

> [t]he purpose of this laboratory was the study of all phenomena which take place during the flow of fluids and gases, of the form of the flow, of the pressure on different objects, of friction phenomena, of the transmission of waves, of jets of fluid and gas.[12]

He was excited by the possibility of developing new fundamental theory, alongside experimental investigations and applications to real-world problems. In his inaugural address he tried to convince his audience that theory had to be taken into account. As he said, however, 'on the whole topics governed by nonlinear equations had not yet come into fashion'.[13] And at Delft he was confronted by old-fashioned, uncomprehending engineers, much more used to traditional rule-based methods. The new laboratory was not opened until 1921, even then only as a temporary wooden construction, and with only limited space, equipment and support for researchers. His chief assistant and experimentalist, Theodorus van der Hegge Zijnen, was appointed in 1921, and they worked together for many years, though they were never close personally. They installed an Eiffel-style wind tunnel and an Ahlborn towing tank, but times were difficult throughout the interwar years and during the Second World War, so Burgers spent much of his time trying, often fruitlessly, to find funds to extend the labs, install equipment and support students and colleagues. He was too diffident to enjoy his teaching, he was uncomfortable as an experimentalist, he disliked administration, and at times he was disappointed at the extent of applied work. When he left Delft in 1955 he reflected,

> I was more of a theoretician than experimenter … as a leader of people, actually I didn't bring along anything. … I was too young and in many senses I have remained like that. … I never properly learned how to attract people and to keep them with me.[14]

Nevertheless, Burgers was very successful scientifically. Experimentally, helped by van der Hegge Zijnen, he developed streamline visualisation methods and the use of hot-wire anemometry to measure flow velocity, which were then applied extensively to explore the complex behaviour of boundary layers, aerofoils and turbulent flow. Theoretically he derived equations to describe vortices, viscous flow, boundary layers, turbulence, rheology, suspensions and compressible flow. He was much exercised for most of his career with developing novel statistical theories of fluid turbulence. At its most general, fluid flow is governed by the *Navier-Stokes equations*, a set of notoriously complex non-linear partial differential equations that are extremely difficult to solve except in the simplest, often trivial, cases. He developed his famous *Burgers equation* to provide a much more convenient (though still complex) one-dimensional version of Navier-Stokes, easier to use, but capturing much of the basic physical behaviour. According to the American mathematician, John von Neumann, commissioned just after the end of the Second World War by the US government to report on worldwide advances in understanding turbulence,

> Burgers found various 1-dimensional problems … which appear, nevertheless, to be able to imitate the essential properties of turbulence. Because of its one dimensionality, and also because of various other mathematical traits, Burgers managed to discuss these examples much more completely than one can do it with those that control true turbulence (the 3-dimensional Navier-Stokes equations).[15]

Burgers also worked on a wide variety of important applications, including: the flow of cooling water for power plants; the mixing of clay and water to manufacture bricks; ventilation in drying ovens; vibration effects in mixing concrete; pneumatic grain elevators; and liquid flow round bends in pipes. He contributed to two of the biggest construction projects of the mid-20th century: the building of the dams to create the Zuiderzee and the ventilation of the road tunnel under the River Maas in Rotterdam. He collaborated extensively with major scientists such as: Aikitsu Tanakadate in Japan, who worked on seismology and military aviation; Maurice Biot in the United States, who provided the first analysis of poro-elasticity, i.e. the elasticity of porous media such as soil and sand beds; Ludwig Prandtl in Germany, who pioneered the mathematical analysis of aerodynamics; G. I. Taylor in England, who demonstrated that individual photons were wavelike and produced interference fringes, and also studied shock waves and turbulence; Abram Ioffe in Russia, who made important contributions to radar, photoelectricity and thermoelectricity; and Richard von Mises in Austria, who developed theories of aerodynamic boundary layers and the strength of materials. He was particularly friendly with Theodore von Kármán, the Hungarian scientist who made important advances in supersonic and hypersonic flow, and whom he visited in TU Aachen in Germany and then later at the California Institute of Technology (Caltech) in the United States. He was active in the Koninklijke Nederlandse Akadamie van Wetenschappen (Dutch Academy of Sciences or KNAW), helped establish a series of influential international conferences on applied mechanics, hosting the first in Delft in 1924, and helped set up the International Union of Theoretical and Applied Mechanics in 1946.

In the 1930s Burgers became interested in the rheology, viscosity and plastic deformation of solids. The KNAW established a Viscosity Committee to understand these related phenomena, and Burgers became secretary of the Committee. As part of this, he began to collaborate with his younger brother Willy, who studied chemistry in Leiden, and would later become Professor of Crystallography at TU Delft in 1940. Plastic flow was thought to be associated with movement of line defects called *dislocations*, but at that time, only G. I. Taylor's edge dislocation in crystal lattices was known. Burgers recalled in 1955,

> Taylor's work stimulated many minds to occupy themselves also with this topic. The problems posed by my brother induced me to consider the mathematical equations which had been given by Volterra for dislocations in elastic media. This made us see that beside the dislocation described by Taylor another type had to exist, and that by combination of both types a very general form could be constructed.[16]

Burgers had discovered the screw dislocation and the general form of mixed dislocations in crystals, including the key displacement vector describing the topological and crystallographic nature of each defect. He published his paper in the proceedings of the KNAW in 1939,[17] and later his Bristol lecture on the same topic was published in the *Proceedings of the Physical Society* in 1940.[18] This led to the general adoption of the term *Burgers vector* to describe the structure of any dislocation. The significance of dislocations in plastic flow remained speculative until their direct observation by the (then) innovative technique of transmission electron microscopy by Peter Hirsch and co-workers in 1956.[19,20]

For most of the inter-war years Burgers, no doubt initially heavily influenced by Ehrenfest, was a committed socialist and communist. He was an idealist and had a fierce belief in the importance of science being not just to understand the world, but also to help improve the lot of his fellow human beings. He was impressed by the Russian Revolution in 1917, and joined the Communist Party of Holland shortly after its foundation in 1918. As he wrote much later in his application to emigrate to the United States,

> My student years coincided with the war of 1914–1918 in Europe, which had cruelly burst in on all the ideals we had cherished in my parental home…the war brought atrocities and the beginnings of suppressions of freedom, which later on have become more and more fierce…in the beginning of 1917 a regime of oppression was overthrown in Russia…the programme of the men who tried to find some solution out of the terrible plight in which the Russian people found itself, was the best approximation available to the ideals of freedom from want and freedom from oppression which always stood before me.[21]

He visited Russia in 1925 at the invitation of Ioffe and fell in love with the beauty of the country, though he was shocked by its poverty and poor social conditions. He was invited in 1926 to succeed Alexander Friedmann as Professor of Physics in Leningrad, and almost accepted, but in the end declined because of Nettie's poor health, which he thought would not withstand the transition from one country to another. Nevertheless he visited Russia regularly and gave lectures there throughout the 1920s and 1930s, made a significant contribution to the Great Soviet Encyclopedia of 1929, published regularly in the *Journal of Applied Physics*, edited by Ioffe, conducted joint research with Russian scientists, and introduced Russian teaching programmes in Delft. During the mid- to late 1930s, however, he lost his love of Russia, lost sympathy with communism and left the Communist Party, as he watched horrified at Stalin's increasing use of convict labour and show trials, and support for Hitler and the Nazis.

In the 1940s and 1950s, Burgers idealism turned increasingly towards the divergence between science and the ethics of its application to society, and then even more philosophically towards the very meaning of science and knowledge. He began to write learned papers on these topics. He was horrified by the way science was being used by big corporations to create massive wealth for the few at the same time as impoverishing the many, and even more appalled at the growing use of science in warfare, culminating of course with the development of the atomic bomb. He was horrified by the 20th-century growth of Nazism, Stalinism, racism, intolerance and totalitarianism. In 1954 he wrote,

> I am convinced that the only way to protect human society against the spread of disruptive trends of thought, as nazism [*sic*], the present form of communism, race conflicts ... must be found in: (a) help to those peoples and to those social groups which suffer from want; (b) extensive research into man's reactions to the changes in his environment, to the increase of knowledge and to the tremendous increase of technical power.[22]

In 1937 he became one of the KNAW's two delegates to the International Council of Scientific Unions (ICSU), the other being the Dutch physical chemist Hugo Rudolph Kruyt. Together, they helped set up ICSU's Committee for Science and Social Relations to 'survey the most important results obtained ... in the physical, chemical and biological sciences, with reference to their interconnection, the development of the scientific picture of the world and their practical application in the life of the community'.[23] In 1939 and 1940 he tried to co-found a similar society locally in Holland, but this was interrupted by the war, and only came to fruition in the late 1940s as the Research Centre for Social Problems.

After dallying with the philosophies of Martin Spinoza, the Indian Vedanta and Henri Bergson, he discovered and became a strong advocate of the English mathematician and philosopher Alfred North Whitehead, working hard to extend his thoughts. He followed Whitehead in rejecting the common Western materialist view that reality consists of independent material objects, and that science just discovers and describes the relationships between them. Instead, reality in Whitehead's *process philosophy* consists of a series of complex, overlapping processes that are intermingled in a complex way. This is an active, interventionist view of the world, rather than a passive, mechanistic one. The scientist cannot just step back, dispassionate and unconcerned with social implications, but must become an active player in building a good society, applying ethical principles alongside scientific ones. As Whitehead put it, 'There is urgency in coming to see the world as a web of interrelated processes, of which we are integral parts, so that all of our choices and actions have consequences for the world around us'.[24] And as Burgers said,

> There is a meaning to life which is directed towards the expression of what we regard as valuable, which we realise to be our freedom. . . . There is a process of creation in the shaping of our thoughts [that] is not the result of sequences of processes which are only ruled by statistical laws. Scientific research can never predict the shape of the future.[25]

Burgers first visited the United States in 1930, invited by the American naval officer and mechanical engineer William F. Durand at Stanford University and his old friend von Kármán, who was in the process of moving from TU Aachen to Caltech in Pasadena. He visited the Mount Wilson Observatory and Yosemite Valley, and was much impressed by the landscape and the people he met: 'When coming to the new continent I have kept open the "eyes of discovery" as an explorer who is elated by the landscapes. . . . This is a constant joy to me which makes me feel at home here'.[26] He visited again after the end of the Second World War, in 1949, giving lectures at Caltech and at Brown, Cornell and Harvard Universities. This deepened his love of the country and the freedom and pioneering spirit of its people and society. He resolved to accept one of many offers for him to move to the United States, but it was the McCarthyite period, and he could not get a visa approved because of his previous communist sympathies. Finally, in 1955, he was invited to become a Visiting Research Professor at the University of Maryland at College Park, just outside Washington, DC, by Professor Munroe Martin, the Director of the Institute for Fluid Dynamics and Applied Mathematics, and this time he was able to get his visa. Later that year, he gave his final address at TU Delft and, on 4 November, he and his second wife, Annie, embarked by ship for the New World. He gave his reasons for leaving as including: disliking the increased engineering rather than scientific work in Delft; difficulties in making progress with his statistical theories of turbulence, and the need for a fresh approach; and the continuing excess bureaucracy and poor experimental support at the University.

At Maryland, he continued to write on social and philosophical topics, but he also regained his enthusiasm for fluid dynamics, working on familiar topics such as non-linear differential equations, magnetohydrodynamics, statistical turbulence and shock waves, and breaking out into new fields such as hypersonic aerodynamics and the structure of matter. He was to stay at the University of Maryland for 26 years, until his death aged 86 on 7 June 1981, following a prolonged period of deteriorating health and the onset of severe Parkinson's disease. When he retired to become an Emeritus Professor, aged 70 in June 1965, he said,

> The way in which we were received at the University of Maryland . . . surpassed anything which we could have imagined when we came to America. . . . We felt at home immediately and a deep love for this country has grown in us.[27]

Burgers was kind, modest, quiet, non-demonstrative, shy and almost anti-social. He didn't drink, didn't smoke, didn't like small talk, didn't engage in organised sports and didn't like bourgeois society. He was a committed Unitarian, which helped him integrate into life in the New World. But his intellectual interests were wide: he loved flowers; he had a sea aquarium; he collected stones and minerals; he loved poetry and literature; he was a good pianist; he loved walking and hiking; he was prolific at sketching and drawing; he collected seaweed; and he was an enthusiastic photographer. He corresponded with scientists in other fields, such as astronomy, geology and biology, discussing such diverse topics as the construction of radio telescopes, cosmic gas dynamics, the rotation of the earth, the deformation of rocks, the lungs of birds and the proteins in muscles. According to his son Herman,

> There was something intense and passionate in many things he undertook.... [He] stood out by the exceptional force of his personality, in at least three respects: by the power of his intellect, the intensity of his feelings, and the strength of his will and energy.[28]

He shared with his mentor Ehrenfest a passion for science and a passion for the advancement of society, but unlike Ehrenfest, fortunately, his mild temperament helped him to control these passions more effectively, through to a happy and well-deserved old age.

3   Dislocation energy

Atoms or molecules near to a dislocation are displaced from their expected lattice positions. The extent of the displacement decreases rapidly with distance away from the dislocation. The region where the displacement of atoms or molecules is large, close to the dislocation line, is referred to as the *dislocation core*. Displaced atoms or molecules have higher energy than in the perfect crystal. In other words, there is a *dislocation energy* or *line tension* that depends on the material and the dislocation type. Outside the core, displacements are small, and the excess energy can be calculated from the associated elastic strain energy. Inside the core, displacements are too large to be treated as elastic.

Consider cutting a material and displacing it by b parallel to the line of the cut, to create a screw dislocation, as shown in Figure 11.4. Consider a thin cylindrical ring of material surrounding the dislocation at a distance r, with a thickness dr and a length l, i.e. a volume $V = 2\pi r dr l$. The shear strain ε throughout the ring is equal to the step height b divided by the circumference of the ring $2\pi r$,

$$\varepsilon = \frac{b}{2\pi r},$$

and the corresponding shear stress τ throughout the ring is, therefore,

$$\tau = \frac{Gb}{2\pi r},$$

where G is the shear modulus.

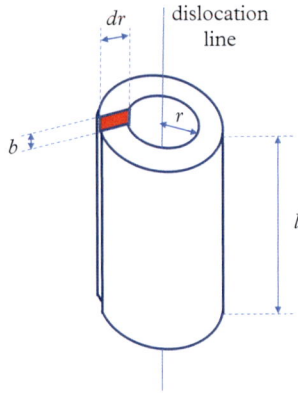

Figure 11.4 *Strain energy around a dislocation*

The elastic strain energy of a stressed material is given by $\frac{1}{2} \times$ stress $\times$ strain $\times$ volume, so the elastic strain energy dE in the ring per unit length of dislocation is

$$dE = \frac{1}{2} \frac{\tau \varepsilon V}{l} = \frac{1}{2} \left(\frac{Gb}{2\pi r} \right) \left(\frac{b}{2\pi r} \right) \frac{(2\pi r dr l)}{l} = \frac{Gb^2}{4\pi r} dr.$$

Integrating from the core radius r_c to the outside edge of the material, at a radius R, gives the energy per unit length of a screw dislocation, i.e. the *dislocation energy* or *line tension*:

$$E_{\text{screw}} = E_c + \int_{r_c}^{R} \frac{Gb^2}{4\pi r} dr = E_c + \frac{Gb^2}{4\pi} \ln \frac{R}{r_c},$$

where E_c is the non-elastic energy in the core of the dislocation, where the atomic or molecular displacements are large. The stress and strain fields around edge and mixed dislocations are more complex, and the equivalent results are

$$E_{\text{edge}} = E_c + \frac{Gb^2}{4\pi (1 - v)} \ln \frac{R}{r_c}$$

$$E_{\text{mixed}} = E_c + \frac{Gb^2 (1 - v\cos^2\theta)}{4\pi (1 - v)} \ln \frac{R}{r_c},$$

where v is the Poisson's ratio and θ is the angle between b and l. In practice, most materials contain quite large numbers of dislocations, with strain fields that overlap and cancel out, so the effective upper limit for integration is the dislocation separation, much smaller than the external radius of the material.

The factors containing v and θ, the logarithmic term and the core energy, are all relatively small, so the line tension can be written approximately as

$$E \approx Gb^2.$$

The dislocation energy or line tension is smallest for small values of the Burgers vector, which explains why common dislocations have Burgers vectors equal to the interatomic spacing. The dislocation energy or line tension causes dislocations to resist curvature wherever possible.

4   Dislocation motion

Dislocation slip

Dislocations can sometimes move rather easily through a crystal when a shear force is applied, as shown for an edge dislocation in Figure 11.5. Effectively, the planes of atoms above and below the dislocation line roll over each other as the extra half plane of atoms moves. This is called *dislocation slip* or *dislocation glide*. The requirements for easy dislocation slip are threefold. The first is a mechanical requirement: the applied forces must produce a sufficiently large shear stress on the *slip plane*. The second is a structural requirement: the *slip plane* must be parallel to both *b* and *l*. The third is a material requirement: the atomic or molecular structure must not be too complex, minimising the barrier to atomic or molecular reorganisation in the dislocation core. This is usually satisfied in materials with relatively close-packed face-centred cubic (fcc), hexagonal close-packed (hcp) and body-centred cubic (bcc) structures, but not often otherwise.

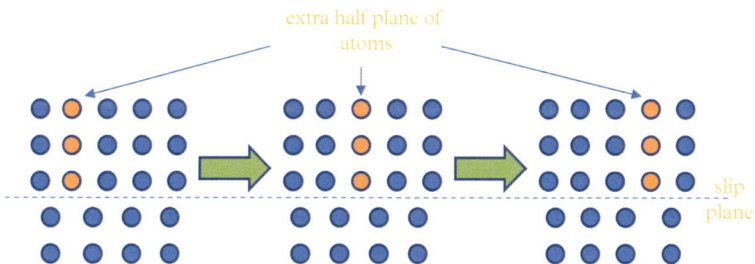

Figure 11.5 *Dislocation slip*

The slip plane

A *glissile dislocation* is one that can slip; and a *sessile dislocation* is one that can't. Dislocations can usually slip easily, i.e. at relatively low shear stresses, only on particular, usually close-packed, crystal planes. In an fcc material, for instance, $a/2<110>$ dislocations slip easily on {111} planes and, in a bcc material, $a/2<111>$ dislocations slip easily on {110} planes. The natural slip planes for some common dislocations in simple crystal structures are included in Table 11.1. Screw dislocations have their Burgers vector and line vector parallel, so they can slip on many planes, but edge and mixed dislocations have their Burgers vector and line vector non-parallel, so there is only one possible slip plane, defined by the cross product of the two vectors.

Peierls stress

The resistance to dislocation motion oscillates as the atomic or molecular planes roll over each other, with maximum resistance when the atoms or molecules are midway between lattice positions. The *Peierls stress* is the stress required to move a dislocation through the crystal lattice, from one lattice position to another, i.e. the stress required to overcome the maximum resistance at the mid-point between adjacent lattice positions. The Peierls stress is small in fcc and hcp materials, a bit larger in bcc materials, and much larger in more complex crystal structures, which is why slip is easiest in fcc, hcp and bcc.

Critical resolved shear stress

In general, a material is subject to a complex set of applied forces. These produce an overall *resolved shear stress* on any given plane. The *critical resolved shear stress* is the resolved shear stress required to move dislocations on that plane, overcoming the Peierls stress and any restraining forces from barriers such as impurities or grain boundaries. The yield point is reached when the critical resolved shear stress has been reached for at least one plane, initiating dislocation slip and plastic flow.

Partial dislocations

A *perfect dislocation* is when the Burgers vector is a lattice-translation vector, so the displacement reproduces the crystal perfectly. In some cases a perfect dislocation can separate into two *partial dislocations*, which are not lattice-translation vectors, as shown for an fcc material in Figure 11.6. The perfect dislocation with a lattice-translating Burgers vector $\boldsymbol{b} = a/2 < 110 >$ dissociates into two *partial dislocations*, with non-lattice-translating Burgers vectors $\boldsymbol{b} = a/6 < 112 >$, which are called *Shockley partials*. This can be represented as a reaction:

$$\frac{a}{2}[110] \rightarrow \frac{a}{6}[211] + \frac{a}{6}[12\bar{1}]$$

Because neither partial dislocation reproduces the lattice, there is a *stacking fault* between the two partials. This means that the stacking sequence on either side of the two partials

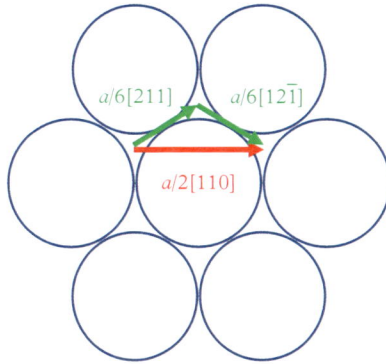

Figure 11.6 *Separation of a perfect a/2<110> dislocation into two partial a/6<112> dislocations in a face-centred cubic material*

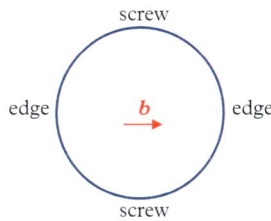

Figure 11.7 *Dislocation shear loop*

is ABCABCABCABC, but between the two partials is ABCABCBCABC, with a plane of hcp in the middle of the fcc structure. Because the Burgers vectors of the two partials are different, dissociated dislocations are less mobile than perfect dislocations.

Dislocation loops

Dislocations are topological line defects in a crystal, and the Burgers vector is conserved. A dislocation cannot end in the crystal. It must extend from one side of the crystal to the other, or form a closed *dislocation loop*. A *shear loop* is when the Burgers vector is parallel to the plane of the loop, and it can move within this plane, as shown in Figure 11.7. Because the line vector varies along the length of a shear loop, so does the dislocation character, from edge to mixed to screw to mixed and finally back to edge, as shown in Figure 11.7. A *prismatic loop* is when the Burgers vector is perpendicular to the plane of the loop, and it cannot move. Prismatic loops correspond to a disc of extra atoms or vacancies.

Plastic strain

The displacement associated with a single dislocation, i.e. its Burgers vector b, is carried with it as it moves. This means that, if a dislocation moves from one side of a block of material to the other, it leaves a permanent set b in the material when it reaches the surface, as shown in Figure 11.8. This is equivalent to producing a plastic displacement of magnitude b parallel to the slip direction, and a *plastic strain* ε_p of

$$\varepsilon_p = \frac{b \sin \varphi}{x},$$

where φ is the angle between the slip direction and the applied stress, and x is the size of the block of material.

Plastic flow

The ability of dislocations to slip and produce a permanent set is the root cause of the ability of many materials to exhibit *plastic flow*. Only a tiny plastic set is caused by the migration of a single dislocation from one side of a material to the other, as shown in Figure 11.8. Measurable plasticity requires the creation and motion of very large numbers of dislocations. The moving dislocations then interact with each other in complex ways, and encounter many barriers, such as grain boundaries and impurity particles. For instance, when dislocations cross they produce *jogs* and *kinks* in the dislocation lines, which are often sessile, and act as pinning points and barriers to further dislocation motion. There is an extensive science of *dislocation mechanics*, which is devoted to understanding dislocation behaviour during plastic flow.

Dislocation cross slip

When dislocations encounter a barrier, they can move around it by the process of *dislocation cross slip*. Cross slip is not usually possible for an edge dislocation with $b \perp l$, which can only, therefore, slip on a single, fixed slip plane; but cross slip is much easier for a screw dislocation with $b \parallel l$, which can, therefore, readily slip on a variety of slip planes. Cross slip is also difficult for dissociated partial dislocations, which have to recombine first.

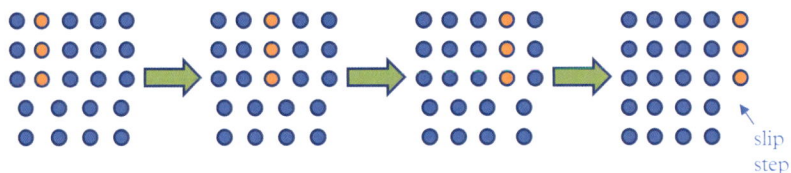

Figure 11.8 *Slip step produced when a dislocation reaches the surface*

Dislocation climb

Dislocations can also move by another mechanism, called *dislocation climb*. This is when a vacancy migrates to a dislocation, allowing it to move out of its slip plane, as shown in Figure 11.9. Dislocation climb is particularly important as a mechanism for dislocations to move around obstacles, such as sessile jogs or impurity particles, which are blocking their motion on the slip plane. Dislocation climb is not significant at low temperatures, when there are few vacancies with only low mobility, but it is particularly effective at high temperatures, when there is a ready supply of mobile vacancies.

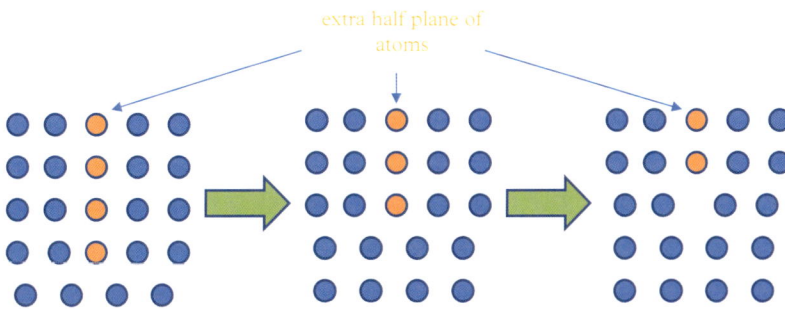

Figure 11.9 *Dislocation climb*

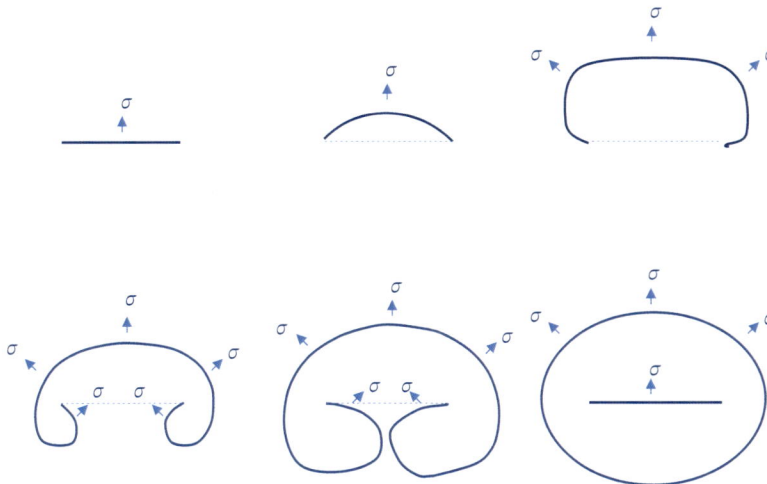

Figure 11.10 *A Frank-Read source for generating dislocations*

Dislocation generation

Dislocations can be generated in many ways. An important mechanism is the operation of a *Frank-Read source*, as shown in Figure 11.10. A dislocation is pinned at two end points. Under an applied stress, it bows between the two pinning points and wraps around them until its two arms interact to create a dislocation loop and re-create the original dislocation.

Dislocation density

Continued application of stress to a material leads to extensive plastic flow, and the creation and motion of large numbers of dislocations. The *dislocation density* ρ is defined as the number of dislocation lines intersecting unit cross sectional area. With no plastic strain, i.e. in the *annealed* state, typical dislocation densities are $\rho \sim 10^7$ lines/cm^2. After heavy deformation, i.e. with extensive plastic strain, dislocation densities can increase up to $\rho \sim 10^{12}$ lines/cm^2.

5   The yield point and work hardening

Figure 11.11 shows a typical tensile stress–strain curve for a material that shows extensive plasticity. There are several different regions, as follows.

Elastic region

Initially, the material extends elastically, obeying Hooke's law with stress proportional to strain, until it reaches its *yield point*.

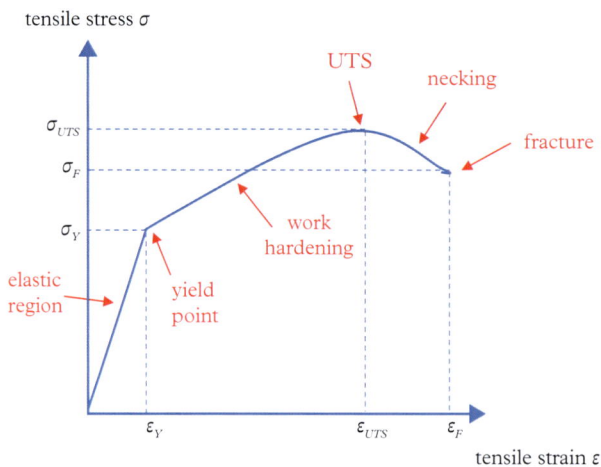

Figure 11.11 *Typical stress–strain curve for a ductile material*

Yield point

The *yield stress* σ_Y is reached at the *yield point*, which is sufficient to release existing dislocations from pinning points, create new ones in large numbers, and drive their motion throughout the material. The resulting dislocation motion leads to *plastic strain*, i.e. non-recoverable extension beyond the *yield strain* ε_Y.

Work hardening

As dislocations move through the material, they interact with each other, and with grain boundaries, precipitates and impurity particles. These interactions create barriers to continued dislocation motion. Increased stress is needed to continue to drive dislocation motion and produce further plastic strain. This process is called *work hardening*. There are various components of work hardening: dislocation–dislocation interactions produce *strain hardening*; dislocation interactions with precipitate particles produce *precipitation hardening*; and dislocation interactions with grain boundaries produce *grain-size hardening*. In strain hardening, the dislocations form extensive tangles.

Ultimate tensile strength

Initially, work-hardening mechanisms increase the tensile stress needed to drive dislocation motion. The material reaches its ultimate tensile strength (UTS), i.e. the maximum stress it can withstand σ_{UTS} at a strain ε_{UTS} as work-hardening mechanisms become exhausted, and obstacles to dislocation motion are overcome.

Ductile fracture

Beyond the UTS, work hardening reduces even further, decreasing the tensile stress needed to drive dislocation motion. Dislocation motion is concentrated intensely in particular regions in the material, a process called *necking*, finally leading to *failure* by *ductile fracture*. The stress in the material at this point is called the *failure stress* or *fracture stress* σ_F, and the strain is called the *failure strain, fracture strain* or *ductility* ε_F. Table 11.2 shows typical values of yield strength, ultimate tensile strength and ductility.

Table 11.2 *Typical Values of the Strength and Ductility of Some Common Materials*

Material	Yield strength σ_Y (MPa)	Ultimate tensile strength σ_{UTS} (MPa)	Failure strain ε_F (%)
Mild steel	250	850	20
High-strength steel	2,200	2,500	5
Aluminium	15–20	50	40
Polyethylene	30	37	100
Glass	—	30	0.1

6 Deformation and annealing

Deformation is often used to manufacture a material into its final shape. There are many different kinds of such *forming* or *deformation processing* methods. In *forging*, a material is squeezed between two dies; in *rolling*, it is turned into sheet between two counter-rotating rolls; in *extrusion*, it is turned into a shaped section by forcing it through a die; and in *drawing*, it is turned into wire by pulling it through a die. Forging and rolling are shown in Figure 11.12. In all cases, the material changes shape by the application of intense shear forces, which are largely compressional to prevent the material from fracturing during manufacture. To achieve the final shape often requires repeated processing steps, e.g. repeated forging with different dies to manufacture complex shapes, or repeated rolling to reduce, gradually, the thickness of a sheet product.

At low temperatures, the shear forces applied during deformation processing cause extensive plastic flow, with a sharp rise in dislocation density and associated work hardening. The material rapidly becomes too hard for further deformation processing. To allow continued deformation processing, the material is *heat treated* or *annealed* to relieve its internal stresses and to remove the excess dislocation density. Alternatively, the deformation processing is conducted at high temperature to prevent this problem.

During heat treatment or annealing, a material undergoes processes of *recovery* and then *recrystallisation*. During *recovery*, the dislocations are gradually re-organised by cross slip and climb into lower energy configurations, with the material hardness decreasing gradually. During *recrystallisation*, new undeformed crystal grains are nucleated and grow into the deformed material, fully replacing it after sufficient time, with the material hardness decreasing sharply.

7 Other forms of plasticity

Some materials can deform plastically by *deformation twinning*. Parts of the crystal shear to produce a mirror image of the crystal structure, called a *twin*. This process is called

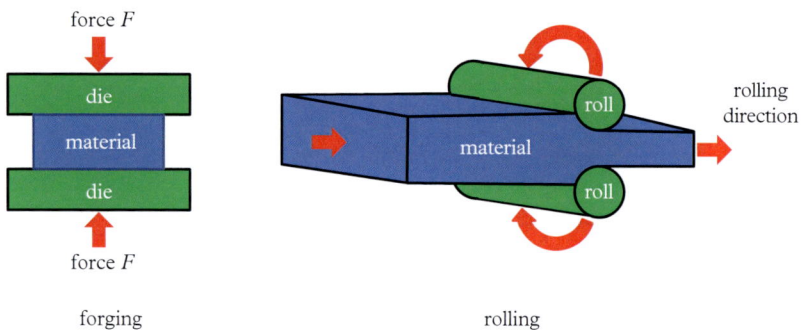

Figure 11.12 *Forging and rolling*

TWIP or *twinning-induced plasticity*. Some other materials can deform plastically by parts of the crystal shearing to produce a new crystal structure. This is called *TRIP* or *transformation-induced plasticity*.

At high temperatures, materials can deform plastically by *creep*, i.e. by large-scale diffusion of atoms or molecules. This can take place in several ways: directly by vacancy migration, in *vacancy creep* or *Nabarro-Herring creep*; by assisted dislocation motion through cross slip and climb, in *dislocation creep*; or by grain boundary sliding in *grain boundary creep* or *Coble creep*.

Some materials are *superplastic*, i.e. can deform uniformly by up to 100% to 200% without developing any instability leading to necking and ductile fracture. Superplasticity is found when deformation takes place at relatively high temperatures in metals and ceramics with a very fine grain size stabilised by small second-phase particles. Superplastic deformation takes place by a process of *grain boundary sliding*, with grains rotating and moving past each other to create extensive overall change of shape in the material.

Most non-crystalline materials are brittle, and fracture without showing any plasticity. However, some non-crystalline polymers can show plasticity at high temperature when the polymer molecules become disentangled. Similarly, most geological materials have complex crystal structures, and are again brittle, and fracture without showing any plasticity. However some rocks can show plasticity through slippage at internal cracks.

....

8 REFERENCES

1. Dirk van Delft. 'Paul Ehrenfest's final years'. *Physics Today* **67** (2014), 41.
2. Ibid.
3. Paul Halpern. 'The tragic fate of physicist Paul Ehrenfest'. 2015. https://medium.com/starts-with-a-bang/the-tragic-fate-of-physicist-paul-ehrenfest-93c946b05d0c (accessed 20 February 2020).
4. F. T. M. Nieuwstadt and J. A. Steketee. *Selected Papers of J. M. Burgers* (Kluwer, Dordrecht, 1995), xii.
5. Ibid. xiv.
6. Ibid.
7. Ibid. xv.
8. Ibid.
9. van Delft, n 1.
10. See ibid.
11. Nieuwstadt and Steketee, n 4, xix.
12. See ibid. xxi.
13. Ibid. xix.
14. Ibid. xxi.
15. Ibid. xxxviii.
16. Ibid. xl.
17. J. M. Burgers. 'Some considerations on the fields of stress connected with dislocations'. *Proceedings of the Koninklijke Akademie van Wetenschappen I and II* **XLII** (1939), 293–325, 378–399.
18. J. M. Burgers. 'Geometrical considerations concerning the structural irregularities to be assumed in a crystal'. *Proceedings of the Physical Society* **52** (1940), 23–33.

19. P. B. Hirsch, R. W. Horne, and M. J. Whelan. 'Direct observations of the arrangement and motion of dislocations in aluminium'. *Philosophical Magazine* **1** (1956), 667–684.
20. P. B. Hirsch, R. W. Horne, and M. J. Whelan. 'Direct observations of moving dislocations: reflections on the thirtieth anniversary of the first recorded observations of moving dislocations by transmission electron microscopy'. *Materials Science and Engineering* **84** (1986), 1–10.
21. Nieuwstadt and Steketee, n 4, liv.
22. See ibid. lviii.
23. Ibid. lix.
24. C. Robert Mesle. *Process–Relational Philosophy: An Introduction to Alfred North Whitehead* (Templeton Foundation Press, West Conshohocken, 2009), 9.
25. Nieuwstadt and Steketee, n 4, lxxxi.
26. See ibid. lxvii.
27. Ibid. lxxi.
28. Ibid. lxxv.

....

9 BIBLIOGRAPHY

Elementary Dislocation Theory. Johannes Weertman and Julia R. Weertman (Oxford University Press, Oxford, 1992).

Introduction to Dislocations (5th edition). D. Hull and D. J. Bacon (Butterworth-Heinemann, Oxford, 2011).

Selected Papers of J. M. Burgers. F. T. M. Nieuwstadt and J. A. Steketee, eds. (Kluwer, Dordrecht, 1995).

12

Griffith's Equation

Fracture

---

1 Fracture

If you drop a cup or a saucer on a tiled floor it shatters. If you drop a tumbler or a wine glass on the floor, the same thing happens. The polycrystalline ceramic materials we use to make crockery and the amorphous soda-lime glass materials we use to make domestic glassware are examples of *brittle solids*, which fail by a mechanism called *brittle fracture*. On the other hand, if you drop a knife or a fork on the floor, it may be dented or bent, but it does not usually break. The metallic stainless steels we use to make cutlery are examples of *ductile solids*, which exhibit *plasticity* or *ductility*, and do not fail by brittle fracture.

What is the key difference between brittle and ductile materials? Ductile materials contain *dislocation*s, line defects in the crystal structure that can move and thus allow the crystal planes to roll over each other rather easily, causing local *plastic flow* and preventing any build-up of stress. The material bends or deforms, but it does not fracture. In brittle materials, on the other hand, either there are no dislocations (such as in glasses) or they are immobile (such as in ceramics), so the stress can build up until it is large enough to fracture the material. When a crack in a brittle material begins to propagate, it weakens the material and the crack accelerates, so the brittle fracture process is catastrophic and the material shatters. Most metals are ductile materials, whereas most ceramics, glasses, rocks and plastics are brittle materials.

2 The theoretical strength of materials

What is the *theoretical strength* of a material? Figure 12.1 shows tensile stress σ versus distance x as two neighbouring planes of atoms or molecules are pulled apart. There is no tensile stress when the planes are undisturbed at their equilibrium separation a_o, i.e. at $x = 0$. As the planes are forced apart, the stress rises initially to overcome the attractive forces between the atoms or molecules. Interatomic or intermolecular attractive forces are only short range, so the stress reaches a maximum and then falls back again towards zero.

The Equations of Materials. Brian Cantor. Oxford University Press (2020). © Brian Cantor.
DOI: 10.1093/oso/9780198851875.001.0001

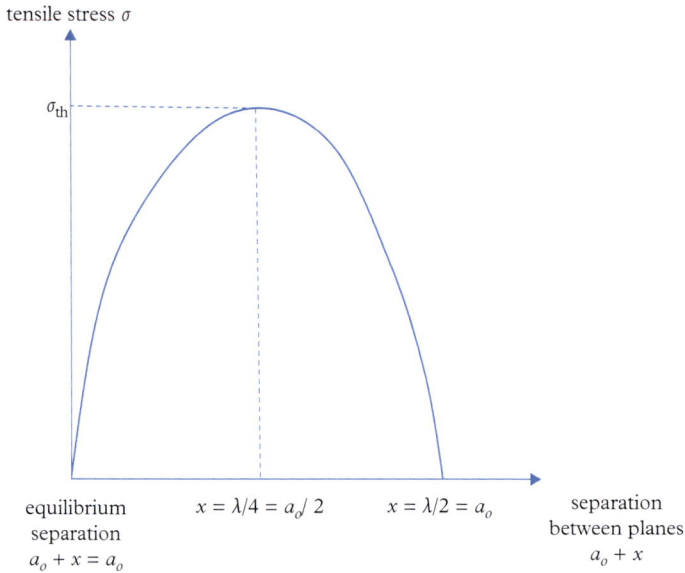

Figure 12.1 *Tensile stress versus separation of neighbouring planes of atoms or molecules*

The maximum stress is the theoretical strength σ_{th}. For simplicity assume the variation of stress with distance is sinusoidal, with a wavelength $\lambda \approx 2a_o$,

$$\sigma = \sigma_{th} \ \sin\left(\frac{2\pi x}{\lambda}\right),$$

with $\sigma = 0$ at $x = 0$ and $x = \lambda/2 = a_o$, and $\sigma = \sigma_{th}$ at $x = \lambda/4 = a_o/2$.

The Young's modulus E is given by the slope of stress $d\sigma$ versus strain dx/a_o for small strains near the equilibrium separation a_o, i.e. as $x \rightarrow 0$:

$$E = a_o\left(\frac{d\sigma}{dx}\right)_{x\rightarrow 0} = a_o\left[\frac{2\pi\sigma_{th}}{\lambda} \ \cos\left(\frac{2\pi x}{\lambda}\right)\right]_{x\rightarrow 0} = a_o\frac{2\pi\sigma_{th}}{\lambda} = \pi\sigma_{th}$$

$$\therefore \sigma_{th} = \frac{E}{\pi}.$$

More accurate calculations, in cases when the interatomic or intermolecular forces are known, lead to a range of values for the theoretical strength, between $E/2$ and $E/5$. In practice, however, measured failure strengths of different materials are almost always at least two orders of magnitude smaller, more like $E/100$ or $E/1{,}000$. Why is this?

The reason is different for ductile and brittle materials. For ductile materials, measured failure strengths are much lower than σ_{th} because they contain mobile *dislocations*, which begin to move at low stresses, causing plastic flow and preventing any build-up of stress.

For brittle materials, measured failure strengths are much lower than σ_{th} because they contain small *cracks*, which begin to propagate at low stresses, causing catastrophic brittle fracture.

3 Griffith's equation

Griffith investigated the brittle fracture of glass. He found that the fracture strength was inversely proportional to the size of the glass rod he tested. He surmised that fracture was initiated from small cracks and flaws in the glass, with large glass rods more likely to contain a large crack or flaw. To test this hypothesis, he inserted controlled surface cracks in a series of glass rods and showed that the fracture strength σ_F was inversely proportional to the square root of the crack length c:

$$\sigma_F \propto \frac{1}{\sqrt{c}}.$$

This relationship can be explained as follows. Consider a thin plate of material of thickness t subjected to an applied tensile stress σ, as shown in Figure 12.2. The elastic energy per unit volume stored in the material because of the applied stress is $\frac{1}{2}$ × stress × strain $= \frac{1}{2} \times \sigma \times \sigma/E = \sigma^2/2E$, where E is the Young's modulus.

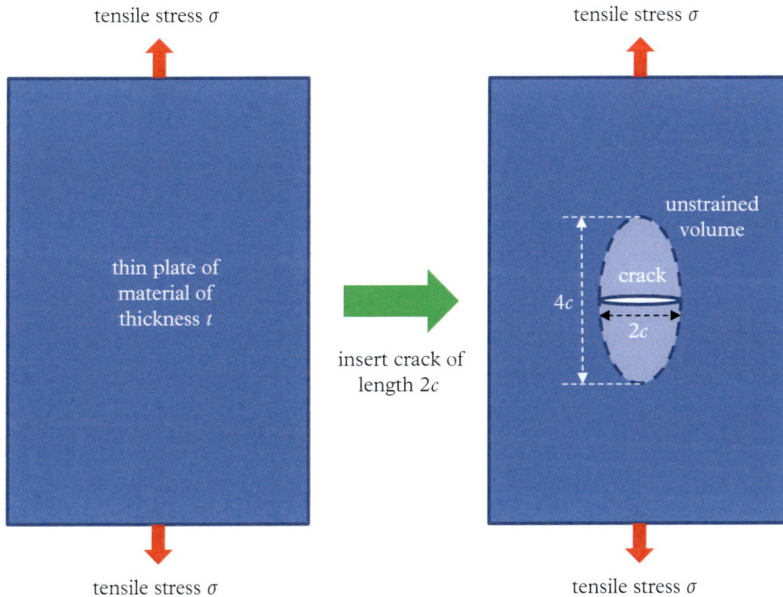

Figure 12.2 *Griffith's equation for the energy of a crack*

energy change ΔU

$4ct\gamma$

ΔU_{crit}

c_{crit}

crack length c

$\dfrac{-\pi c^2 t \sigma^2}{E}$

ΔU

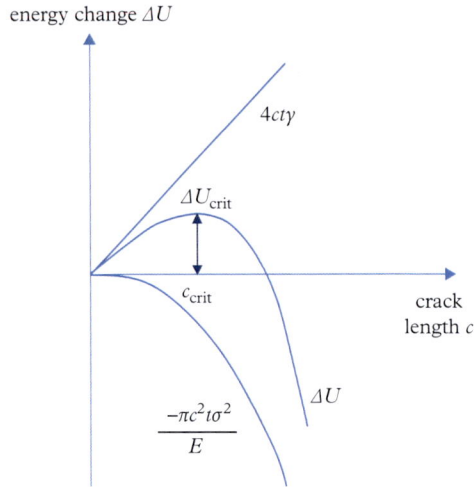

Figure 12.3 *Variation of energy with crack length*

Now consider introducing a small, thin crack of length $2c$, as shown in Figure 12.2. The total energy of the material is reduced because strain energy is relieved in the vicinity of the crack, but is also increased because of the creation of the crack surface. The unstrained volume surrounding the crack is approximately elliptical with axes $2c$ and $4c$, as shown in Figure 12.2, so the reduction in strain energy is given by the elliptical volume $2\pi c^2 t$ multiplied by the strain energy per unit volume $\sigma^2/2E$. The increase in surface energy is given by the crack area $4ct$ multiplied by the surface energy of the material γ. In other words, the change in energy ΔU associated with introducing the crack is given by

$$\Delta U = -\frac{2\pi c^2 t\sigma^2}{2E} + 4ct\gamma.$$

The surface energy term increases linearly with crack length, and the strain energy term reduces quadratically with crack length, as shown in Figure 12.3. This means that the change in energy is positive for small cracks, increases up to a maximum, and then reduces for large cracks. In other words, very small cracks are stable and do not propagate because their energy increases with increasing length, but larger cracks are unstable and propagate because their energy decreases with increasing length.

The condition for crack propagation, i.e. for fracture, can be determined by finding the maximum in the variation of energy ΔU:

$$\frac{d\Delta U}{dc} = -\frac{4\pi ct\sigma^2}{2E} + 4t\gamma = 0$$

$$\therefore \quad \frac{4\pi ct\sigma^2}{2E} = 4t\gamma.$$

This means that, for a given applied stress σ, the critical size of crack that will propagate is

$$c_{\text{crit}} = \frac{2E\gamma}{\pi\sigma^2}.$$

Any crack of length $c > c_{\text{crit}}$ will propagate. Any crack of length $c < c_{\text{crit}}$ will not. Or, for a given crack of length c, the applied stress required for it to propagate leading to fracture is

$$\sigma_F = \sqrt{\frac{2E\gamma}{\pi c}}.$$

This is known as *Griffith's equation* for crack propagation and brittle fracture. All brittle materials contain a range of cracks of different lengths. When the stress reaches that given by Griffith's equation for the largest crack, it begins to propagate and the material fails by brittle fracture.

If we put $\gamma \approx 0.01Ea_o$ and $c \approx 10,000a_o \approx 1\ \mu$m, $\sigma_F \approx E/1,000$, i.e. approximately two orders of magnitude smaller than the theoretical strength, as discussed previously.

4   A. A. Griffith

A. A. Griffith

At the beginning of the Second World War, the German Army had overrun much of Europe and was being opposed only by Britain. The war effort was taxing, and Britain was struggling to grow its own food and manufacture war and other goods. German naval U-boats were lurking everywhere, blocking imports and sinking cargo ships faster than they could be replaced. American public opinion was instinctively non-interventionist, and the United States did not enter the war until later, when the Japanese bombed Pearl Harbor. Instead, as part of his lend-lease defence policy, the US President, Franklin Delano Roosevelt, instituted a programme of building so-called Liberty ships to help supply Britain and the rest of Europe. Conrad Black, one of Roosevelt's biographers, commented, 'If there was no practical alternative there was certainly no moral one either. Britain and the Commonwealth were carrying the battle for all civilization, and the overwhelming majority of Americans … wished to help them'.[1] Roosevelt himself said, more pithily, 'There can be no reasoning with incendiary bombs'.[2,3]

Between 1941 and 1945, a total of 2,710 Liberty ships were built, based on a British design, 441 feet (134 m) long, made from welded instead of the more conventional riveted steel plates, with speeds of up to 13 miles per hour (20 km/h), a range of 23,000 miles (37,000 km), and a capacity of more than 10,000 long tons (a long ton is 2,240 pounds or 1,016 kg). The welded steel construction was new and fast. Riveted ships took months to build. By comparison, the average time to build a Liberty ship was just 42 days, with one ship, the *SS Robert E Peary*, built in less than five days. The ships were not pretty. Roosevelt called the first one 'a dreadful looking object'[4,5] and *Time* magazine called it 'an ugly duckling',[6,7] but the programme was successful, making a major contribution to Britain's war effort. Lend lease overall provided a total of $50 billion of goods (equivalent to almost $1 trillion today), 13% of the total US war effort. However, cracks in the ship hull and deck were commonplace, with more than 1,500 brittle fractures reported, and 12 ships breaking completely in half with no warning. For instance, the *SS Schenectady* completed its sea trials in January 1943 and returned to harbour in almost freezing water. Suddenly, with a crack that could be heard a mile away, the deck, sides, bulkhead and bottom girders fractured instantaneously, the ship jack-knifed with its centre lifted out of the water, and its bow and stern settled into the riverbed. Later that year, on 24 November, the *SS John P Gaines* broke in half and sank with the loss of 10 lives. Able seaman Paul Tatman of Spokane, one of the last to leave the *Gaines*, said,

> She [the sea] was pounding heavily and green ones were coming over the foc's'le.... At 2.41 in the morning there was a loud pop and then a tremendous tearing grinding noise.... She [the ship] just seemed to bend in the middle a couple of times as she rode the seas, and then she just parted, the aft and bow ends separating about a third of the way back from the stern. The bow end then disappeared in the darkness.[8]

The US Coast Guard established a Board of Investigation, which published its final report in 1946. More than 1,000 ships had developed weld cracks. Constance Tipper, Cambridge Professor of Engineering, showed that the steel hulls had become *embrittled*. Poor steel-making had left sulphur levels too high and manganese levels too low, so that low-temperature seas caused the normally ductile steel plates to undergo a *ductile–brittle transition*, and the weld cracks propagated catastrophically.

On 10 January 1954, the world's first-ever production jet airliner, the deHavilland Comet G-ALYP, took off from Rome's Ciampino airport on the last leg of the British Overseas Airways Corporation (BOAC) route 781 from Singapore to Heathrow, London. Twenty minutes later, it broke up in mid-air and crashed into the Mediterranean Sea near the Italian island of Elba, killing all 35 passengers and crew. The accident was instantaneous. The flight commander, Captain Alan Gibson, was in contact with a nearby plane G-ALHJ, and was saying, 'George How Jig [G-HJ] from George Yoke Peter [G-YP], did you get my—',[9] when the line went dead. The *New York Times* reported the following day: 'there was [only] a slight hope that there were any survivors ... [but] the search continued in the sub-zero weather and rising seas'.[10]

BOAC immediately grounded its entire Comet fleet. Speculation as to the cause of the crash ranged from sabotage to control flutter, from wing failure to explosion in an empty fuel tank. The possibility of cabin failure was ruled out because the design strength was much higher than was considered necessary. The hastily convened Abell Committee of Inquiry concluded, as it later transpired erroneously, that fire was the most likely cause of the problem. Changes were made to all serving Comets to protect their wings and fuselage from possible future fires.

BOAC was desperate to resume Comet flights as soon as possible, and it did so on 23 March, with their chairman commenting, 'We obviously wouldn't be flying Comets with passengers if we weren't satisfied conditions were suitable.'[11] Nevertheless. two weeks later, on 8 April, Comet G-ALYY was flying from Rome to Cairo as part of South African Airways Flight 201 from London to Johannesburg when it too crashed into the Mediterranean, this time near Naples, again killing all 21 passengers and crew. The Comet fleet was again immediately grounded. The Cohen Committee of Inquiry tasked Sir Arnold Hall, Director of the Royal Aircraft Establishment (RAE) at Farnborough, to conduct a detailed investigation. By this time, much of G-ALYP had been recovered from the sea near Elba, aided by the pioneering use of underwater TV cameras. BOAC donated an entire airframe G-ALYU to be tested in a specially constructed large, pressurised water tank. Hall's team demonstrated conclusively that both crashes had been caused by catastrophic failure of the fuselage initiated by small *fatigue cracks* emanating from the corners of the square-shaped windows. (Fatigue is the continual stressing and re-stressing of a material many times over.) The Committee concluded, 'Probable Cause: We have formed the opinion that the accident at Elba was caused by structural failure of the pressure cabin, brought about by fatigue'.[12]

The evidence was overwhelming and decisive. Window corner stresses were shown to be up to five times higher than the design stress; tears were found to have started from window corners and rivets in the crashed aircraft; no other defects were found anywhere else in the aircraft; and in the water tank simulation G-ALYU burst open from a window corner crack after about 3,000 flight cycles. All Comets were withdrawn from service, and later Comet series were substantially re-designed, including replacing all windows with an oval shape, as used on all subsequent aircraft. Comet flights did not resume until

1958, after which they continued to fly passenger services until 1981 without significant problems.

The Liberty ships failed ostensibly because of weld cracks and poor-quality steel. The Comets failed ostensibly because of fatigue cracks at sharp window corners. These engineering failures led to improvements in welding, steel manufacture and aircraft window design. However, the key underlying problem was that nominally ductile materials (such as steel ship hulls and aluminium alloy airframes) suddenly became susceptible to brittle fracture. Thus the field of *fracture mechanics* was born in the 1950s, based on Alan Griffith's earlier investigations of brittle fracture in glass. Fracture mechanics extends Griffith's description of brittle fracture to include strong materials with limited ductility. The key concept is the catastrophic extension of a crack when stresses at the crack tip become sufficient to overcome its surface energy and any localised plastic flow.

Alan Arnold Griffith was born on 13 June 1893, the eldest of three children of George Chetwynd Griffith and his wife Elizabeth Brierly. George Griffith was a colourful, rumbunctious, larger-than-life character, a prolific science fiction writer and a buccaneering explorer, poet, schoolmaster and journalist, who barnstormed his way around the world with strong-minded if somewhat incoherent views, anti-monarchist, anti-republican, socialist and communist, yet fiercely pro-British. He was born in Plymouth in 1857, the son of a vicar, and went to private school in Southport when the family moved to Ashton under-Lyne in Manchester. Aged 16 he was apprenticed to a Liverpool merchant ship sailing to Australia, where he deserted, before later re-joining and then sailing several times round the world. He said, '[I]n the 78 days between Liverpool and Melbourne I learnt more of the world than in the previous 14 years'.[13] He claimed to have turned down the offer from a Polynesian king of marriage to one of his daughters. Just short of his 20th birthday, he returned to Britain to become a schoolmaster at Worthing College, where he began to write freelance articles and poems. He moved to Bolton, where he met and married Elizabeth Brierly in 1887, then on to London where his children were born. He lost all his money in a failed newspaper venture, before starting work as an office boy for Cyril Arthur Pearson, who later went on to found the *Daily Express*. George Griffith began writing so-called *marvel tales* in the style of Jules Verne, fantasy stories dealing with 'heavier-than-air flying machines, compressed air guns, submarines … spectacular aerial, land or undersea combat'.[14] Many of his tales appeared first in *Pearson's Weekly* or *Pearson's Magazine* before being published as novels, including his most famous work: *The Angel of the Revolution*. Other, rather lurid, titles included *The Syren of the Skies*; *The Gold Finder*; *The Virgin of the Sun*; and *A Honeymoon in Space*.[15] He was replaced by H. G. Wells at *Pearson's*, a better writer who overshadowed him in the United States, though Griffith remained popular in the United Kingdom. Pearson urged George to circumnavigate the globe again, which he did in a record time of just 65 days, writing up his adventures for the magazine. For a while he was a special correspondent for the *Daily Mail*, reporting on extended trips to Peru and South Africa, where he covered many of Cecil Rhodes' exploits, before moving briefly to Littlehampton on the south coast of England, and then back to Australia. In 1904 George's health began to deteriorate and finally he settled down with his family in Douglas on the Isle of Man. He died of cirrhosis of the liver on 4 June 1906.

Alan Griffith was just seven years old when his father died, leaving his mother an impoverished widow with three young children. His early education was somewhat variable, given his father's peripatetic life. He was taught privately at home until 1906, when he went to Douglas Secondary School. Unlike his father, he was studious and gifted intellectually. He passed his Cambridge Senior exams in 1910, achieving first-class passes with distinctions in physics, chemistry and maths. He won a Sir W. H. Tate Science Scholarship the following year, giving him £35 per annum to study mechanical engineering at Liverpool University. In 1914 he graduated with first-class honours, winning the Rathbone Medal, as well as a university scholarship, which allowed him to do research on the heat flow between metals and gases. He received his Master's and then Doctorate degrees in 1917 and 1921 respectively for the continuation of his research after he had left the university. In 1925 he married Constance Vera Falkner, daughter of R. T. Falkner of the Royal Engineers. They had three children: Betty, June and John. Betty, the eldest, died suddenly and tragically in January 1946, aged just 19, from a horse-riding accident. Griffith was, naturally, much affected. Afterwards, he always wore a small lapel badge and, when asked what it was, replied quietly, 'It was Betty's'.[16]

After leaving Liverpool in 1915, Griffith joined the Royal Aircraft Factory, later known as the Royal Aircraft Establishment or RAE Farnborough. He started as a draughtsman in the Physics and Instrument Department, but gradually built his career through the different technical grades, becoming a Senior Scientific Officer in April 1920. During this period, he published an important paper with the famous mathematician and physicist G. I. Taylor on the use of soap bubbles as a model system to solve torsional stress problems,[17] which was read at the Institution of Mechanical Engineers and won the Thomas Hawksley Gold Medal. In 1920 he produced his famous paper on the fracture of glass: 'The Phenomena of Rupture and Flow in Solids' published in *Philosophical Transactions*.[18] This paper demonstrated that the strength of a brittle material such as glass is inversely proportional to its size, and that brittle fracture is caused by pre-existing cracks. On the one hand this explained the discrepancy between the measured and theoretical strengths of materials, and on the other hand it showed that extremely high strengths can be obtained in very fine whiskers of a brittle material. Robert Glenny of the National Gas Turbine Establishment said, 'The importance of Griffith's work was in pointing out that cracks could cause rupture. All subsequent theories of fracture strength took into account the existence of micro-cracks, dislocations etc., either pre-existing in the material or generated in the process of deformation'.[19]

Alongside his fundamental work on the strength of materials, Griffith was also working on the efficiency of propeller blades and turbines. In 1926 he wrote another famous paper, report no. H1111 for the RAE: 'An Aerodynamic Theory of Turbine Design'.[20] This paper showed that high efficiency could be achieved if turbine and propeller blades were designed as aerofoils, based on aerodynamic theory. He also showed that compression was much more efficient if designed as a multi-stage axial process. A test unit was built under Griffith's supervision in 1927 incorporating these features, and was tested in detail the following year by W. C. Clothier, achieving efficiencies as high as

91%. The Rolls-Royce aeroengine designer, Arthur Rubbra, described Griffith as 'the true originator of the multi-stage, axial engine'.[21] Multi-stage, axial compressors are now used in almost all high-output, high-efficiency industrial gas and aerojet turbines.

Early aeroplanes were powered by *internal combustion engines* with *reciprocating pistons*, similar to those used in a car. The piston energy was transferred to a rotating shaft that drove the propeller blades, which in turn drove the aeroplane forwards. In the 1920s and 1930s, Frank Whittle, Griffith and others began to consider the use of *gas turbine engines* instead of piston engines because they were, in principle, much simpler and, therefore, more reliable. A *gas turbine engine* consists of four stages: compression, combustion, expansion and exhaust. The working gas, usually air, is sucked into the *compressor* at the front of the engine by one or more rings of rotating *compressor blades*, increasing the gas pressure and causing its temperature to rise; fuel is then injected and ignited in the *combustor*, creating a blast of hot gas; the hot gas plume expands explosively and strikes one or more rings of rotating *turbine blades* in the *turbine*; and the gas then exhausts at the back of the engine. The compressor and turbine are connected by a shaft, so that some of the rotation energy created in the turbine is used to drive the compressor. The remainder of the energy is available as usable power. This excess energy can be used to drive a shaft with rotating propeller blades and thus drive the plane forwards, in a similar way to a piston engine. This is called a *turboprop engine*. Alternatively, however, the excess energy can be used to create a jet exhaust to power the plane directly. This is called a *turbojet engine* or just a *jet engine*. In a turboprop, as much energy as possible should be transferred to the rotating shaft, with as little as possible left in the exhaust gas, i.e. the exit gas pressure and temperature should be as close as possible to ambient. In a turbojet, by contrast, as little energy as possible should be transferred to the rotating shaft, and as much as possible should be retained in the gas to provide the strongest jet plume and thus maximise the direct forward thrust. There is a halfway house engine called a *turbofan* in which some of the excess energy drives a propeller fan at the front of the engine and some drives a jet exhaust. The *bypass ratio* is the ratio of cold air going through the outer part of the fan, bypassing the compressor and turbine and driving the plane forward by a propeller effect, to the hot air going through the central part of the fan and on into the compressor and turbine, driving the plane forward by the resulting jet exhaust.

Frank Whittle was born in a terraced house in Earlsdon, Coventry, in 1907, eldest son of Moses Whittle and Sara Alice Garlick. His father, Moses, was an engineer and mechanic, owner of the Leamington Valve and Piston Ring Company. Frank was intelligent, adventurous and rebellious, interested in astronomy, engineering and aviation, and largely self-taught. He left school in 1923, aged 15, and applied to join the Royal Air Force (RAF) as an aircraft apprentice, finally succeeding after failing the physical twice. He was specially selected for training as a pilot officer at RAF Cranwell, where he was something of a fish out of water amongst mostly upper class former private schoolboys. He was a strange mixture of charismatic pilot and radical engineer. He developed a reputation for daredevil low flying and aerobatics, for which he was nearly court-martialled. He considered deserting, but in the end completed the course. In 1928 he wrote his cadet's dissertation, entitled 'Future Developments in Aircraft Design',[22,23] proposing that 'the turbine is the most efficient prime mover known so it is possible it will be developed for aircraft'.[24] He graduated as a Pilot Officer, winning the Andy Fellowes Memorial Prize. He was described as an exceptional to above-average pilot, but was warned about showboating and overconfidence. He joined 111 Squadron in Hornchurch and was rapidly posted to Wittering for a flying instructor course, where he was selected to perform the so-called crazy flying routine in the 1930 RAF Air Display. He managed to destroy two aircraft in accidents during rehearsals, and Flight Lieutenant Harold Raeburn said, 'Why don't you take all my bloody aeroplanes, make a heap of them … and set fire to them—it's quicker'.[25]

Whittle was particularly interested in novel engine designs. In 1929 he realised that a gas turbine could be used directly to produce propulsive power via a jet exhaust instead of via a propeller shaft. Thus was born the idea of the *jet engine*. Whittle was still only 22 years old. His commanding officer arranged for him to discuss his ideas with W. L. Tweedie from the Department of Scientific and Industrial Research and Griffith from RAE. Whittle described the outcome as depressing. He subsequently received a letter from Griffith saying that his engine was impracticable because no material could withstand the high temperatures and high stresses that would be set up in the turbine:

> [T]he internal combustion turbine [i.e. turbojet] will almost certainly be developed into a successful engine, but before this can be done the performance of both compressors and turbines will have to be greatly improved. However, it has been of real interest to investigate your scheme and I can assure you that any suggestion submitted by people in the Service is always welcome.[26,27]

Andrew Nahum, in his biography of Whittle, quotes an RAE turbine worker describing this letter as kind, but it was unmistakeably a case of don't call us, we'll call you. Tweedie and Griffith's negative assessment put back the development of jet engines, probably by as much as a decade or two. In fact, Griffith was technically correct. No material at that time could withstand the high temperatures and high stresses in a jet engine. But this problem was about to be solved. That very year, 1929, Bedford, Pilling and Merica added small amounts of aluminium and titanium to increase the strength of high-temperature nickel–chromium alloys, effectively creating the first high-temperature nickel-based superalloys, the material of choice ever since for hot jet-engine components.[28]

The RAF sent Whittle to Cambridge University, from where he graduated in 1936 with a first-class degree in mechanical sciences. The same year he set up a company, Power Jets, in partnership with RAF colleagues Rolf Dudley-Williams and James Collingwood Tinling, funded by the investment bank O.T. Falk. Falk's engineer, Lancelot Law Whyte commented,

> The impression he made was overwhelming. I have never been so quickly convinced.... This was genius not talent. Whittle expressed his ideas with superb conciseness: 'Reciprocating engines are exhausted. They have hundreds of parts jerking

to and fro. The engine of the future must produce 2,000 horsepower with one moving part: a spinning turbine and compressor'.[29,30]

Falk put in £2,000, with the option of a further £18,000. Whittle, Williams and Tinling took a 49% share of the company. Griffith was again asked for his views, but was again critical, dampening Falk's ardour. Throughout the late 1930s and early 1940s Power Jets struggled with funding while trying to construct and test various prototypes, involving at different times companies such as Gloster Aircraft, BTH, Vauxhall, Rover and Rolls-Royce. Whittle was promoted to Flying Officer then Flight Lieutenant, Squadron Leader and Wing Commander, but he was suffering badly. According to Nahum, he had headaches, indigestion, insomnia and eczema, took Benzedrine to help him work 16 hours a day and then tranquillisers to help him sleep, lost weight to just 9 stone (57 kg), became irritable with an explosive temper, and suffered a nervous breakdown. Finally, however, his W1 jet engine, installed on a Gloster E28/39, took off from Cranwell on 15 May 1941 and flew for 17 minutes, reaching a top speed of 340 miles per hour (500 km/h). Whittle's long-standing colleague Pat Johnson said, 'Frank it flies'.[31] And he replied, 'Well that's what it was bloody well designed to do'.[32] Within days, it was outperforming the Spitfire, a top performance plane with one of the best-available piston engines. The W1 was the precursor of the Rolls-Royce Welland, Britain's first production jet engine. Unfortunately, Whittle was not first. Because of all the delays, he was beaten by the German Hans von Ohain, whose HeS3 jet engine powered a Heinkel He 178 on its first flight on 27 August 1939. In the end Power Jets was nationalised and absorbed into the UK National Gas Turbine Establishment after the war. Whittle retired from the RAF in 1948 on medical grounds.

In the meantime, in 1928 Griffith became Principal Scientific Officer in charge of the Air Ministry Lab in South Kensington, where he continued to develop new engine geometries, but he remained wedded to a turboprop configuration, driven by a multi-stage axial compressor. In 1929 he developed a contra-flow design, with nine stages rotating alternately clockwise and counter-clockwise, each carrying a ring of compressor blades concentric with a ring of turbine blades. In 1931 he moved back to Farnborough to take charge of the Engine Department, and in 1938 and 1939 an experimental contra-flow engine rig was built and tested, but it was not successful. According to his colleague at RAE, the aeronautical engineer Hayne Constant,

> When designing turbomachinery it always pays to be a pessimist. This Griffith was not; he always aimed for the stars.... He was never a man for whom half measures or pragmatic solutions based on empiricism had any attractions.... His biggest contribution was the inspiration he gave to lesser men.[33]

During this period, he made a series of other innovations and took out patents for components as diverse as flame traps, ice indicators, carburettors, superchargers and fuel feeds. In early 1939 E. W. Hives (later Lord Hives), General Manager of Rolls-Royce, heard of his work and offered him a job as a research engineer, which he accepted and took up on 1 June. His instruction from Hives was to 'go on thinking'.[34] He continued to work on multi-stage axial and contra-flow designs. Major problems included: accurate forming of the aerodynamic blades; the complexity of alternate counter-rotating stages; and the difficulty of linking compressor and turbine. By 1944 it was clear that contra-flow engines were too difficult to manufacture, and the decision was taken to suspend contra-flow tests and shift at last onto simpler jet engine designs similar to that proposed by Whittle. Between 1943 and 1946, as well as continuing to develop Whittle's ideas, leading to the development of the Rolls-Royce Welland engine, Griffith also worked extensively on the use of compact jet engine designs, which could be mounted at the rear of the fuselage, leading ultimately to the Rolls-Royce Avon engine, and on axial bypass turbofan engines, culminating in the development of the Rolls-Royce Conway engine.

In 1945 Griffith conceived the idea of using the greater power and stability of jet engines to develop vertical take-off and landing (VTOL) planes, developing ideas from an earlier report of his in 1941:

> This suggests the possibility of a jump take-off, a flap being provided behind the power unit which could be lowered so as to deflect the slipstream downwards.... An obvious further development is to use the flap for landing also, whereby ground speed might be brought practically to zero before touching down.[35]

Griffith called this *Thistledown landing*.[36] A series of different designs led to the famous *flying bedstead*[37] test rig, a wingless platform with two Rolls-Royce Nene engines back-to-back, with cascades of deflecting vanes to provide vertical lift. Its first flight was on 3 August 1954. It rose about 9 feet (3 m) in the air and was then manoeuvred successfully using compressed air jets. This led directly to the Rolls-Royce RB108 engine, the first engine specifically designed for VTOL, which was first used on the Short Brothers and Harland SC1 aircraft. The flying bedstead and RB108 were ultimately precursors to the successful VTOL Harrier *jump jet* and its Rolls-Royce Pegasus engines. It is hard not to believe that Griffith got the idea of the flying bedstead subliminally from his father who, in his flight-of-fancy romances, described many weird and wonderful flying machines, including 'strange looking aeroplanes ... with vertical lift provided by five airscrews on vertical shafts'.[38]

Griffith was tall, slim, serious and taciturn, calm and somewhat aloof, unlike Whittle, the very epitome of a grey and dour engineer, but to friends and close associates, amusing and engaging, with a dry sense of humour. According to Hayne Constant,

> Griffith was a delightful man to work for. Not once during our time together did he utter a single word of reproof—not even when I was running an engine on vapourized [*sic*] petrol, a leak developed and I blew up the whole test bed.... He was a remarkable colleague; I have never had a better one.[39]

He published nothing after 1928 because his work was subject to national and/or commercial security. Most of his work was issued as RAE or Rolls-Royce internal documents. Nevertheless, he was widely respected and influential. He became a Fellow of the Royal Society in 1941 and a Commander of the British Empire in 1948. He received the Silver Medal of the Royal Aeronautical Society in 1955. It is a supreme irony that he is famous for his work developing aeroengines, yet he turned down Whittle's ground-breaking invention of the jet engine. And it is another supreme irony that he invented the theory of brittle fracture, which only became famous as a consequence of the terrible brittle fracture-induced crashes of the first jet airliners.

After his retirement in 1960, he went into hospital in Chertsey and then in Farnborough with a hip problem. He resumed work as a consultant for Rolls-Royce briefly in late 1961 and early 1962, but the following year returned to hospital, where he passed away, aged 70, on 13 October 1963. The Secretary of the Aeronautical Research Council, R. W. Gandy wrote in an obituary in *The Times* on 30 October 1963,

> It has been said that he was outwardly a reserved man but he had extraordinary charm in the presentation of his ideas and, among his fellow scientists, was of an attractively convivial disposition. He will be greatly missed by all who knew him.[40]

5　Fracture toughness

Griffith's equation says that a crack will propagate in a brittle material when the release of strain energy is big enough to overcome the corresponding creation of new surface energy. This happens more easily for high applied stresses and large cracks. Brittle materials contain small cracks and flaws, so their fracture strength is the applied stress at which the largest crack just begins to propagate,

$$\sigma_F = \sqrt{\frac{2E\gamma}{\pi c}},$$

where $2c$ is the largest crack length.

Cracks do not propagate so easily in ductile materials because localised plastic flow takes place, relieving the high stresses generated near any crack tip. In other words, a crack will propagate in a ductile material only when the release of strain energy is big enough to overcome the work of plastic flow γ_p as well as the creation of surface energy:

$$\sigma_F = \sqrt{\frac{2E(\gamma + \gamma_p)}{\pi c}}.$$

This is called *Irwin's modified Griffith's equation*. For high-strength materials with only limited ductility, the extent of plastic flow at a crack tip is relatively small, so $\gamma_p \approx \gamma$ and fracture is catastrophic, similar to brittle materials. For highly ductile materials such as pure metals, $\gamma_p \gg \gamma$, crack tips are easily blunted by extensive plastic flow so they cannot propagate, and fracture takes place by ductile tearing of the material.

Table 12.1 *Typical Values of Fracture Toughness*

Material	Fracture toughness K_c (MPa$\sqrt{m}$)
Aluminium alloy	24
Steel	50
Alumina	4
Silicon carbide	4
Glass	0.8
Polystyrene	1

The *stress intensity factor* $K = \sigma\sqrt{\pi c}$ is the increase in stress at the crack tip. The *critical stress intensity factor* K_c, when a crack begins to propagate and fracture takes place, is called the *fracture toughness*:

$$K_c = \sigma_F\sqrt{\pi c} = \sqrt{2E\left(\gamma + \gamma_p\right)}.$$

Table 12.1 shows typical fracture toughness values for a number of different materials.

The *critical strain energy release rate* $G_c = 2\left(\gamma + \gamma_p\right)$ is the strain energy release needed for a crack to propagate and initiate fracture. Griffith's equation then becomes

$$K_c = \sqrt{EG_c}$$

or

$$G_c = \frac{K_c^2}{E}.$$

In other words, the critical strain energy release needed for a crack to propagate is proportional to the square of the critical stress intensity factor at the crack tip (which is the same as the square of the fracture toughness), and the constant of proportionality is the Young's modulus.

There are three major fracture modes that describe different ways a crack can propagate. *Mode I* is tensile fracture, *mode II* is shear fracture, and *mode III* is tearing fracture, as shown in Figure 12.4. The fracture toughness is different for each fracture mode. The most common is the tensile mode, and we often refer to the tensile fracture toughness K_{1c}.

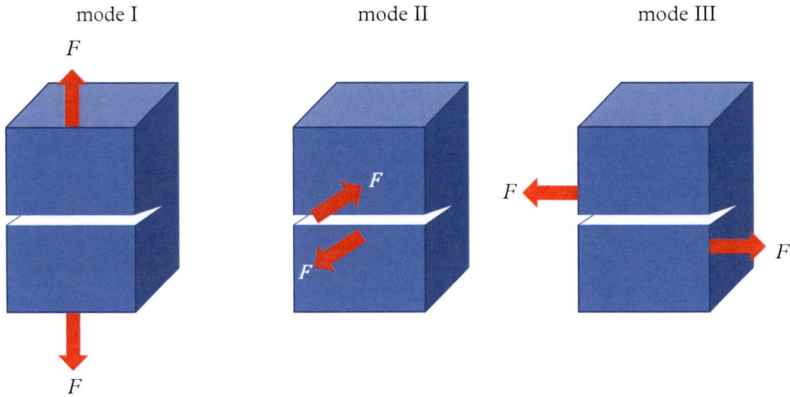

Figure 12.4 *Fracture modes*

6   Fractography

The structure of a fracture surface indicates the mechanism of fracture, and the study of fracture surfaces is called *fractography*.

Brittle cleavage fracture

Brittle crystalline materials exhibit *cleavage fracture*. The fracture surface consists of a near atomically flat *facet* across each crystal in the material. The facet plane is a low-energy surface plane, the nearest one normal to the maximum applied tensile stress. Once a crack begins to open up, it is highly constrained to one atomic or molecular plane, and runs catastrophically along that plane, because any deviation would represent an increase in energy. In a single-crystal material, there is one single flat facet; in a polycrystalline material, there are differently orientated flat facets in each crystal.

River fracture

Brittle non-crystalline materials exhibit *river patterns* on their fracture surface. The fracture plane is, again, near atomically flat, but without any crystallographic constraint. There is no particular low-energy surface plane, so a crack runs normal to the maximum applied tensile stress. The so-called river patterns represent unstable attempts to deviate from the fracture plane.

Ductile fracture

Ductile materials fracture from voids that open up around impurity particles, with ductile tearing between the voids. This produces a distinctive dimpled pattern on the fracture surface. The fracture surface is very rough, with the average fracture plane normal to the maximum applied tensile stress.

....

7 REFERENCES

1. Conrad Black. *Franklin Delano Roosevelt: Champion of Freedom* (Public Affairs, New York, 2003), 603.
2. Franklin Delano Roosevelt. 'Address is spur to British hopes; confirmation of American aid in conflict is viewed as heartening, a joining of interests, discarding of peace talks is regarded as a major point in the speech'. *New York Times*, 30 December 1940.
3. Stephen W. Stathis. *Landmark Debates in Congress* (Sage Publishing, London, 2009), 471.
4. Franklin D. Roosevelt. *Public Papers and Addresses of Franklin D. Roosevelt*, vol. IX (Random House, New York, 1941), 647.
5. Greg H. Williams. *The Liberty Ships of World War II: A Record of the 2,710 Vessels and Their Builders, Operators and Namesakes, with a History of the Jeremiah O'Brien* (McFarland, Jefferson, 2014), 15.
6. *Time*, 13 January 1941, 14.
7. Fredric Chapin Lane. *Ships for Victory: A History of Shipbuilding under the US Maritime Commission in World War II* (Johns Hopkins University Press, Baltimore, 1951; reprint 2001), 67.
8. 'Liberty ship breaks in two'. *The Log* **39** (1944), 72.
9. Graham M. Simons. *Comet! The World's First Jet Airliner* (Pen and Sword Books, Barnsley, 2013), 133.
10. 'Jet crash off Italy kills 35'. *New York Times*, 11 January 1954, https://www.nytimes.com/1954/01/11/archives/jet-crash-off-italy-kills-35-chester-wilmot-among-29-passengers-on.html (accessed 21 February 2020).
11. Kari Fay. 'The de Havilland Comet'. 1 February 2019. https://www.greatdisasters.co.uk/the-de-havilland-comet/ (accessed 2 March 2020).
12. Lord Lionel Cohen. *Report of the Court of Inquiry into the Accidents to Comet G-ALYP on 10 January 1954, and Comet G-ALYY on 8 April 1954* (HMSO, London, 1955), 22.
13. Ailise Bulfin. *Gothic Invasions: Imperialism, War and Fin-de-Siècle Popular Fiction* (University of Wales Press, Cardiff, 2018), 182.
14. Auguste Nemo, ed. *Essential Novellists: George Griffith: Edwardian Science Fiction* (Tacet Books, Brazil, 2018), 1.
15. Goodreads. 'George Chetwynd Griffith'. n.d. https://www.goodreads.com/author/show/3000622.George_Chetwynd_Griffith (accessed 21 February 2020).
16. A. A. Rubbra. 'Alan Arnold Griffith 1893–1963'. *Biographical Memoirs of Fellows of the Royal Society* **10** (1964), 133.
17. A. A. Griffith and G. I. Taylor. 'The use of soap films in solving torsion problems'. *Proceedings of the Institution of Mechanical Engineers* **93** (1917), 755–809.
18. Alan Arnold Griffith. 'The phenomena of rupture and flow in solids'. *Philosophical Transactions of the Royal Society A***221** (1921), 582–593.
19. Rubbra, n 16, 120.
20. A. A. Griffith. *An Aerodynamic Theory of Turbine Design*. Royal Aircraft Establishment report H1111 (Royal Aircraft Establishment, Farnborough, 1926).
21. Rubbra, n 16, 122.
22. F. Whittle. 'Future developments in aircraft design'. PhD diss., Royal Air Force College Cranwell, 1928. [Science Museum Group archives].

23. F. Starr. 'A commentary on "Future Developments in Aircraft Design"'. *Journal of Aeronautical History* (2019), paper 01.
24. Ibid. 35.
25. John Golley. *Jet* (Datum, Fulham, 1996), 42.
26. Andrew Nahum. *Frank Whittle: Invention of the Jet* (Icon Books, London, 2004), 19.
27. Golley, n 25, 36.
28. Chester T. Sims. *A History of Superalloy Metallurgy for Superalloy Metallurgists: Superalloys 1984* (TMS-AIME, Warrendale, 1984), 401.
29. Nahum, n 26, 27.
30. Peter Pugh. *The 50 Most Influential Britons of the Past 100 Years* (Icon Books, London, 2015), 2.
31. 'Sir Frank Whittle'. *Daily Telegraph*. 10 August 1996. [obituary].
32. Ibid.
33. Rubbra, n 16, 124.
34. See ibid. 125.
35. Ibid. 128.
36. Ibid. 129.
37. Ibid. 131.
38. Ibid. 117.
39. Ibid. 125.
40. Ibid. 132.

· ·

8 BIBLIOGRAPHY

'Alan Arnold Griffith 1893–1963'. A. A. Rubbra. *Biographical Memoirs of Fellows of the Royal Society* **10** (1964), 117–136.
Comet: The World's First Jet Airliner. Graham M. Simons (Pen and Sword Books, Barnsley, 2013).
Fracture Mechanics: Fundamentals and Applications (4th edition). Ted L. Anderson (CRC Press, Boca Raton, 2017).
Frank Whittle: The Invention of the Jet. Andrew Nahum (Icon Books, London, 2004).
Fundamentals of Fracture Mechanics. A. F. Knott (Butterworths, London, 1976).
Liberty: The Ships That Won the War. Peter Elphick (Naval Institute Press, Annapolis, 2001).
Strong Solids (3rd edition). A. Kelly and N. H. Macmillan (Clarendon Press, Oxford, 1987).

13

The Fermi Level

Electrical Properties

---

1   Electrical conduction

A light bulb is switched on and lights up when it is connected by copper wires to the two terminals of a battery, as shown in Figure 13.1. However, the same light bulb is not lit up when the copper wires are replaced by glass fibres, as also shown in Figure 13.1. This is because copper conducts electricity and glass does not. Electrons move easily within a piece of copper, but cannot move within a piece of glass. We say that copper is a *conductor* and glass is an *insulator*.

We can describe more accurately what happens when a light bulb is connected to a battery as follows. A battery is a device that produces electrons (e^-) at its negative terminal or *anode* ($\ominus$) and absorbs electrons at its positive terminal or *cathode* ($\oplus$). When the two battery terminals are connected with a conductor such as copper, electrons are produced at the anode and move along the copper to be absorbed at the cathode. A light bulb is a device that turns a flow of electrons into light. Interspersing the light bulb between the two battery terminals, as in Figure 13.1, provides a flow of electrons to the bulb and switches it on. When the two battery terminals are connected instead with an insulator such as glass, electrons cannot move along the glass from the anode to the cathode. Interspersing the light bulb between the two battery terminals, as in Figure 13.1, does not, therefore, provide a flow of electrons to the bulb, so it stays unlit.

The strength of a battery is measured by its *voltage V*. This is technically the difference in energy between electrons at the two terminals, i.e. it measures the strength with which the electrons want to move from the anode to the cathode through a conductor such as copper. The flow of electrons produced, i.e. the number of electrons moving per second, is called the *electric current* or just the *current I*. The current produced in a conductor is proportional to the voltage applied across its ends:

$$V \propto I = IR.$$

The Equations of Materials. Brian Cantor. Oxford University Press (2020). © Brian Cantor.
DOI: 10.1093/oso/9780198851875.001.0001

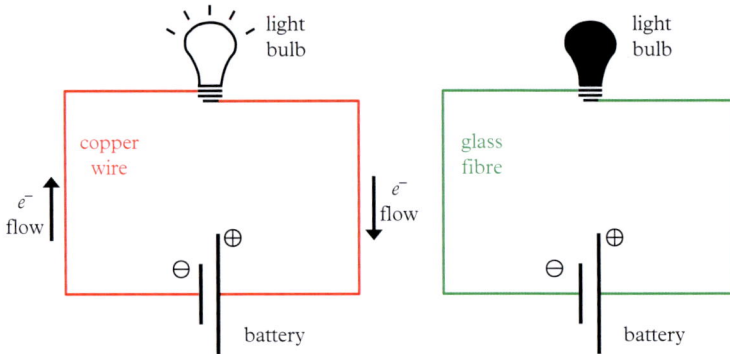

Figure 13.1 *Electrical conductors and insulators*

This is *Ohm's law*, and the constant of proportionality R is called the *resistance* of the conductor. Since $R = V/I$, the resistance is the voltage required to produce unit current. Electrons flow less easily in a conductor of high rather than low resistance, and we need to apply a higher voltage to achieve unit current.

The resistance of a conducting wire depends on its size and shape, as well as the material it is made from, increasing with increasing length l, and decreasing with increasing cross sectional area A,

$$R = \rho \frac{l}{A},$$

where $\rho = RA/l$ is a material property called its *resistivity*. Sometimes it is convenient to use the inverse of the resistivity, or the *conductivity* $\sigma = \rho^{-1} = l/RA$.

The voltage V is the difference in energy ΔE for each electron at the anode compared to the cathode, i.e. $V = \Delta E/q$, where q is the charge on the electron. We define the *electric field* $\mathcal{E}$ in a material as the force F on each electron caused by this difference in energy, $\mathcal{E} = F/q$. Energy is force × distance moved, so for an electron moving along a length of wire l across which there is a voltage V, the difference in energy is $\Delta E = Fl$, so $V = Fl/q = \mathcal{E}l$ or $\mathcal{E} = V/l$, i.e. the electric field is the voltage per unit distance along the wire. Similarly, the *current density* i is the current per unit cross sectional area of the wire $i = I/A$. Ohm's law can then be re-written as

$$V = IR = I\rho \frac{l}{A},$$

or, rearranging,

$$\mathcal{E} = \rho i.$$

In other words, when an electric field is applied to a material, the resulting current density is proportional to the applied electric field, with the resistivity as the constant of proportionality. Since $\rho = \mathcal{E}/i$, the resistivity is the electric field required to produce unit current density.

Table 13.1 shows typical resistivity values and their temperature variation for some common materials. Metals such as copper, aluminium and iron are good conductors with very low resistivity, typically $<10^{-7}\,\Omega$m, so they are highly amenable to electron flow. Ceramics and polymers such as silica, glass and polythene are insulators with very high resistivity, typically $>10^{10}\,\Omega$m, so they are effectively impermeable to electron flow. Materials such as silicon and germanium are an intermediate category called *semiconductors* with intermediate resistivity, typically 10^{-2} to $10^{2}\,\Omega$m, so electrons can flow, but only sluggishly.

The resistivity of a conductor increases with increasing temperature, because increasing atomic or molecular vibrations interfere with the motion of the electrons. The increase is not very large and is typically proportional to temperature T according to

$$\rho = \rho_o (1 + \alpha \Delta T),$$

where ρ_o is the resistivity at room temperature, α is the *temperature coefficient of resistivity*, and ΔT is the increase in temperature above room temperature. Insulators do not conduct at any temperature, so their temperature coefficient is effectively zero. The resistivity of a semiconductor decreases with increasing temperature, because the number of available electrons increases as the temperature rises. In other words α is positive for conductors, zero for insulators and negative for semiconductors, as shown in Table 13.1.

Table 13.1 *Typical Values of Room-temperature Resistivity and Temperature Coefficient of Resistivity for Some Common Materials*

Material	Resistivity ρ (Ωm)	Temperature coefficient α (C^{-1})
Copper	1.7×10^{-8}	0.04
Aluminium	2.6×10^{-8}	0.04
Iron	10^{-7}	0.006
Germanium	10^{-3}–0.5	−0.05
Silicon	0.1–60	−0.07
Glass	10^{13}	0
Polythene	10^{14}	0
Silica	10^{17}	0

2   Electrons in atoms

Why do different materials show such different conduction behaviour? The answer is that it depends on the availability of electrons to detach from individual atoms and molecules so they can move throughout the material. It depends on the number of *free electrons*. In metals, there are many free electrons; in semiconductors, there are only a few; and in insulators, there are none, because they are all closely bound to the atoms or molecules of the material. To understand this in more detail, we have to consider first how electrons are distributed within individual atoms, before considering their behaviour in a material consisting of a large number of atoms or molecules.

Electrons in atoms are governed by the laws of *quantum mechanics*. The number of electrons in an atom is called the *atomic number z*. The electrons are described by four *quantum numbers*:

1. *The principal quantum number n* defines the different electron shells that electrons can occupy within the atom according to their distance from the atomic nucleus. It can take values of $n = 1, 2, 3, 4$ etc. in increasing distance from the nucleus.

2. *The orbital quantum number l* describes the different kinds of orbital that electrons can occupy as a subshell within each shell. It can take values of $l = 0, 1, 2 \ldots n - 1$, usually labelled respectively $s, p, f, d \ldots$ orbitals. There are only s orbitals for the first shell, $n = 1$; there are s and p orbitals for the second shell, $n = 2$; there are s, p and f orbitals for the third shell, $n = 3$; and so on.

3. *The magnetic quantum number m* describes the particular orbital and corresponding angular momentum that electrons can have within a given subshell. It can take integer values between $-l$ and $+l$, i.e. $m = 0$ for an s orbital; $m = -1, 0$ or $+1$ for a p orbital; $m = -2, -1, 0, +1$ or $+2$ for an f orbital; and so on.

4. *The spin quantum number s* describes the magnetic spin that electrons can have, and can take only two values, either $+\frac{1}{2}$ or $-\frac{1}{2}$ (sometimes referred to as *spin up* and *spindown*).

According to the *Pauli exclusion principle*, each of the z electrons has its own unique set of quantum numbers called the *electron state*. No two electrons can have the same set of quantum numbers, i.e. no two electrons can be in the same state. Moreover, each electron has a discrete energy defined by its set of quantum numbers, and the electron energy increases as the quantum numbers increase. This means that the z electrons in an atom fill up the different quantum numbers, i.e. the different shells and subshells, starting with the lowest quantum numbers, which correspond to the lowest energy, until all the z electrons have been accommodated. Electrons in the outer shell are called *valence electrons*, and electrons in the inner shells are called *core electrons*. The number of valence electrons is called the *valency* of the atom.

For instance, an atom of sodium has 11 electrons, and its electronic structure is $1s^2 2s^2 2p^6 3s^1$. This means the following:

- $1s^2$: there are two electrons in the first shell, both in s orbitals, with quantum numbers $n = 1, l = 0, m = 0$ and $s = +\frac{1}{2}$ and $-\frac{1}{2}$.

- $2s^2 2p^6$: there are eight electrons in the second shell, of which two are in s orbitals, with quantum numbers $n = 2, l = 0, m = 0$ and $s = +\frac{1}{2}$ and $-\frac{1}{2}$, and six are in p orbitals, with quantum numbers $n = 2, l = 1, m = -1, 0$ and $+1$ and $s = +\frac{1}{2}$ and $-\frac{1}{2}$.

- $3s^1$: there is one electron in the third shell, in an s orbital, with quantum numbers $n = 3, l = 0, m = 0$ and $s = +\frac{1}{2}$.

The two electrons in the 1s shell have the lowest energy; the two electrons in the 2s shell have the next lowest energy; the six electrons in the 2p shell have the next lowest energy; and the single electron in the 3s shell has the highest energy. This is shown more clearly in Figure 13.2 and Table 13.2. There is only one valence electron, the one in the 3s shell, the others are all core electrons, and the valency of sodium is one.

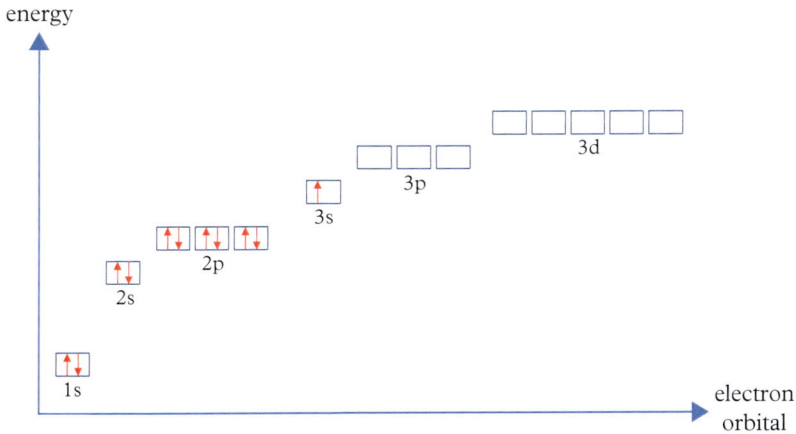

Figure 13.2 *Energy of electron states 1s, 2s, 2p, 3s, 3p and 3d, and the electronic structure of a sodium atom (each electron represented by a red arrow)*

Table 13.2 *The Electronic Structure of a Sodium Atom*

Shell	No. of electrons	Quantum numbers			
		n	l	m	s
1s	2	1	0	0	$\pm\frac{1}{2}$
2s	2	2	0	0	$\pm\frac{1}{2}$
2p	6	2	1	$-1, 0, +1$	$\pm\frac{1}{2}$
3s	1	3	0	0	$+\frac{1}{2}$

An atom can gain excess energy, for instance when it is heated, or when it is irradiated with a beam of light or X-rays. Its outer valence electrons can absorb thermal energy or radiation energy by being promoted to a higher set of quantum numbers. The energised electron and atom are then said to be in an *excited state*. For example, the 3s valence electron in a sodium atom can absorb excess energy and be promoted into a 3p state, producing an excited sodium atom.

3   The Fermi level

When a large number of atoms or molecules is brought together to make a material, the discrete energy levels associated with the different states with their different quantum numbers spread into energy bands. Consider one mole of the material, i.e. N_o atoms or molecules, where N_o is *Avogadro's number*. As a collection of individual, separated atoms, there are N_o electrons with each set of quantum numbers, all with the same energy. When they are brought together in a material, they cannot all have the same energy, so instead they fill an *energy band* of very closely spaced energies, as shown in Figure 13.3. There is a different energy band corresponding to each of the different sets of quantum numbers, i.e. corresponding to each of the different shells and subshells in the atoms and molecules, with *band gaps* of disallowed energies between the bands. The bands are of different types. The inner core electrons remain bound to their individual atoms or molecules and form core bands. The valence electrons and any excited states are less well bound, however, and make bands that permeate through the material and are called, respectively, the *valence band* and the *conduction band*.

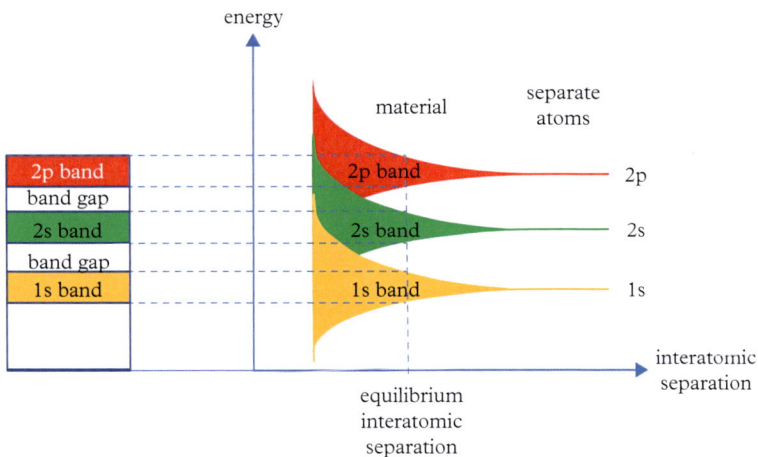

Figure 13.3 *Electron energy band structure*

The *Fermi level E_F* is defined as the highest energy an electron can have at absolute zero. Effectively there is a *Fermi equation* for the maximum electron energy:

$$E_F = E_{max}.$$

For an individual atom, the Fermi level is the energy of the outermost valence electron. For an assembly of atoms or molecules in a material, the Fermi level is the highest energy of the N_o valence electrons in the valence band. In a metal conductor, the Fermi level is in the middle of the valence band, as shown in Figure 13.4. This means that the highest energy electrons, i.e. valence electrons with the Fermi energy E_F, can easily gain a small amount of excess energy and move readily throughout the material. In other words, the valence electrons are *free electrons*. In a ceramic or polymer insulator, on the other hand, the Fermi level is at the top of the valence band, as also shown in Figure 13.4. This means that the highest energy electrons, i.e. valence electrons with the Fermi energy E_F, cannot gain extra energy and are immobile because of the *band gap* of disallowed energies. Effectively, all the electrons are bound to their individual atoms or molecules and there are no free electrons. In a semiconductor, as in an insulator, the Fermi energy is again at the top of the valence band, as also shown in Figure 13.4, but in these materials the band gap of forbidden energies is only small. This means that, under some circumstances, when extra energy is available, the highest energy electrons at the top of the valence band can gain enough extra energy to jump across the band gap into the conduction band and become mobile. In other words there are some free electrons, depending on the circumstances and the amount of available extra energy.

The *Fermi function $f(E)$* is the probability of occupation of an energy level as a function of the energy. This is shown in Figure 13.5 for a metal conductor. At absolute zero,

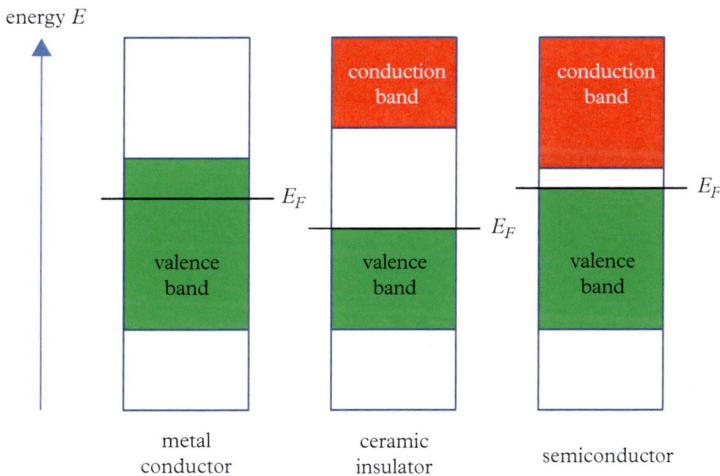

Figure 13.4 *Fermi level in a metal conductor, ceramic insulator and semiconductor*

Fermi function
$f(E)$

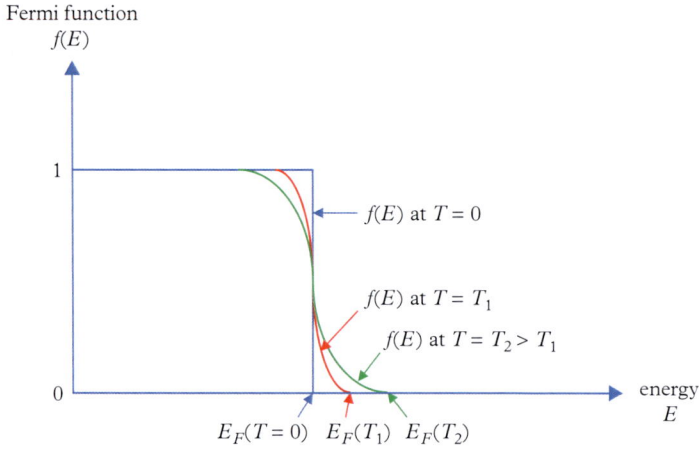

Figure 13.5 *The Fermi function f(E) for a metal conductor at different temperatures*

$f(E) = 1$ for all energies equal to or below the Fermi level E_F and $f(E) = 0$ for all energies above the Fermi level E_F. In other words, at absolute zero all states at or below the Fermi level are occupied, but all states above the Fermi level are empty:

$$f(E) = 1 \text{ for } E \le E_F$$
$$f(E) = 0 \text{ for } E > E_F.$$

In a metal conductor at temperatures above absolute zero, thermal energy allows some of the electrons to gain excess energy and have energies somewhat higher than the Fermi level, as shown in Figure 13.5. This is because of the *Boltzmann function*, which says that the probability p of gaining excess energy $E - E_F$ is proportional to the exponential of minus the excess energy $E - E_F$ divided by the thermal energy kT, where k is Boltzmann's constant and T is the temperature,

$$p = \exp \left(\frac{-(E - E_F)}{kT} \right),$$

and the Fermi function is

$$f(E) = \frac{1}{\exp \left(\frac{(E - E_F)}{kT} \right) + 1}.$$

In a ceramic insulator, however, $f(E) = 1$ for all energies equal to or below the Fermi level E_F and $f(E) = 0$ for all energies above the Fermi level E_F, but in this case at all temperatures, because of the band gap of disallowed energies so that no states are available just above the Fermi level. In a semiconductor, the band gap is very small, as shown in Figure 13.4, so that at temperatures above absolute zero, a small number of electrons can gain excess energy to jump up from the valence to the conduction band and become mobile, i.e. there is a small number of thermally excited *free electrons*.

4   Enrico Fermi

Enrico Fermi

At 3 a.m. on the morning of 6 August 1945, Colonel Paul Warfield Tibbets Jr took off from North Field military airport on the island of Tinian in the Northern Mariana Islands in the Pacific Ocean, flying a specially converted B29 Superfortress bomber named *Enola Gay* (after Tibbets' mother), heading for Japan 1,500 miles (2,500 km) away. He was accompanied by two other back-up B29s and, after making a rendezvous with weather reconnaissance, photography and instrumentation planes over Iwo Jima, they all (a total of seven planes) flew on to Japan. At 8.09 a.m. Tibbets started his bomb run, and at 8.15 a.m. his bombardier, Major Thomas Ferebee, pushed a lever to actuate the bombing system, dropping a single 9,000-lb (4,082-kg) bomb on its intended target, the Aioi Bridge in the western Japanese city of Hiroshima. The bomb was called *Little Boy* after the character Wilmer Cook played by Elisha Cook Jr in director John Huston's famous film noir *The Maltese Falcon*. It was a gun-triggered uranium device containing about 141 lb (64 kg) of enriched uranium, enriched to contain 80% of fissile radioactive uranium U^{235}. It was the second-ever atomic bomb to be exploded and the first to be used in combat (the first to be exploded was the so-called Trinity test bomb). The bomb dropped from the airplane's flying height of 31,000 feet (9,400 m) to its detonation height of about 1,900 feet (580 m), missing its target, the bridge, due to a crosswind, and exploding instead immediately above the Shima Surgical Clinic. Although only about 1.5% of the uranium exploded, the blast was equivalent to 16,000 tons (14.5 Gg) of TNT. There was a massive boom, the sky erupted in a dazzling bluish light, and the earth seethed like a boiling liquid. The mission commander, Captain William S. Parsons said later, 'The whole thing was tremendous and awe-inspiring . . . the men aboard with me gasped "My God"'.[1-3] Everything within a radius of 2 miles (3.2 km) was vaporised. About 70% of Hiroshima's buildings were destroyed. About 80,000 people were killed instantly or would die later from radiation, and another 70,000 were injured, making a total of almost half of Hiroshima's population of 340,000. In preparation for the Hiroshima bomb, a US scientific panel had decided against using just a demonstration bomb (i.e. without a real target), and against dropping a warning leaflet in advance, because of the unproven and therefore uncertain success of the bomb, and because of a desire to shock the Japanese leadership (somewhat conflicting reasons). Later that day, the US President Harry Truman issued a public statement: 'If they [the Japanese] do not now accept our terms [of surrender] they may expect a rain of ruin from the air, the like of which has never been seen on this earth'.[4,5]

Three days later, another atomic bomb was dropped at 11.02 a.m. over another western Japanese city, Nagasaki. It exploded at a similar height of about 1,800 feet (548 m) with a blast equivalent to 21,000 tons (19 Gg) of TNT, midway between the Mitsubishi steelworks and the Nagasaki arsenal, killing about 40,000 people and injuring about 60,000 more, a total of about 40% of Nagasaki's population of 260,000, and destroying up to 80% of the city's industrial production capacity. The Nagasaki bomb was called *Fat Man* after the character Kaspar Gutman played by Sidney Greenstreet in *The Maltese Falcon*. Unlike *Little Boy*, it was an implosion-triggered plutonium device, containing 11 lb (5 kg) of fissile plutonium Pu^{239}. It was the second and the last (so far) atomic bomb to be used in combat. Truman issued another public statement: 'We shall continue to use it [the atomic bomb] until we completely destroy Japan's power to make war. Only a Japanese surrender will stop us'.[6-8] The Potsdam Declaration, issued earlier in the year by Truman on 26 July during a high-level conference in the German city of Potsdam with the UK Prime Minister Winston Churchill and the Russian Premier Joseph Stalin, had outlined America's terms for Japan's unconditional surrender to end the Second World War (Germany had already surrendered), but had been roundly rejected by the Japanese Emperor Hirohito. The Japanese War Minister Korechiki Anami, General Yoshijiro Umezo and Admiral Teijiro Toyoda said the terms were 'too dishonourable',[9] and the Prime Minister Kantaro Suzuki said their policy towards the declaration was one of *mokusatsu* (killing with silence),[10] equivalent to ignoring it as contemptible. On 15 August, however, six days after the bombing at Nagasaki, Hirohito changed his mind and broadcast his capitulation statement to the nation:

[T]he enemy now possesses a new and terrible weapon with the power to destroy many innocent lives and do incalculable damage. Should we continue to fight, not only would it result in an ultimate collapse and obliteration of the Japanese nation, but also it would lead to the total extinction of human civilization.[11-13]

Dropping atomic bombs on Hiroshima and Nagasaki was one of the defining events of the 20th century. It brought to an end the Second World War and initiated the Cold War era. Ever since 1945, preventing the proliferation and use of atomic weapons has continued to be a key, though difficult, policy objective for the world's politicians and diplomats.

Atomic bombs were developed through a remarkable secret US government programme code-named the *Manhattan Project*. Many scientists, technologists, engineers and politicians were involved, but the key scientific driving force was provided by the Italian scientist Enrico Fermi. Fermi was born on 29 September 1901 in Rome. His father, Alberto Fermi, was Division Head in the Italian government-run railway system, and his mother, Ida de Gattis, was a primary school teacher. He was the youngest of three children: his sister, Maria, was two years older, born in 1899; and his brother, Giulio, was one year older, born in 1900. All three children were extremely bright at school, but Enrico in particular excelled: 'easily the best in his class'.[14] He loved maths and physics, and aged just seven began to buy physics books from the *Campo del Fiori* (Field of Flowers) market in Rome. He was baptised as a Roman Catholic, but his family was not particularly religious, and he became and remained an agnostic throughout his life. Sadly, his brother Giulio died when Enrico was just 14 years old, when an operation to remove an abscess in his throat went wrong. This was, of course, a massive blow to the family. The three children had been very close. Mrs Fermi was desolate and never really recovered: 'Giulio had been her favourite'.[15]

Enrico's father had a work colleague, Adolfo Amidei, who was a university-trained engineer, and he began to act as Enrico's mentor, discussing with him topics such as advanced geometry, trigonometry, calculus and mechanics. From an early age Enrico had a truly prodigious memory and was able to recite verbatim long poems as well as tracts of calculus. He was also very active physically, playing soccer, going swimming, and hiking in the mountains. Amidei persuaded him and his family that he should apply to the Scuola Normale Superiore, an elite institution associated with the University of Pisa and founded by Napoléon in 1810 (following the success of the *École Normale* in Paris), limited to an enrolment of no more than 40 students each year, the best and brightest across all Italy. In 1918 Enrico took the highly competitive scholarship exam and came top. To quote his biographer Dan Cooper: 'he creamed it...the examining professor could hardly believe his eyes'.[16] Within a year of starting at Pisa, Enrico was ahead of his teachers in understanding newly emerging fields such as Einstein's relativity and Planck's quantum theory, which were not yet taught in Italy. He published a number of theoretical papers in the scientific literature during his period as an undergraduate, on topics such as the dynamics and gravitational field of electrical charges, including relativistic effects. His doctoral studies were quite different, concerning the experimental formation of X-ray images, and his thesis was modestly entitled '*Un Teorema di Calcolo delle Probabilità ed Alcune sue Applicazioni*' ('A Theorem on Probability and Some of Its Applications').[17] In fact, Fermi was very unusual in being equally outstanding as a theoretician and an experimentalist. He received his *laurea* (doctorate) in 1922 at the age of just 20.

In 1922 Fermi won a prestigious scholarship to study abroad with Max Born at the University of Göttingen in Germany; in 1923 he returned to a fellowship at the University of Rome; in 1924 he won a Rockefeller Fellowship to study abroad, this time with Paul Ehrenfest at the University of Leiden in the Netherlands; in 1924 he moved to a teaching position at the University of Florence; and in 1926, aged just 25, he won the competition for the newly created Chair of Theoretical Physics back at the University of Rome, set up by the influential Director of the Institute of Physics and ex-Minister of Education, Orso Mario Corbino. These were tumultuous years for Fermi. In 1924 both his parents died. The following year he met his future wife, Laura Capon. And in 1926, aged just 24, he published his seminal paper on what are now called *Fermi-Dirac statistics*[18], discovered the same year independently by the English physicist Paul Dirac. This was probably the most important theoretical work of his career, and it immediately projected him onto the world stage.

In 1925, Laura Capon, aged 16, was a student of science at the University of Rome. She met Enrico, aged 23, in a streetcar, on a group outing to the country. They played soccer, with Laura as a novice in goal helping to win the game. She said later, 'That was the first afternoon I spent with Enrico Fermi and the only instance in which I did better than he'.[19] Two years later they met again, on a hiking vacation in the Dolomites, and began to go out together regularly in Rome and on trips to the country. Fermi had bought a tiny, bright-yellow sports car he called the *Bébé Peugeot*, which frequently went wrong and required him to demonstrate his engineering abilities, but also, without doubt, helped with his courtship of Laura. On 19 July 1928 they were married, even though, according to Dan Cooper: 'Laura had planned to follow a career and not marry . . . [and] Fermi had said he could afford the Peugeot or a wife but not both'.[20] They subsequently had two children, Nella and Giulio, born in 1931 and 1936 respectively.

In the mid-1920s, it was becoming clear that there are two different kinds of fundamental particle, those with spin, such as electrons and protons, and those without spin, such as photons and α-particles. The former are now called *fermions*, after Fermi, and the latter are now called *bosons*, after the Indian physicist Satyendra Nath Bose. Fermions and bosons behave in a totally different way. Fermions obey the *Pauli exclusion principle* (discovered by the Austrian physicist Wolfgang Pauli in 1925), which means that no two particles can be in the same quantum state, and each particle must have a different set of quantum numbers and therefore different energy. On the other hand, bosons can all be in the same quantum state, with the same set of quantum numbers and, therefore, the same energy. Fermi's paper was entitled '*Sulla Quantizzazione del Gas Perfetto Monoatomic*' ('On the Quantisation of the Perfect Monatomic Gas').[21] It describes the theoretical statistical distribution of groups of electrons across the different quantum states and energies, the so-called Fermi-Dirac statistics. Technically, the Fermi-Dirac distribution gives the probability that particular quantum states are occupied by a group of particles, and this is the same as the probability distribution of the different particle energies. For any group of particles, the highest quantum state occupied and the corresponding highest energy is referred to as the *Fermi level*. At almost the same time, Bose and Einstein were independently deriving the *Bose-Einstein statistics* for bosons.

In 1934, Fermi published an important paper on β-decay, when the nucleus of an atom emits an electron.[22] The measured energies did not fit theoretical calculations, and he solved the problem by postulating that *neutrinos* (Italian for *little neutral ones*) were being emitted at the same time. The neutrinos carried off some of the energy, but were difficult to detect because they were so small and were not charged. Effectively, his theory confirmed a previous speculation by Pauli about the existence of the neutrino as a new fundamental particle, and was the first description of the *weak interaction*, one of the four fundamental forces of nature. He developed his theory in 1933 and submitted his paper to the prestigious journal *Nature*, but it was rejected as 'abstract speculations remote from physical reality'.[23] Unabashed, he published elsewhere the following year.[24]

When Rutherford first split the atom in 1911, he fired high-energy α-particles at nitrogen atoms and found that they emitted positively charged protons. In 1932, James Chadwick, working in Rutherford's lab, showed that when α-particles were fired at beryllium atoms they emitted neutral particles that he called *neutrons*. And in 1934, Irène Joliot-Curie, the daughter of Marie Curie, and her husband, Frédéric Joliot-Curie, showed that beams of α-particles could be used to realise the alchemist's dream and transmute one element into another, initially creating radioactive nitrogen from boron, and then going on to produce radioactive phosphorus from nitrogen and radioactive silicon from magnesium. This demonstrated that radioactivity could be produced artificially, and was not only emitted naturally from heavy elements such as uranium, providing a method for creating radioactive materials quickly and cheaply for use in medicine. Chadwick won the 1935 Nobel Prize in Physics for the discovery of the neutron, and the Joliot-Curies won the 1935 Nobel Prize in Chemistry for the discovery of artificial radioactivity. In fact, Rutherford's experiment was an example of an (α, p) nuclear reaction, $\alpha + \mathrm{N}^{14} \rightarrow \mathrm{O}^{17} + p$, i.e. when struck by an α-particle, a nitrogen nucleus N^{14} absorbs it, emits a proton p and is transformed into the nucleus of a radioactive isotope of oxygen O^{17}. Chadwick's and the Joliot-Curies' experiments were examples of (α, n) nuclear reactions, for example $\alpha + \mathrm{B}^{10} \rightarrow \mathrm{N}^{13} + n$, i.e. when struck by an α-particle a boron nucleus B^{10} absorbs it, emits a neutron n and is transformed into the nucleus of a radioactive isotope of nitrogen N^{13}.

Fermi quickly saw the significance of these results and began a major series of experiments to study the behaviour of beams of neutrons. He discovered a much more effective way of producing neutrons, using a mixture of beryllium powder and radon gas. Natural α-particles from radioactive radon impinge on beryllium, causing an (α, n) reaction and creating a beam of neutrons. This proved to be a stronger neutron source, more convenient and controllable than the previous use of polonium as the primary source of α-particles. He and his colleagues in Rome began to investigate the behaviour of beams of neutrons when they strike different target materials, working their way through the periodic table in a typically rigorous and organised fashion. Very rapidly they had published many new papers, identifying different reactions and new radioactive isotopes. They expected that higher energy neutrons would be more effective at producing nuclear reactions, but they found exactly the opposite. According to Fermi's first student, Emilio Segrè, 'there were certain [wooden] tables ... which had miraculous properties, since silver irradiated on these tables became much more active than when it was irradiated on other marble tables in the same room'.[25] The neutrons slowed down when they interacted with carbon atoms in the wood and, while high-energy neutrons can pass more or less straight through a silver target, slow neutrons are captured much more easily. Similarly, neutrons were found to be hundreds of times more effective when passed through paraffin wax, because hydrogen atoms act like carbon atoms and slow the neutrons down. To prove this theory, Fermi and his colleagues decided to repeat the experiment under water, using the goldfish pond in Corbino's garden. They did indeed get the same result: fortunately 'the goldfish retained their calm and dignity despite the neutron shower'.[26] Overall, Fermi's group produced 25 major papers on neutron beams by the end of 1934. The same year, Fermi took out a patent on the uses of slow neutrons, for which the US government later paid $40,000, in 1953.

In 1938, the influential Danish scientist Niels Bohr asked Fermi confidentially if he would accept the Nobel Prize that year, and Fermi said yes. The reason for doubt was that the fascist regimes in Germany and Italy had banned receipt of Nobel Prizes. Benito Mussolini and his Fascist Party had come to power in Italy in 1922, following a mass demonstration and a march on Rome. Fermi was appointed as a member of the Royal Academy of Italy by Mussolini in 1929, and joined the Fascist Party later the same year. In the 1930s, however, Mussolini increasingly began to follow Hitler in enacting racist laws, restricting and then persecuting non-Aryan and especially Jewish scientists. Fermi's wife, Laura, was Jewish and, with the promulgation of new, even-tighter racist laws in 1938, Fermi lost patience with the fascists. He decided to leave Italy secretly immediately after the Nobel Prize ceremony in Sweden, and accept one of many offers to work in the United States. On 10 November, he received the call from the Swedish Academy of Sciences confirming he had won the prize; on 10 December, he received the prize from King Gustav V in Stockholm; and he and his family immediately embarked on the ocean liner *Franconia* for the United States, arriving in New York on 2 January 1939. On arrival, he was offered professorships at five US universities, and he accepted one at Columbia University in New York.

In 1934, Fermi and his colleagues in Rome had bombarded thorium and uranium with slow neutrons, but their results were difficult to interpret because of the confusing effect of the natural radioactivity of the target materials. They ruled out all elements lighter than uranium down to lead, and concluded that they had produced heavier, previously unknown, so-called *transuranic* elements, which they named *hesperium* and *ausonium*. This was wrong, and they were criticised by the German chemist Ida Noddack, who argued that they might, instead, have produced even lighter elements, i.e. that the neutrons might have broken the heavy target nuclei into smaller pieces. Noddack was not, however, taken seriously by the scientific community, partly because she did not present any new experimental results, but mainly because such *nuclear fission* was widely (though erroneously) thought to be completely impossible. On 16 and 17 December 1938, ironically at almost exactly the same time as Fermi was receiving his Nobel Prize and fleeing Italy for America, the German radiochemist Otto Hahn and his student Fritz Strassmann at the Kaiser Wilhelm Institute for Chemistry in Berlin were doing the experiments which showed definitively that the lighter element barium is the product when uranium is bombarded with neutrons, confirming that fission is indeed possible. Noddack's criticism of Fermi's interpretation was, after all, correct.

On 22 December 1938, Hahn and Strassmann submitted a manuscript on their ground-breaking results to *Naturwissenschaften*, and it was published on 6 January 1939.[27] As a radiochemist, Hahn was chary of proposing a major physics breakthrough. He did not refer to fission (or *bursting*, as he called it) in the January paper, and he did not discuss his results with physicists in his Institute. Instead he wrote about the results to his longstanding friend and colleague, the Austrian Jewish scientist Lise Meitner, whom he had advised and then helped to escape from Berlin to Stockholm earlier that year: 'Perhaps you can provide some fantastic explanation. We ourselves realize that it [uranium] can't really burst into barium'.[28] Meitner, then at the Nobel Institute for Physics in Stockholm, and her nephew Otto Frisch at the Niels Bohr Institute in Copenhagen, provided the underlying theoretical explanation of the new results, showing why there were no stable elements beyond uranium (because of the repulsion of so many protons), and how Einstein's famous equation $e = mc^2$ explains the tremendous release of energy associated with nuclear fission. Frisch confirmed Hahn and Strassmann's experimental results in January 1939, and it was he who coined the term *fission*. Hahn and Strassmann published two other papers in 1939,[29,30] including a discussion of the possibility of a *chain reaction*, because of the production of more neutrons than used in the initial bombardment. Meitner and Frisch published their theoretical justification for the results in *Nature* with two papers later that year.[31,32] Hahn and Strassmann received the 1944 Nobel Prize for Chemistry for the discovery of nuclear fission. Many scientists and historians have subsequently suggested that Meitner and Frisch should have shared in the prize, but not Meitner herself. As she wrote in 1945 in a letter to one of her friends, 'Surely Hahn fully deserved the Nobel prize in chemistry. There is really no doubt about it. But I believe that Otto Robert Frisch and I contributed something not insignificant to the clarification of the process of uranium fission'.[33]

Towards the end of 1938, Hahn, Meitner and Frisch discussed their emerging results on nuclear fission with Niels Bohr, when Hahn gave a lecture at Bohr's Institute in Copenhagen in November 1938. The news of the discovery of nuclear fission travelled across the Atlantic very rapidly, because Bohr was visiting the United States at the beginning of 1939, giving lectures at Princeton University. Fermi commented,

> I remember very vividly the first month, January 1939, that I started working at the Pupin Laboratories [at Columbia] because things began happening very fast. Niels Bohr was on a lecture engagement at Princeton and I remember one afternoon Willis Lamb came back very excited and said that Bohr had leaked out great news…the discovery of fission and at least the outline of its interpretation.[34]

Fermi was embarrassed because he had missed the discovery of nuclear fission, but he was quick to see the implications and immediately began a feverish period of experimentation. In fact the news spread like wildfire, and physicists everywhere rushed to study the fission of uranium by slow neutrons. The possibility of a chain reaction had first been suggested in 1933 by the Hungarian physicist Leo Szilard, who had taken out a patent on the idea and assigned it to the British Navy (he was working in London at the time). By a remarkable coincidence, Szilard had decided in late 1938 to emigrate to the United States because he believed that war in Europe was imminent, and he had just moved to Columbia University a month before Fermi. Fermi himself had first mentioned that Einstein's famous equation $e = mc^2$ carried the possibility of enormous energy being released from a nuclear reaction, when he wrote an appendix in 1923 to the Italian edition of the astronomer August Kopf's book on relativity:

> It does not seem possible, at least in the near future, to find a way to release these dreadful amounts of energy – which is all to the good because the first effect of an explosion...would be to smash into smithereens the physicist who had the misfortune to find a way to do it.[35]

Despite such a stark warning, Fermi and Szilard set about with gusto to work on this very project.

In August 1939, Szilard persuaded Albert Einstein to write his famous letter to the US President Franklin Delano Roosevelt:

> Some recent work by E. Fermi and L. Szilard, which has been communicated to me in manuscript, leads me to expect that the element uranium may be turned into a new and important source of energy in the immediate future...it may become possible to set up a nuclear chain reaction in a large mass of uranium, by which vast amounts of power...would be generated.... This new phenomenon would also lead to the construction of bombs.... A single bomb of this type, carried by boat and exploded in a port, might very well destroy the whole port.[36]

How prophetic! Einstein's letter prompted the formation of a US government Advisory Committee on Uranium, which amongst other things provided grants for the research at Columbia.

Natural uranium consists of two isotopes: 0.7% U^{235}, which is radioactive and fissile; and 99.3% U^{238}, which is stable. Every time a neutron strikes a U^{235} nucleus and causes nuclear fission, it produces at least two new neutrons and releases enormous amounts of energy. The problem facing Fermi and Szilard was twofold: how to have enough fissile material, the *critical mass*, to create a self-sustaining chain reaction releasing energy continuously; yet how to prevent a premature explosion. Otto Frisch and Rudolf Peierls at Birmingham University in the United Kingdom had shown that the critical mass of U^{235} was about 10 kg. To prevent premature explosion, Fermi hit on the idea of separating individual pieces of uranium or uranium oxide with graphite. The graphite would make sure that the neutrons were slowed down and therefore not lost, but it would also act as a moderator by absorbing some of them. Non-fissile U^{238} also acted as a powerful moderator. Fermi and his team began to build *atomic piles*: blocks of graphite, interspersed with lumps of uranium oxide, initially 3 feet (0.9 m) across and 8 feet (2.4 m) tall, with a radon–beryllium neutron source at the base to kick-start the reaction, and rhodium foils at different heights to measure the induced radioactivity. Fermi said, '[it was] the first time that I started to climb on top of my equipment'.[37] They built several of these piles to study and understand their behaviour, initially in the Pupin Labs at Columbia University in New York and then in the Metallurgical Lab at the University of Chicago.

Fission research was boosted dramatically in 1941 when the Japanese bombed Pearl Harbor and America joined the war. Fermi, however, was immediately classified as an enemy alien, with severe restrictions on his activities: he could not own a camera or shortwave radio; he was not allowed to fly; and he had to notify the authorities if he moved outside his district. Commuting between New York and Chicago became irksome in the extreme. In the end he was given a special travel pass; better still, in late 1942 Italians were taken off the enemy alien list; and in 1944 he became a US citizen.

Fermi's team succeeded in building the world's first nuclear reactor, Chicago Pile Number One, or CP1, in the abandoned squash courts under the University of Chicago's football stadium, Stagg Field. For safety reasons, the original plan had been to use a site at Argonne Woods Forest Preserve, 20 miles (32 km) from the city, but construction at Argonne was stopped by an industrial dispute, so the work reverted to the city. CP1 was a squashed sphere, 25 feet (7.6 m) wide and 20 feet (6.1 m) high, consisting of 400 tons (360 Mg) of graphite in the form of 57 machined layers, interspersed with 40 tons (36 Mg) of uranium oxide and 6 tons (5.4 Mg) of uranium metal in the form of 22,000 small lumps or 'eggs'. Slots in the graphite allowed the insertion of cadmium control rods to absorb neutrons rapidly and damp out any runaway reaction.

Throughout November 1942, Fermi and his team gradually built and tested this pile, until they were close to a self-sustaining reaction. On 2 December 1942, the full team of 49 scientists and others assembled, and one of the cadmium rods was gradually inched out of its controlling place. They worked all morning, broke for lunch, reconvened in the afternoon to continue and, finally, Fermi smiled and said, 'The reaction is self-sustaining'.[38,39] One of the scientists, Eugene Wigner, opened a bottle of Chianti, and the group all sipped red wine from paper cups and signed the straw wrapping on the bottle. Wigner later wrote, 'For some time we had known we were about to unlock a giant: still, we could not escape an eerie feeling when we knew we had actually done it, with far-reaching consequences we could not foresee'.[40] Another of the scientists, Arthur Compton, reported in code to the government's head of research, James Conant in Washington, 'The Italian navigator has just landed in the new world'.[41,42] Conant responded, 'Were the natives friendly?' To which, Compton replied, 'Everyone landed safe and happy'.[43,44] Szilard stayed after the crowd had left, shook Fermi's hand and said, 'This day will go down as a black day in the history of mankind'.[45] The structure of CP1 proved to be the blueprint for subsequent post-war development of nuclear energy power plants. A myth later grew up that all those present at the first reactor had died young of cancer, but Argonne National Laboratory (as the Argonne site later became) tracked everyone to demonstrate that this was not true.

In the meantime, the US government and military establishment were setting up research and development activities to explore the possibility of building an atomic bomb. They were worried that, because fission had been discovered in Germany, Hitler might be well ahead. On 9 October 1941 Roosevelt approved the atomic programme, set up the top-level S1 Committee to oversee it, and asked the US Army to take the lead. In the wake of the attack on Pearl Harbor on 7 December 1941, the Committee proposed pursuing a range of nuclear technologies at different sites across the United States, on topics such as the manufacture of uranium, enrichment of U^{235}, heavy-water moderation, carbon moderation (including Fermi and Szilard's work at Columbia), plutonium production and nuclear fusion (later to become the hydrogen bomb). A key problem was again how to reach critical mass without premature explosion. Unlike a nuclear reactor such as CP1, it was essential to trigger the reaction instantaneously. In the end, two options were pursued, effectively prototypes for the two bombs at Hiroshima and Nagasaki: fission of U^{235}, with critical mass reached rapidly by firing one plug of uranium into another one; and fission of Pu^{239}, with critical mass reached rapidly by imploding a plug of plutonium by detonating surrounding explosive material.

In June 1942 the project head, Colonel (later General) James C. Marshall established project headquarters on the 18th floor of 270 Broadway in mid-town Manhattan, New York, close to Stone and Webster, the principal project contractor, and to Columbia University. The original code-name for the project was *Substitute Materials*, but this was rapidly superseded by *Manhattan District* and then *Manhattan Project*. The initial budget was $34 million. In the end, the Manhattan Project would employ a total of 130,000 people, with a budget of almost $2 billion (equivalent to $22 billion today). In September 1942, Colonel Leslie Groves was promoted to General and took over leadership of the project. One of his first problems was who to appoint as Director of *Project Y*, the group that would design and build the bomb. On the recommendation of Arthur Compton, he selected Robert Oppenheimer, Professor of Physics at the University of California, Berkeley. There were concerns that Oppenheimer had little administrative experience, insufficient scientific standing, and too many links to friends and family members who were communists. In fact, he was regarded as a major security risk and was put under intensive investigation and surveillance, but Groves knew who he wanted and insisted on clearing him personally. His appointment was confirmed on 20 July 1943.

The Los Alamos Ranch School was founded in 1917 on a 7,200-foot (2,200-m) high plateau in the Sangre de Cristo (Blood of Christ) Mountains above the Rio Grande Valley, 35 miles (56 km) northwest of Santa Fe in the state of New Mexico. The school's programme combined a college curriculum with a rigorous outdoor life. Notable alumni include the novelists William Burroughs and Gore Vidal; the President of Sears Roebuck, Arthur Wood; and the founder of the Santa Fe Opera, John Crosby. In 1928, Robert Oppenheimer had visited Paul Ehrenfest's laboratory at the University of Leiden in the Netherlands, and on his return had accepted the offer of a professorship at Berkeley. Before moving to Berkeley to take up his professorship, he was diagnosed with a mild case of tuberculosis and spent some weeks recovering with his brother, Frank, at a ranch in New Mexico near Los Alamos, which he liked so much that he later bought it. In 1943, knowing the area well, he persuaded the US Army to purchase the nearby Los

Alamos Ranch School to be the site of the laboratory for Project Y, i.e. to be the place where the first atomic bombs would be designed and built. It was an ideal location for a top secret laboratory: isolated, with access via one steep and winding road; and with the surrounding countryside full of natural beauty, with trails for hiking and skiing and rivers for fishing, attractive to young scientists and their families.

Oppenheimer was a tall, thin chain-smoker, an enthusiastic and charismatic scientist, but intense and prone to fits of depression verging on self-harm, particularly when he was deep in thought. The physicist Murray Gell-Mann worked with him after the war at the Institute for Advanced Studies at Princeton and said,

> He didn't have sitzfleisch, 'sitting flesh', when you sit on a chair. As far as I know he never wrote a long paper or did a long calculation.... He didn't have patience.... But he inspired other people to do things, and his influence was fantastic.[46]

Oppenheimer wrote to a friend, 'my two great loves are physics and desert country, it's a pity they can't be combined'.[47] For the first scientists at Los Alamos it was a peculiar pioneer life, living in a chaotic jumble of dude ranches, trailers and prefabs in a boom town, with unpaved streets and mud everywhere, poor-quality power and water supplies, surrounded for security reasons by fences and barbed wire, and with all telephone calls and mail monitored and censored.

Fermi attended the first meeting of the Los Alamos scientists in April 1943, assembled to discuss the research programme. He continued to visit on a consultancy basis, and the following year was appointed Associate Director and Head of F Division (F for Fermi), in charge of the theory of fission, thermonuclear fusion and aqueous reactors. Overall, Fermi was working on a dizzying array of projects, shuttling between Columbia in New York, Argonne in Chicago, the first plutonium reactor at Hanford in Washington state, where he was also a consultant, and Los Alamos. Enrico, Laura and the family moved to Los Alamos, taking their boots to cope with the mud, and looking forward to the excitement of living in a young and vibrant, rapidly growing scientific community.

In February 1945 there was sufficient scientific and technological progress for General Groves to call a design freeze on the two prototypes. The first-ever explosion of an atomic bomb was then set up 210 miles (338 km) south of Los Alamos at the Alamogordo bombing and gunnery range in the Jornada del Muerto (Journey of Death) Desert. The site and test were both code-named *Trinity*. In advance, in May, 100 tons (91 Mg) of conventional high explosive were detonated at the Trinity site, at that time the largest man-made explosion ever, as a rehearsal to assess the test plans. The atomic bomb itself, code-named *The Gadget*, consisted of a ball of about 10 lb (4.5 kg) of Pu^{239} (sub-critical mass) corresponding to about 20,000 tons (18 Gg) of TNT, surrounding a small polonium–beryllium initiator to kick-start the neutron flux, and surrounded by a natural uranium tamper to prevent neutron escape, which itself was further surrounded by 5,000 lb (2,270 kg) of conventional high explosive to create a spherical inward shock to cause implosion of the plutonium, instantaneously pushing it beyond its critical mass at the same time as triggering the initiator. The bomb was assembled on the top of a

100-foot (30-m) steel tower, on four legs that went 20 feet (6.1 m) into the ground on concrete footings.

The detonation was set for 5.29 a.m. on Monday 16 July 1945, and the automated countdown started with 45 s to go. More than 400 people stayed at a base camp specially built for the test that weekend. The VIPs watched from Campania Hill, about 20 miles (32 km) northwest of the tower. There was a sweepstake on how strong the blast would be, ranging from 0 to 18,000 tons (16.3 Gg) of TNT (the nearest answer). General Groves wrote in his autobiography, 'As we approached the final minute the quiet grew more intense . . . I thought only what I would do if the countdown got to zero and nothing happened'.[48] His worst fears were not realised. The bomb completely vaporised the steel tower, melted the sand at the base of the tower, turning it into a light-green radioactive glass, and created a 300-foot (91-m) crater in the desert, with a shockwave that was felt more than 100 miles (160 km) away. Fermi himself wrote, 'After a few seconds, the rising flames lost their brightness and appeared as a huge pillar of smoke with an expanding head like a gigantic mushroom that rose rapidly beyond the clouds'.[49] Emilio Segrè wrote, 'I thought the explosion might set fire to the atmosphere and thus finish the earth, even though I knew it was not possible'.[50] More poetically but more darkly, Oppenheimer remembered lines from the *Bhagavad-Gita* in Hindu scripture: 'Now I am become Death, the destroyer of worlds'.[51,52] How prescient. Just 21 days later, Colonel Tibbets would take off from the Marianas with an atomic bomb onboard, bound for Hiroshima.

After the war, brought to a precipitate close in 1945 by the atomic bombs dropped on Hiroshima and Nagasaki, Fermi accepted the Charles H. Swift Professorship of Physics at the University of Chicago. He was elected to the US National Academy of Sciences later that year. The Argonne site became the Argonne National Laboratory in 1946, and the Manhattan Project became the US Atomic Energy Commission in 1947, with Fermi joining its General Advisory Committee. He continued to study at Chicago, Argonne and Los Alamos, working on a variety of projects: *pions*, which he predicted were composite particles, prefiguring *quark* theory; *cosmic rays*; *neutron scattering* with Leona Marshall; and *spin–orbit coupling* with Maria Mayer. He espoused the *Fermi paradox*: the contradiction between our lack of contact with extra-terrestrial life, despite its high probability of existence. He opposed the development of the hydrogen bomb, more on technical than moral grounds: he thought it wouldn't work. In the summer of 1954, he went to Europe and lectured on pions at an advanced summer school run by the Italian Physical Society on Lake Como. He went hiking in the mountains, but was lacking in energy. On returning to Chicago, he was discovered to have late-stage cancer, which had already spread to several parts of his body. He died a few months later, aged 53, on 16 November 1954.

To many people, Fermi will always be a dark, at best equivocal, figure, given his significant contribution to the development of nuclear weapons. As a scientist, however, he was extraordinary. He discovered neutrinos, Fermi-Dirac statistics, slow neutrons, nuclear fission (almost) and atomic piles. He was the key scientist behind nuclear energy and the atomic bomb. Many science artefacts have been named after him, including: Fermilab, the US National Accelerator 35 miles (56 km) west of Chicago; the Enrico

Fermi Institute for Nuclear Studies at the University of Chicago; the Enrico Fermi International School of Physics, the summer school run by the Italian Physical Society at Lake Como, where he spent his last summer; fermium, the radioactive element with atomic number 100; fermions, fundamental particles with spin, which obey Fermi-Dirac statistics; the Enrico Fermi Prize of the Italian Physical Society, awarded to those who have particularly honoured physics with their discoveries; the Enrico Fermi Award, given by the US Department of Energy for lifetime achievement in the development, use or production of energy; the Fermi constant, which is a measure of the strength of the weak interaction; Fermi's golden rule, which describes transition rates between different quantum states; the fermi, a unit of length equal to one femtometre, or 10^{-15} m; and, of course, the Fermi level, which determines the electrical properties of materials. The website history.com called him 'the architect of the nuclear age'[53]; the *New York Times* called him 'the architect of the atomic bomb'[54]; his colleagues at the University of Rome called him 'the Pope of physics'[55]; and his graduate students said he 'had an inside track to God'.[56] The scientist and novelist C. P. Snow said 'anything one says about Fermi is hyperbole'[57]; Robert Oppenheimer said 'he was simply unable to let things be foggy. Since they nearly always are, this kept him pretty active'[58]; and his daughter Nella said 'it wasn't that he lacked emotions, but that he lacked the ability to express them'.[59] Fermi himself said, 'I could never learn to stay in bed late enough in the morning to be a theoretical physicist',[60] and 'If I could remember the names of all these particles, I'd be a botanist'.[61]

5 Conductors

In a conductor, the Fermi level is in the middle of the valence band, so the valence electrons can easily acquire excess energy from the application of a voltage and move throughout the material. The valence electrons are not bound to their individual atoms and are *free electrons*. If the material is a single crystal at absolute zero, the free electrons can accelerate without restriction in response to the force produced by an applied voltage. However, real crystals contain defects such as vacancies, dislocations, grain boundaries and impurity particles, and at realistic temperatures the atoms or molecules are vibrating and rotating. The defects and the atomic or molecular motions obstruct and deflect the moving electrons, so that at any given time some electrons are moving forwards and some backwards. This is what causes the resistance of the material to electron flow.

At any given time, the free electrons in a conductor are all moving in different directions and with different velocities, but there is an overall average *drift velocity v* in the direction of the force produced by an applied voltage and its associated electric field. The drift velocity of the electrons is proportional to the applied electric field:

$$v = \mu \mathcal{E},$$

where the constant of proportionality μ is called the *mobility* of the electrons, determined by the extent of atomic and molecular vibrations and the number of defects in the material to obstruct and deflect the electrons. Since $\mu = v/\mathcal{E}$, the mobility is the velocity produced by unit applied electric field. The current density i depends on the density of free electrons n, their charge q, and their average velocity v:

$$i = nqv = nq\mu\mathcal{E}.$$

Inserting this into Ohm's law gives

$$\mathcal{E} = \rho i = \rho nqv = \rho nq\mu\mathcal{E}.$$

Dividing by $\mathcal{E}$,

$$1 = \rho nq\mu$$
$$\therefore \sigma = \frac{1}{\rho} = nq\mu.$$

Not surprisingly, the conductivity and resistivity are, respectively, proportional and inversely proportional to the electron mobility.

The resistivity of a metal conductor arises from the deflection of moving electrons by defects and by atomic or molecular motions, and can be written as the sum of these different contributions,

$$\rho = \rho_{vib} + \rho_{vac} + \rho_{disloc} + \rho_{gb} + \rho_{imp},$$

where ρ_{vib}, ρ_{vac}, ρ_{disloc}, ρ_{gb} and ρ_{imp} are contributions from atomic or molecular vibrations, vacancies, dislocations, grain boundaries and impurities respectively. The best metal conductors with the lowest resistivity are materials such as face-centred cubic copper and silver, which have low ρ_{vib}, at high purity to reduce ρ_{imp}, in single-crystal or large-grain-size form to reduce ρ_{gb}, and annealed to reduce ρ_{vac} and ρ_{disloc}. However, these conditions make conductors very soft, so it is common to add a controlled amount of alloying additions to enhance conductor strength, ideally without damaging conductivity too much.

Conductors are used to bring electricity from power stations where it is generated, to houses and factories where it is used. Conductors are also used to distribute electricity within houses to provide power for many of our familiar utilities and devices, such as lights, heaters, fridges, ovens, TVs, radios and computers, and in factories to provide power for manufacturing and production machines.

6   Superconductors

There are a number of materials, such as niobium–titanium alloys and yttrium barium copper oxides (or YBCO), which are called *superconductors*. In these materials, the resistivity drops to zero at a critical temperature that is well below room temperature.

Superconductivity is a quantum mechanical effect, whereby at low temperatures the electrons form *Cooper pairs*, with wave functions that permeate throughout the material, so they are not scattered by atomic or molecular vibrations or defects in the material. Having zero resistance makes superconductors potentially attractive in many applications but, unfortunately, the need for low temperature usually makes it relatively difficult to use them.

7   Insulators

Insulators do not conduct electrons because their valence band is full. Electrons at the Fermi level, i.e. at the top of the valence band, cannot gain extra energy because of the band gap of disallowed energies between the valence and conduction bands. Effectively, therefore, all the valence electrons remain bound to their individual atoms or molecules, like the core electrons. When a battery is used to apply a voltage to an insulator, as shown in Figure 13.6, electrons cannot move from one end of the material to the other, despite the electric field and the associated force on the electrons. Instead an electric charge builds up between the two ends of the material, with an excess of electrons at one end and a deficiency at the other, as shown in Figure 13.6. This is called the *dielectric effect*, and the insulator is said to act as a *dielectric*. An insulator or dielectric between two parallel plates, as shown in Figure 13.6, is called a *capacitor* or *condenser*.

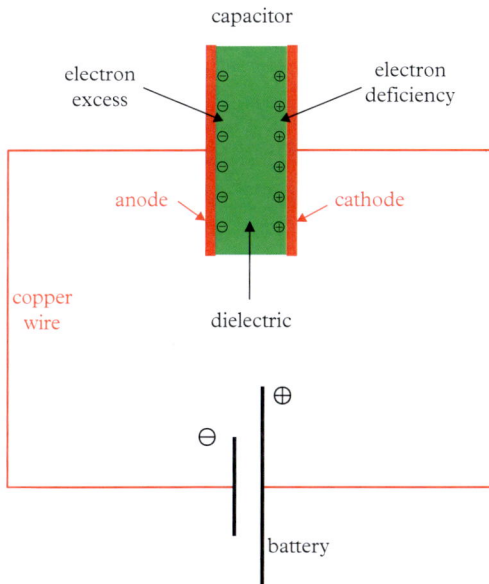

Figure 13.6 *Parallel-plate capacitor*

The charge Q built up in a parallel-plate capacitor such as shown in Figure 13.6 is proportional to the applied voltage V:

$$Q = CV,$$

where the constant of proportionality C is called the *capacitance*. Since $C = Q/V$, the capacitance is the charge built up per unit applied voltage. Charge builds up more readily in an insulator of high capacitance. The capacitance of a parallel-plate capacitor depends on its size and shape as well as the material between the plates, increasing with increasing cross sectional area of the plates A, and decreasing with increasing separation d between them,

$$C = \epsilon \frac{A}{d},$$

where ϵ is a material constant called its *permittivity*. The *charge density* on the capacitor's plates is the charge built up per unit area, $D = Q/A$, and the applied electric field is $\mathcal{E} = V/d$, so the charge build-up can also be described by

$$Q = \epsilon \frac{A}{d} V$$
$$\therefore D = \epsilon \mathcal{E}.$$

In other words, when an electric field is applied to a material, the resulting charge density is proportional to the applied electric field, with the permittivity as the constant of proportionality. Since $\epsilon = D/\mathcal{E}$, the permittivity is the charge density built up per

Table 13.3 *Typical Values of Dielectric Constant for Some Common Insulators*

Material	Dielectric constant κ
Alumina	10
Silica	5
Bakelite	5–22
Glass	5–10
Glass fibre-reinforced nylon	3.7
Polyethylene	2.5
Rubber	2–4
wood	1.4–2.9

unit applied electric field. If the material is replaced by a vacuum, the charge density is $D = \epsilon_o \mathcal{E}$, where ϵ_o is the *vacuum permittivity*. So, in general,

$$D = \epsilon \mathcal{E} = \epsilon_o \kappa \mathcal{E},$$

where $\kappa = \epsilon/\epsilon_o$ is the ratio of the permittivity to the vacuum permittivity, and is called the *relative permittivity* or, more often, the *dielectric constant*. Table 13.3 shows typical values of dielectric constant for some common insulator materials.

Insulators are used as sheaths and outer casings to protect conductors, preventing us from getting electric shocks when we handle them, and as capacitors when we want to build up a charge, before discharging to power a device.

8   Semiconductors

Semiconductors conduct electricity, but only weakly. Like insulators their valence band is full, but the band gap of disallowed energies between the valence and conduction bands is small. This means that, at temperatures above absolute zero, some of the electrons at the top of the valence band can gain enough thermal energy to jump across the gap into the conduction band, as shown in Figure 13.7. To do so, an electron must gain excess energy equal to the band gap energy E_B, and the likelihood of a jump is given by the Boltzmann probability:

$$p = \exp\left(\frac{-E_B}{kT}\right).$$

The probability of electrons jumping across the band gap only becomes significant when the band gap energy is approximately equal to or less than the thermal energy, $E_B \leq kT$, which only happens at room temperature in a few materials.

Every time an electron jumps from the valence band to the conduction band, it creates an *electron–hole pair*, as shown in Figure 13.7. A free electron is created near the bottom of the conduction band, and the electron is able to move freely throughout the material. In addition, a hole is left behind near the top of the valence band, and the hole is also able to move freely (and independently) throughout the material, by the mechanism of adjacent electrons jumping into the hole. The electrons and the holes are both *charge carriers*. Each electron carries a negative charge, and responds to an applied voltage by moving from the anode to the cathode, i.e. from the negative terminal to the positive terminal. Each hole carries a positive charge, because it is an absence of an electron, with an absence of negative charge, and responds to an applied voltage by moving in the reverse direction from the cathode to the anode, i.e. from the positive terminal to the negative terminal.

Elemental semiconductors such as pure silicon or germanium are called *intrinsic semiconductors*. In an intrinsic semiconductor, the number of free electrons equals the

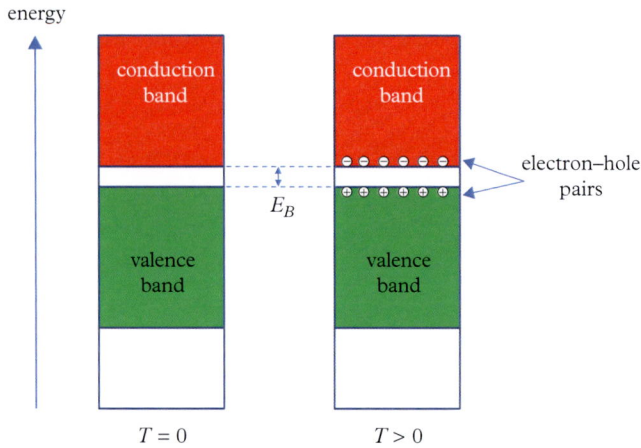

Figure 13.7 *Creation of electron–hole pairs*

number of holes, as is obvious from the mechanism by which electron–hole pairs are created. The number of free electrons and holes remains constant, determined by a dynamic equilibrium between: (1) continuous creation of electron–hole pairs as electrons randomly acquire sufficient excess thermal energy to jump into the conduction band; and (2) continuous destruction of electron–hole pairs as electrons and holes randomly meet and recombine as they move around the material. In one mole of a univalent metal conductor, there are Avogadro's number $N_o \sim 6 \times 10^{23}$ of atoms, and each atom contributes one electron to the valence band, so there are also Avogadro's number $N_o \sim 6 \times 10^{23}$ of free electrons to act as charge carriers. By comparison, in one mole of an intrinsic semiconductor such as pure germanium at room temperature, there are approximately 10^{14} electrons that have enough thermal energy to jump into the conduction band, leaving behind the same number of holes in the valence band, i.e. a total of only about 2×10^{14} charge carriers, many orders of magnitude less than the metal conductor. This is why semiconductors are only poor conductors of electricity. The overall conductivity of an intrinsic semiconductor depends on the mobility of the free electrons μ_e and the holes μ_h,

$$\sigma = \frac{1}{\rho} = nq(\mu_e + \mu_h).$$

The rate at which free electrons can move through the conduction band is usually faster than the rate at which valence electrons can jump into holes in the valence band, i.e. $\mu_e > \mu_h$.

The number of charge carriers in an intrinsic semiconductor such as pure silicon or germanium can be increased enormously by *doping*, i.e. by adding a small amount of an impurity element with a different valence. This creates an *extrinsic semiconductor*.

Doping with an element of greater valence injects extra electrons into the conduction band, leading to an excess of free electrons over holes: this is called *n-type doping* (negative-type doping), creating an *n-type semiconductor*. On the other hand, doping with an element with a lower valence injects excess holes into the valence band, leading to an excess of holes over free electrons: this is called *p-type doping* (positive-type doping), creating a *p-type semiconductor*. For instance, pure silicon has a diamond cubic structure and a valency of 4. It can be made n-type by doping with a small amount of 5-valent nitrogen or phosphorus, and it can be made p-type by doping with a small amount of 3-valent aluminium or gallium. In each case the doping leads to a dilute substitutional solid solution, with dopant atoms replacing occasional silicon atoms on the diamond cubic lattice. Each n-type dopant atom has one extra valence electron, which has to go into the conduction band, and each p-type dopant atom has one fewer valence electron, leaving a hole in the valence band. Even doping a semiconductor with a few parts per million (1 ppm $= 10^{-6}$) of an n-type or p-type impurity enhances the conductivity dramatically, because each part per million corresponds to an extra $10^{-6} \times 6 \times 10^{23} = 6 \times 10^{17}$ charge carriers, swamping the number of intrinsic electron–hole pairs.

There are intrinsic *compound semiconductors* as well as intrinsic elemental semiconductors. For instance, equiatomic compounds between 3-valent and 5-valent elements, such as gallium arsenide GaAs or indium antimonide InSb, have an average valency of 4, the same as in pure silicon and germanium, and have the zinc blende structure, similar to diamond cubic silicon and germanium. They have a similar band structure, with a small energy gap between the valence and conduction bands, and are known as *III-V semiconductors*. Equiatomic compounds between 2-valent and 6-valent elements, such as cadmium sulphide CdS and zinc telluride ZnTe also have similar band structures, and are known as *II-VI semiconductors*. Table 13.4 shows typical band gap values for some intrinsic elemental and compound semiconductors. Compound semiconductors can also be doped to increase their conductivity in a similar way to elemental semiconductors.

Table 13.4 *Typical Band Gap Values for Some Intrinsic Elemental and Compound Semiconductors*

Material	Band gap E_B (eV)
Silicon	1.11
Germanium	0.66
Gallium arsenide	1.47
Indium antimonide	0.17
Cadmium sulphide	2.59
Zinc telluride	2.26

9   Electronic devices

Semiconductors are used to manufacture miniature components that are the basis of *electronic devices* for processing information in *computers, telecommunications* and *intelligent systems*. We will briefly examine one or two basic *semiconductor devices*.

Diodes

An important component of electric circuits is the *rectifier* or *diode*, which only lets electric current flow in one direction. Miniature diodes can be manufactured by joining p-type and n-type semiconductors to form a *p–n junction*. In fact this can be achieved in a small region of a single-semiconductor material by doping adjacent regions with p-type and n-type impurities. Doping is achieved by depositing tiny amounts of the dopant on controlled regions of the surface of the semiconductor and heating gently to allow the dopant to diffuse into the semiconductor. The semiconductor is usually a carefully grown single crystal to prevent defects interfering with the device, and the typical size of the dopant regions is in the micron or even nanometre scale.

In a p–n junction with no electrical bias, holes are distributed randomly throughout the p-type region and electrons are distributed randomly throughout the n-type region. When a battery is connected across a p–n junction, its behaviour differs radically depending on the direction of the applied voltage. The junction is *forward biased* when the positive terminal or cathode (⊕) is connected to the p-type region and the negative terminal or anode (⊖) is connected to the n-type region, as shown in Figure 13.8. The anode then supplies a flow of electrons to the n-type region, pushing excess electrons into the conduction band and towards the p-type region. At the same time, the cathode removes a flow of electrons from the p-type region, pushing excess holes into the valence band and towards the n-type region. The excess electrons and holes meet at the interface between the p and n regions, and recombine, allowing the flow of electricity to be maintained. Forward bias produces, therefore, a steady current that increases with increasing applied voltage.

The junction is *reverse biased* when the electrodes are connected the other way round, i.e. when the anode is connected to the p-type region and the cathode is connected to the n-type region, as also shown in Figure 13.8. The anode now supplies a flow of electrons to the p-type region, pulling excess holes away from the junction; and the cathode now removes electrons from the n-type region, pulling excess electrons away from the junction. This creates an insulating region, with no charge carriers (neither electrons nor holes) in the vicinity of the interface between the p and n regions, stifling the continued flow of electricity. Reverse bias, therefore, prevents the flow of electricity, and there is no sustained current. Overall, this produces an asymmetric variation of current versus applied voltage, or *I–V curve*, which only allows electric flow in the forward direction. This allows *alternating current* (or *ac*) to be converted into *direct current* (or *dc*), and can also be used to protect components from unwanted electron flow.

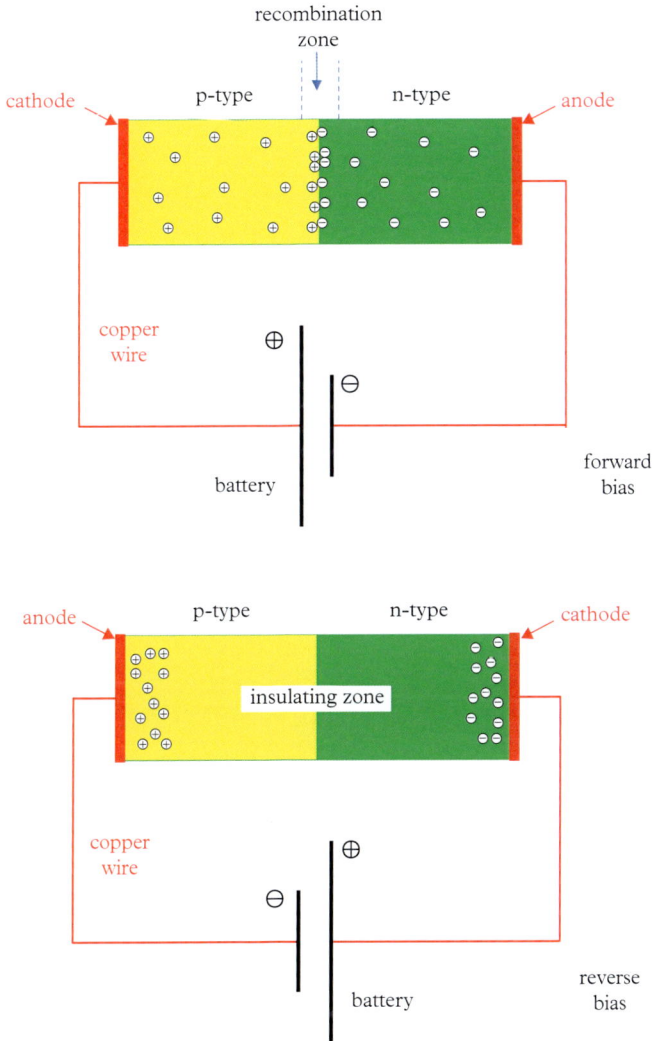

Figure 13.8 *Forward- and reverse-biased p–n junctions*

Transistors

Another important component of electric circuits is the *amplifier*, often achieved by using a *transistor*. A simple *bipolar junction transistor* (or *BJT*) consists of a pair of adjacent p–n junctions, in an n–p–n configuration as shown in Figure 13.9. A thin p region, called the *base* or *gate* separates two n regions, one called the *emitter* or *source* and the other called the *collector* or *drain*. The emitter–base junction is forward biased with a voltage so, as with a diode, the anode supplies a flow of electrons to the n-type emitter, pushing

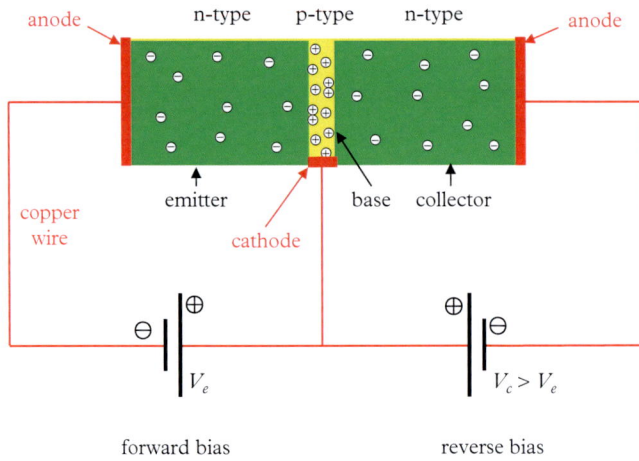

Figure 13.9 *Bipolar junction transistor*

excess electrons towards the junction with the p-type base. In this case, however, the base region is very thin, so the excess electrons are not all consumed by recombination near the junction. Instead, they are driven across the junction and into the n-type collector. The base–collector junction is reverse biased so, again as with a diode, the cathode pulls electrons away from the junction, removing them from the n-type collector. The reverse voltage across the base–collector junction V_c is greater than the forward voltage across the emitter–base junction V_e, and the resulting current in the collector I_c depends exponentially on the emitter voltage,

$$I_c = I_o \exp\ (qV_e/kT),$$

where I_o is a constant. This means that the collector current is very sensitive to emitter voltage, and the transistor acts as an amplifier, with a big increase in current for a small change in voltage. Transistors can also have an p–n–p configuration, and can again be miniaturised by doping adjacent regions of a single semiconductor material. In a *field effect transistor* (or *FET*), the extent of recombination and, therefore, the current through the central base region is controlled by applying an electric field to change the bias. This enhances miniaturisation and control of the device.

Data storage

Diodes and transistors can both be used as switches, to allow binary digital *data storage,* with 1 or 0 corresponding to the switch being open or closed respectively.

Integrated circuits

Semiconductor diodes and transistors can be miniaturised, allowing large amounts of data to be stored and manipulated in a single piece of semiconductor material. A single crystal surface of a semiconductor such as silicon is doped on a micron or nanometre scale, using a variety of *deposition*, *masking* and *etching* techniques, to create a complex architecture of densely packed, interconnected devices. This is called *large-scale integration* (or *LSI*) or *very large-scale integration* (or *VLSI*), and the resulting components are called, colloquially, *silicon chips*. A wide range of different processing methods are used to create complex miniaturised electronic circuits, with many kinds of overall functionality. This technology underpins the so-called *digital revolution* and growth of the *information technology industry*.

··

10   REFERENCES

1. 'Super-fortress crew tell their story'. *The Guardian*, 8 August 1945.
2. 'Atom bomb crew's story'. *Chicago Tribune*, 8 August 1945.
3. Sylvia Engdahl, ed. *The Atomic Bombings of Hiroshima and Nagasaki* (Greenhaven Press, Farmington Hills, 2011), 36.
4. Harry S. Truman. 'Speech at the White House, 9 August 1945'. In: Priscilla Roberts (ed.). *The Cold War: Interpreting Conflict through Primary Documents* (ABC-CLIO, Santa Barbara, 2018), 144.
5. Engdahl, n 3, 33.
6. Harry S. Truman. 'Announcing the bombing of Hiroshima'. 6 August 1945, Truman Library, http://www.pbs.org/wgbh/americanexperience/features/truman-hiroshima/ (accessed 26 February 2020).
7. 'Britain and US to keep atomic bomb secret'. *The Guardian*, 10 August 1945.
8. Engdahl n 3, 31.
9. Dennis Wainstock. *The Decision to Drop the Atomic Bomb* (Praeger, Westport, 1996), 76.
10. Ibid. 77.
11. Emperor Hirohito. Radio speech. 15 August 1945.
12. 'The Emperor's speech: 67 years ago, Hirohito transformed Japan forever'. *The Atlantic*, 15 August 2012.
13. Jeffrey K. Smith. *Fire in the Sky: The Story of the Atomic Bomb* (Authorhouse, Bloomington, 2010), 89.
14. Dan Cooper. *Enrico Fermi: And the Revolutions of Modern Physics* (Oxford University Press, Oxford, 1999), 20.
15. Ibid. 21.
16. Ibid. 24.
17. Luisa Bonolis. 'Enrico Fermi's scientific work'. In: C. Bernardino and Luisa Bonolis (eds.). *Enrico Fermi: His Work and Legacy* (Società Italiana di Fisica, Springer, Bologna, 2001), 321.
18. Enrico Fermi. 'Sulla quantizzazione del gas perfetto monoatomico' ['On the quantisation of the perfect monoatomic ideal gas']. *Rendiconti Lincei* 3 (1926), 145–149.
19. Cooper, n 14, 38.

20. See ibid.
21. Fermi, n 18.
22. E. Fermi. 'Versuch einer theorie der β-strahlen' ['Attempt at a theory of β-rays']. *Zeitschrift für Physik* **88** (1934), 161–177.
23. Cooper, n 14, 43.
24. Fermi, n 22.
25. Cooper, n 14, 47.
26. See ibid. 48.
27. Otto Hahn and Fritz Strassmann. 'Über den nachweis und das verhalten der bei der bestrahlung des urans mittels neutronen entstehenden erdalkalimetalle' ['On the detection and characteristics of the alkaline earth metals formed by irradiation of uranium with neutrons']. *Die Naturwissenschaften* **27** (1939), 11–15.
28. Patricia Rife. *Lise Meitner and the Dawn of the Nuclear Age* (Birkhäuser, Boston, 2010), 183.
29. Otto Hahn and Fritz Strassman. 'Über die bruchstticke beim zerplatzen des urans' ['On the fragments when uranium bursts']. *Die Naturwissenschaften* **27** (1939), 163–164.
30. Otto Hahn and Fritz Strassman. 'Zur frage nach der existenz der "trans-urane"' ['The question of the existence of trans-uranium elements']. *Die Naturwissenschaften* **27** (1939), 451–453.
31. Lise Meitner and O. R. Frisch. 'Disintegration of uranium by neutrons: a new type of nuclear reaction'. *Nature* **143** (1939), 239–240.
32. Lise Meitner and O. R. Frisch. 'Products of the fission of the uranium nucleus'. *Nature* **143** (1939), 472–472.
33. Lise Meitner. 'Letter to B. Broomé-Aminoff, November 1945'. In: Anne Hardy and Lore Sexl. *Lise Meitner* (Rowohlt Taschenbuch Verlag, Reinbeck, 2002), 119.
34. Emilio Segrè. *Enrico Fermi, Physicist* (University of Chicago Press, Chicago, 1995), 222.
35. Bonolis, n 17 321.
36. Albert Einstein. Letter to President Franklin Delano Roosevelt, 2 August 1939. In: Al Cimino. *The Manhattan Project: The Making of the Atomic Bomb* (Arcturus Publishing, London, 2015), 11.
37. Cooper, n 14, 59.
38. Bonolis, n 17, 366.
39. Al Cimino. *The Manhattan Project: The Making of the Atomic Bomb* (Arcturus Publishing, London, 2015), 63.
40. Jagdish Mehra, ed. *The Collected Works of Eugene Paul Wigner: Part B Historical, Philosophical and Socio-Political Papers*, vol. VII (Springer, New York, 2001), 447.
41. Arthur Compton. *Atomic Quest* (Oxford University Press, New York, 1956), 144.
42. Cimino, n 39, 64.
43. Compton, n 41, 144.
44. Cimino, n 39, 64.
45. US Department of Energy. 'The Manhattan Project: an interactive history'. n.d. https://www.osti.gov/opennet/manhattan-project-history/Events/1942-1944_pu/cp-1_critical.htm (accessed 26 February 2020).
46. Kai Bird and Martin J. Sherwin. *American Prometheus: The Triumph and Tragedy of J. Robert Oppenheimer* (Alfred A. Knopf, New York, 2005), 375.
47. Richard Rhodes. *The Making of the Atomic Bomb: 25th Anniversary Edition* (Simon and Schuster, New York, 1986 and 2012), 451.

48. Los Alamos Scientific Laboratory. *Los Alamos: Beginning of an Era 1943–1945* (Bathrub Row Press, Los Alamos, 1986), 51.
49. Ibid. 53.
50. Segrè, n 34, 147.
51. Cimino, n 39, 150.
52. Los Alamos Scientific Laboratory, n 48, 54.
53. Beatriz Guilén. 'Enrico Fermi, architect of the nuclear age'. 28 November 2016. https://www.bbvaopenmind.com/en/science/mathematics/enrico-fermi-the-architect-of-the-nuclear-age/ (accessed 26 February 2020).
54. 'Enrico Fermi dead at 53; architect of atomic bomb'. *New York Times*, 29 November 1954, 1.
55. Gino Segrè and Bettina Hoerlin. *The Pope of Physics: Enrico Fermi and the Birth of the Atomic Age* (Henry Holt, New York, 2016), 3.
56. Ibid. 272.
57. C. P. Snow. *The Physicists: A Generation That Changed the World* (Little Brown, Boston, 1981), 79.
58. Time Magazine/CBS News. *People of the Century: One Hundred Men and Women Who Shaped the Last One Hundred Years* (Simon and Schuster, New York, 1999), 209.
59. Segrè and Hoerlin, n 55, 295.
60. See ibid. 32.
61. Michio Kaku. *Hyperspace: A Scientific Odyssey Through Parallel Universes, Time Warps, and the Tenth Dimension* (Oxford University Press, Oxford, 1994), 118.

• •

11 BIBLIOGRAPHY

Electrical Properties of Materials (*9th edition*). Laszlo Solymar, Donald Walsh, and Richard Syms (Oxford University Press, Oxford, 2014).

Enrico Fermi: And the Revolutions of Modern Physics. Dan Cooper (Oxford University Press, New York, 1999).

The Last Man Who Knew Everything: The Life and Times of Enrico Fermi, Father of the Nuclear Age. David Schwartz (Basic Books, New York. 2017).

The Manhattan Project: The Making of the Atomic Bomb. Al Cimino (Arcturus, London, 2016).

The Pope of Physics: Enrico Fermi and the Birth of the Atomic Age. Gino Segrè and Bettina Hoerlin (Henry Holt, New York, 2016).

Index

.

The manufacturer's authorised representative in the EU for product
safety is Oxford University Press España S.A. of El Parque Empresarial
San Fernando de Henares, Avenida de Castilla, 2 - 28830 Madrid
(www.oup.es/en or product.safety@oup.com). OUP España S.A. also acts
as importer into Spain of products made by the manufacturer.
Printed and bound by CPI Group (UK) Ltd, Croydon, CR0 4YY

01/07/2025
01909155-0002